Physical Methods to Characterize Pharmaceutical Proteins

Pharmaceutical Biotechnology

Series Editor: Ronald T. Borchardt
The University of Kansas
Lawrence, Kansas

Volume 1 PROTEIN PHARMACOKINETICS AND METABOLISM
Edited by Bobbe L. Ferraiolo, Marjorie A. Mohler, and Carol A. Gloff

Volume 2 STABILITY OF PROTEIN PHARMACEUTICALS, Part A: Chemical and Physical Pathways of Protein Degradation
Edited by Tim J. Ahern and Mark C. Manning

Volume 3 STABILITY OF PROTEIN PHARMACEUTICALS, Part B: *In Vivo* Pathways of Degradation and Strategies for Protein Stabilization
Edited by Tim J. Ahern and Mark C. Manning

Volume 4 BIOLOGICAL BARRIERS TO PROTEIN DELIVERY
Edited by Kenneth L. Audus and Thomas J. Raub

Volume 5 STABILITY AND CHARACTERIZATION OF PROTEIN AND PEPTIDE DRUGS: Case Histories
Edited by Y. John Wang and Rodney Pearlman

Volume 6 VACCINE DESIGN: The Subunit and Adjuvant Approach
Edited by Michael F. Powell and Mark J. Newman

Volume 7 PHYSICAL METHODS TO CHARACTERIZE PHARMACEUTICAL PROTEINS
Edited by James N. Herron, Wim Jiskoot, and Daan J. A. Crommelin

Physical Methods to Characterize Pharmaceutical Proteins

Edited by

James N. Herron
University of Utah
Salt Lake City, Utah

Wim Jiskoot
National Institute of Public Health and Environmental Protection
Bilthoven, The Netherlands

and

Daan J. A. Crommelin
Utrecht University, and
Utrecht Institute for Pharmaceutical Sciences
Groningen-Utrecht Institute for Drug Exploration
Utrecht, The Netherlands

Plenum Press • New York and London

Library of Congress Cataloging-in-Publication Data

On file

ISBN 0-306-45026-7

A Division of Plenum Publishing Corporation
233 Spring Street, New York, N. Y. 10013

10 9 8 7 6 5 4 3 2 1

Printed in the United States of America

Contributors

Michael Bloemendal • Department of Protein and Molecular Biology, Royal Free Hospital School of Medicine, London NW3 2PF, England

E. A. Cooper • Department of Bioengineering, University of Utah, Salt Lake City, Utah 84112

Reinout J. Driebergen • Ares Serono, Chemin des Mines, Geneva, Switzerland

Ernesto Freire • Department of Biology and Biocalorimetry Center, The Johns Hopkins University, Baltimore, Maryland 21218

Paul T. Hamilton • Becton Dickinson Research Center, Research Triangle Park, North Carolina 27709

James N. Herron • Department of Pharmaceutics and Pharmaceutical Chemistry, University of Utah, Salt Lake City, Utah 84112

Vladimir Hlady • Department of Bioengineering, University of Utah, Salt Lake City, Utah 84112

Joost J. M. Holthuis • OctoPlus b.v., 2300 AS Leiden, The Netherlands

Wim Jiskoot • Department of Pharmaceutics and Pharmaceutical Chemistry, University of Utah, Salt Lake City, Utah 84112; *present address*: Laboratory for Product and Process Development, National Institute of Public Health and Environmental Protection, BA Bilthoven, The Netherlands

W. Curtis Johnson, Jr. • Department of Biochemistry and Biophysics, Oregon State University, Corvallis, Oregon 97331

K. Knutson • Department of Pharmaceutics and Pharmaceutical Chemistry, University of Utah, Salt Lake City, Utah 84112

Mark C. Manning • School of Pharmacy, University of Colorado Health Sciences Center, Denver, Colorado 80262

James Matsuura • School of Pharmacy, University of Colorado Health Sciences Center, Denver, Colorado 80262

Kenneth P. Murphy • Department of Biochemistry, University of Iowa, Iowa City, Iowa 52242

John J. Naleway • Marker Gene Technologies, University of Oregon, Eugene, Oregon 97403

Peter Roepstorff • Department of Molecular Biology, Odense University, DK-5230 Odense M, Denmark

Tom A. A. M. van de Goor • Eindhoven University of Technology, Laboratory of Instrumental Analysis, 5600 MB Eindhoven, The Netherlands; *present address*: Hewlett Packard Laboratories, Palo Alto, California 94303-0867

David G. Vander Velde • NMR Laboratory, University of Kansas, Lawrence, Kansas 66045

Preface

Proteins are still gaining importance in the pharmaceutical world, where they are used to improve our arsenal of therapeutic drugs and vaccines and as diagnostic tools. Proteins are different from "traditional" low-molecular-weight drugs. As a group, they exhibit a number of biopharmaceutical and formulation problems. These problems have drawn considerable interest from both industrial and academic environments, forcing pharmaceutical scientists to explore a domain previously examined only by peptide and protein chemists.

Biopharmaceutical aspects of proteins, e.g., low oral bioavailability, have been extensively investigated. Although all possible conventional routes of administration have been examined for proteins, no real, generally applicable alternative to parenteral administration in order to achieve systemic effects has yet been discovered. Several of these biopharmaceutical options have been discussed in Volume 4 of this series, *Biological Barriers to Protein Delivery.*

Proteins are composed of many amino acids, several of which are notorious for their chemical instability. Rational design of formulations that optimize the native structure and/or bioactivity of a protein is therefore of great importance when long shelf life is required, as it is for pharmaceutical products. This issue has also been examined in two prior volumes of this series: Volume 2: *Stability of Protein Pharmaceuticals* (Part A) and Volume 5: *Stability and Characterization of Protein and Peptide Drugs.*

Of equal importance for the therapeutic or diagnostic success of proteins is their physical stability. The integrity of their secondary, tertiary, and quaternary structure should be guaranteed during their shelf life (e.g., liquid formulations) or upon administration to the patient (e.g., freeze-dried products). This means that techniques that provide information about various structural aspects of proteins have to be used. These aspects include, but are not limited to, three-dimensional structure, hydrodynamic properties, physicochemical behavior, kinetics, thermodynamic properties, and dynamic behavior.

Unfortunately, no one technique in the present arsenal of structural methods is able to provide 100% of this information. Therefore, a rational strategy is to employ a concerted approach in which the protein is examined using several different structural techniques. The resulting information is cross-correlated to provide a more complete picture of the chemical and physical state and/or bioactivity of the protein under different conditions. The most frequently used techniques for structural analysis of proteins are described in this volume. The impressive recent progress in all these techniques and expected future developments are also discussed.

Several spectroscopic techniques can be employed for studying the structure and function of proteins. These include fluorescence spectroscopy (Chapter 1), circular dichroism (Chapter 2), and infrared spectroscopy (Chapter 3). The first of these, fluorescence spectroscopy, exhibits a level of sensitivity (subnanomolar) unachievable by any other technique described in this volume. For that reason, it is especially well suited for protein function studies where information about ligand–receptor interactions (including antigen–antibody binding) and enzyme kinetics is required. Furthermore, recent advances in time-resolved fluorescence have made it possible to study dynamic processes that occur in proteins in the nanosecond to microsecond time scale.

Circular dichroism (CD) is a spectroscopic technique that has long been part of the concerted approach. Different parts of the spectrum (far UV, near UV/VIS, infrared) provide information on the secondary and tertiary structure of proteins. The degree of structural information is less detailed than with new nuclear magnetic resonance (NMR) or X-ray diffraction techniques, but CD has the advantage that relatively simple equipment is used, scans are rapidly obtained, and the (semi) empirical interpretation is not complicated.

It has long been known that certain transitions in the infrared spectrum of a protein were due to the vibrations of the carbonyl and amide functions of the peptide bond, but only in the last decade have instrumentation and analytical techniques improved to the point where these transitions could be effectively used to predict the secondary structure of a protein. Thus, infrared spectroscopy, or more correctly, Fourier transform infrared (FTIR) spectroscopy, has become a viable technique for examining secondary structure. Although the information obtained is similar to that provided by CD spectroscopy, FTIR offers a greater flexibility in detection that facilitates measurements with specimens such as cells, crystals, tissue slices, and thin films, in addition to aqueous solutions of proteins.

In the last decade, the role of mass spectrometry for protein characterization changed from only a marginal one to a core position. This dramatic change is mainly the result of the development of a number of new ionization methods. Potentials and limitations of these new mass spectrometry approaches and future developments are discussed in Chapter 4.

As with mass spectrometry, the role of NMR spectroscopy in protein characterization has changed dramatically in the last decade and now rivals that of X-ray diffraction, at least in the case of small- to medium-sized proteins (10–50 kDa). This has largely been due to three developments: the availability of high-resolution spectrometers (both high magnetic field and high frequency), the effective usage of spin-coupling relaxation times to provide two-dimensional data, and the labeling of proteins with isotopes such as ^{13}C and ^{15}N to provide data of even higher dimensionality (three- and four-dimensional) than is available with standard proton (^{1}H) NMR experiments. These developments and their application to proteins of pharmaceutical interest, are discussed in Chapter 5.

X-ray diffraction has been and continues to be the definitive method for determining the three-dimensional structure of a protein. However, the technique is usually performed by specialists, and four excellent works reviewing the state of the art in X-ray diffraction analysis of proteins appeared recently. Besides, X-ray crystallography is not a technique that can be easily used in quality control protocols of pharmaceutical proteins because of the problems encountered with crystallization. Therefore, we have not included a chapter on X-ray diffraction in the present volume and refer interested readers to recent works by Ducruix and Giegé (1992), Rhodes (1993), McRee (1994), and Drenth (1994).

Thermodynamic parameters provide valuable information about stability, and differential scanning calorimetry has been used for many years in the characterization of small molecules. However, it has only been in the last 5 to 10 years that microcalorimetry instrumentation has progressed to the point where similar measurements could be obtained for macromolecules. Recent developments in this field and their application to both protein folding and protein–ligand interactions are discussed in Chapter 6.

Chromatographic approaches have been the standard-bearers of the concerted approach mentioned above. As discussed in Chapter 7, several fundamentally different chromatographic separation approaches, each with its own pros and cons, have been developed over the years. Reversed-phase chromatography, hydrophobic interaction chromatography, ion-exchange chromatography, size-exclusion chromatography, affinity and immunoaffinity chromatography, and perfusion chromatography all provide characteristic information about the peptide or protein involved. This tendency to diverge has made chromatography a highly flexible tool for monitoring different aspects of proteins. Equally important for optimization of the separation process is the further improvement of detectors. The reader is therefore informed about, for example, "the state of the art" in hyphenated high-performance liquid chromatography–mass spectrometry configurations (Chapters 4 and 7).

A new branch on the tree of protein characterization methodologies is capillary electrophoresis (Chapter 8). The potential of capillary electrophoresis for the characterization of peptides and proteins has not yet been fully explored, but it is clear that upon "maturation" of the technique in the future, capillary electrophoresis will be recognized as extremely important because it provides data complementary to other approaches.

We would be remiss not to include a chapter about the impact of molecular biology on the characterization of pharmaceutical proteins. First, a number of such proteins owe their existence to recombinant DNA technology and pose a unique set of problems in purification and characterization, particularly in cases where exogenous material of microbial origin has to be removed in order to make a safe product. Second, in development and formulation of recombinant proteins, it is now possible to "go back to the laboratory" to engineer the protein in such a way as to improve its physical properties. An overview of the present state of the art of molecular biology and its application to several specific problems in pharmaceutical biotechnology is presented in Chapter 9.

After reading through the chapters in this volume, it is clear that even a combination of the above-described physical techniques—a concerted approach—does not necessarily guarantee that information about structural stability and reproducibility will be completely accurate. Progress in these techniques should help reduce the level of uncertainty in years to come, but the challenges of obtaining a full physicochemical description of these macromolecules may not be met in the foreseeable future. This means that, in principle, we can be confronted and confounded by unexpected behavior in biological systems, such as changes in immunogenic behavior. In practice, however, employment of the concerted physical approach, together with a thorough chemical examination of the macromolecule, has worked well so far for pharmaceutical proteins used in therapy and as vaccines. This underlines the importance of having a good conceptual understanding of the pros and cons and the potentials and limitations of these techniques. Therefore, the purpose of the present volume is to disseminate existing knowledge, to critically review current progress, and to identify future trends and needs.

James N. Herron
Wim Jiskoot
Daan J. A. Crommelin

REFERENCES

Drenth, J., 1994, *Principles of Protein X-ray Crystallography*, Springer-Verlag, New York.
Ducruix, A., and Giegé, R. (eds.), 1992, *Crystallization of Nucleic Acids and Proteins*, Oxford University Press, New York.
McRee, D. E., 1994, *Practical Protein Crystallography*, Academic Press, New York.
Rhodes, G., 1993, *Crystallography Made Crystal Clear*, Academic Press, New York.

Contents

Chapter 1

Application of Fluorescence Spectroscopy for Determining the Structure and Function of Proteins

Wim Jiskoot, Vladimir Hlady, John J. Naleway, and James N. Herron

Chapter 2

Structural Information on Proteins from Circular Dichroism Spectroscopy: Possibilities and Limitations

Michael Bloemendal and W. Curtis Johnson, Jr.

Chapter 3

Fourier Transform Infrared Spectroscopy Investigations of Protein Structure

E. A. Cooper and K. Knutson

Chapter 4

Mass Spectrometry in Protein Structural Analysis

Peter Roepstorff

Chapter 5

Two-, Three-, and Four-Dimensional Nuclear Magnetic Resonance Spectroscopy of Protein Pharmaceuticals

David G. Vander Velde, James Matsuura, and Mark C. Manning

Chapter 6

Thermodynamic Strategies for Rational Protein and Drug Design

Kenneth P. Murphy and Ernesto Freire

Chapter 7

Chromatographic Techniques for the Characterization of Proteins

Joost J. M. Holthuis and Reinoud J. Driebergen

1

Application of Fluorescence Spectroscopy for Determining the Structure and Function of Proteins

Wim Jiskoot, Vladimir Hlady, John J. Naleway, and James N. Herron

I. INTRODUCTION

The ideal technique to investigate the structure of macromolecules would feature both atomic-level resolution and high sensitivity and would be able to provide this information dynamically. Unfortunately, none of the present-day techniques satisfy all of these criteria simultaneously. For instance, while X-ray diffraction can provide atomic-level information, it requires milligram quantities of material and produces (in most cases) only a time-averaged structure. Likewise, nuclear magnetic resonance (NMR) spectroscopy requires milligram quantities of protein and provides only an incomplete picture of the macromolecule. Furthermore, only relatively small macromolecules ($\leqslant$ 20,000 mol. wt.) can presently be resolved.

Wim Jiskoot and James N. Herron • Department of Pharmaceutics and Pharmaceutical Chemistry, University of Utah, Salt Lake City, Utah 84112. *Vladmir Hlady* • Department of Bioengineering, University of Utah, Salt Lake City, Utah 84112. *John J. Naleway* • Marker Gene Technologies, University of Oregon, Eugene, Oregon 97403. *Present address of W. J.*: Laboratory for Product and Process Development, National Institute of Public Health and Environmental Protection, 3720 BA Bilthoven, The Netherlands.

Physical Methods to Characterize Pharmaceutical Proteins, edited by James N. Herron *et al.*, Plenum Press, New York, 1995.

On the other hand, NMR spectroscopy can provide information in the time domain, such as proton exchange rates and the vibrational motion of atoms.

The topic of this chapter is fluorescence spectroscopy, a technique that is very sensitive (picogram quantities of material can be detected) and well-suited for measurements in the time domain. Its principal shortcoming, however, is that only the structure of the fluorescent probe and the immediate environment is reported. Nonetheless, fluorescence spectroscopy has found wide use in studying the physicochemical properties of proteins, protein–ligand interactions, and protein dynamics. This is because almost all proteins contain naturally fluorescent amino acid residues such as tyrosine and tryptophan. In addition, a large number of fluorescent dyes have been developed that can be used to specifically probe the function and/or structure of macromolecules. In cases where only limited quantities of protein are available (e.g., a recently expressed recombinant protein or a precious protein pharmaceutical), fluorescence spectroscopy is often the method of choice for studying properties such as stability, hydrodynamics, kinetics, or ligand binding, because of its exquisite sensitivity. Even in cases where the structure of the protein is well-known (either from X-ray diffraction or NMR), time-resolved fluorescence spectroscopy can be particularly useful for investigating the dynamic behavior of the macromolecule.

No discussion of modern fluorescence spectroscopy would be complete without mentioning Prof. Gregorio Weber (University of Illinois at Urbana-Champaign), whose pioneering work over the past four decades laid the foundation for using fluorescence techniques to investigate the structure, function, and dynamics of macromolecules. Prof. Weber was the first to recognize that the three factors mentioned above (sensitivity, capacity for time-resolved measurements, and intrinsic protein fluorescence) offered unique opportunities for probing the public lives* of proteins. This is a critical distinction between fluorescence and other techniques such as NMR and infrared spectroscopy, which focus more on the protein's private life. Because proteins are active elements in almost all dynamic processes within an organism, it is important to study their interactions with other biomolecules, i.e., their public life. This is especially true in pharmaceutical applications where almost all relevant proteins are either ligand-binding proteins (e.g., monoclonal antibodies) or enzymes [e.g., tissue plasminogen activator (tPA)].

Most readers will be familiar with ultraviolet/visible absorption spectra,

*When discussing protein dynamics, one can refer to both the private and public life of a protein. The private life includes events that are confined to the protein itself, such as its vibrational and rotational modes. These motions typically take place on the picosecond time scale, which is too short to allow the protein to interact with other molecules. In contrast, the public life includes events that occur over a much longer time scale (nanoseconds to seconds). Such events include interactions of the protein with other molecules in its environment (e.g., ligands, substrates, other proteins) and are often diffusion controlled.

which are often used both qualitatively and quantitatively to analyze proteins, nucleic acids, drugs, and other biomolecules. Such spectra are based on the absorption of light (photons) by molecules, a process in which an electron is excited from the singlet ground state (S^0) to the first singlet excited state (S^1).* As shown in Fig. 1, the transition from ground to excited state is almost instantaneous (10^{-15} sec) and does not allow time for atomic nuclei to move during excitation (Franck-Condon principle). Depending on the configuration of the molecule at the moment of excitation, an electron can populate any one of several different vibrational levels in the S^1 state. Thus, the shape of the absorption spectrum (in which a multitude of electrons are excited) reflects the distribution of electrons in the vibrational levels of the excited state. However, since the higher vibrational levels exhibit energies far in excess of kT, electrons that initially reside in these levels quickly (10^{-12} sec) relax down to the lowest vibrational level of the S^1 state. This process is called internal conversion or thermalization.

Once the lowest vibrational level of the S^1 state is reached, one of several competing processes can ensue. These are shown in Fig. 1 and include: (1) decay to the S^0 state by emission of a photon (fluorescence); (2) decay to the S^0 state by thermal relaxation, quenching, or other nonradiative events (nonradiative processes); and (3) "intersystem crossing" to the first excited triplet state (T^1), followed by decay back to the S^0 state (phosphorescence). Because both fluorescence and phosphorescence result in the emission of a photon, they are collectively referred to as *luminescence*. From an experimental perspective, the main difference between the two resides in their relative emission rates. In fluorescence, the transition from S^1 to S^0 is allowed because the electrons in these states have opposite or "paired" spins. Thus, fluorescence emission is generally shorted-lived (10^{-9} to 10^{-7} sec). In contrast the transition from T^1 to S^0 is forbidden because the electrons in these states have the same spins. As a consequence, phosphorescence emission occurs over much longer times ($\geqslant 10^{-4}$ sec).

As with absorption, fluorescence (or phosphorescence) emission can result in the population of any one of several vibrational levels in the S^0 state. Thus, an emission spectrum reflects the distribution of electrons in the vibrational levels of the ground state. After the S^1 (or T^1) to S^0 transition occurs, electrons in the higher vibrational levels relax back down to the lowest vibrational level, again by internal conversion. In most cases, molecules that are excited return to the same electronic state that they started from; thus, the fluorophore is not consumed in the process. In practice, this is true to a first approximation, although certain fluorophores possess excited sites that can participate in photochemical reactions. In such cases, the fluorophore is converted into photoproducts that may decay radiatively (with a different emission spectrum) or nonradiatively (referred to as "photobleaching").

*In a singlet-to-singlet transition, the spin of the electron does not change. Thus, the electron in the S^1 excited state has a spin opposite to that of its counterpart in the S^0 ground state.

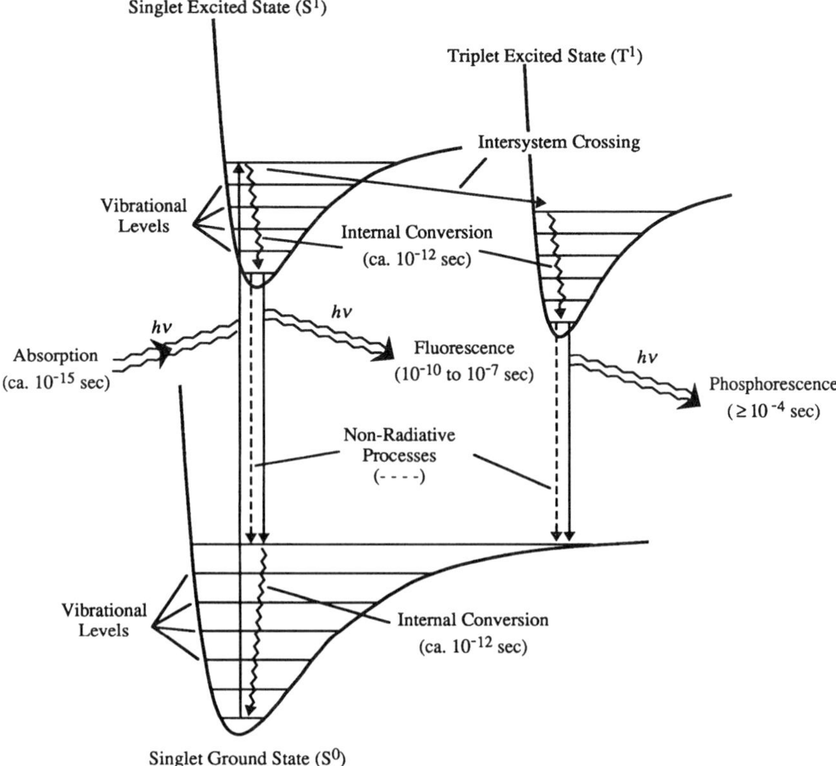

Figure 1. Physical basis of fluorescence and other related phenomena. Many biomolecules and drugs contain π and/or nonbonded electrons that can be excited by light in the ultraviolet or visible regions of the spectrum. When such a molecule is irradiated by light of an appropriate wavelength, an electron is excited from the singlet ground state (S^0) to the first singlet excited state (S^1), a process that is called absorption. Usually one of the upper vibrational levels of the S^1 state is excited rather than the lowest-lying level, and the electron relaxes down to the lowest vibrational level within a few picoseconds of the absorption event; this process is called internal conversion. Once the lowest vibrational level of the S^1 state is reached, one of several competing processes can ensue. These include: (1) decay to the S^0 state by emission of a photon (fluorescence); (2) decay to the S^0 state by nonradiative processes; and (3) "intersystem crossing" to the first excited triplet state (T^1), followed by decay back to the S^0 state (phosphorescence).

In solution, fluorescence emission invariably occurs at a longer, less energetic wavelength than absorption. This red shift in wavelength was first observed by Stokes (1852), and in his honor is referred to as the "Stokes' shift." The decrease in energy can best be understood by comparing the lengths of the absorption and fluorescence emission vectors in Fig. 1. The absorption vector is the longer of the two, denoting a more energetic transition. The emission vector is shorter because a portion of the absorbed energy is converted to waste heat (by internal conversion) in both the S^1 and S^0 states. Finally, there is a third factor (solvent relaxation) that can contribute to the Stokes' shift. In this case, a portion of the absorbed energy is transferred to the surrounding solvent. This phenomenon will be discussed in Section 2.1.4.

With the exception of monoclonal antibodies, the literature is not replete with examples of fluorescence techniques used to study proteins of pharmaceutical interest. However, numerous fluorescence studies do exist for proteins in general. Thus, it is our hope and expectation that the reader will use the information and case studies presented in this chapter as examples of how to apply fluorescence spectroscopy to examine pharmaceutical proteins. To this end we will examine the following three topics in the remainder of this chapter: (1) fluorescence techniques used for examining proteins; (2) intrinsic protein fluorescence and the use of extrinsic fluorescent labels; and (3) applications of fluorescence spectroscopy in studying the structure and function of proteins. These will be discussed in Sections 2, 3, and 4, respectively, followed by a brief set of conclusions in Section 5. It is not our intent to provide a comprehensive review of each topic; rather, we hope to provide the reader with a general knowledge of the field and inform him or her of the options that are available. For more information, the reader is referred to the numerous research articles and reviews that are cited in the text.

2. TECHNIQUES

Although the Stokes' shift was first described almost 150 years ago, and steady-state fluorescence techniques have been applied to studying proteins for almost 50 years, fluorescence spectroscopy is still a rapidly growing field. For example, time-resolved fluorescence methodology has only come into its own in the last decade. In this section, we will discuss the fluorescence techniques (both old and new) most often employed for studying proteins. These include: (1) spectral measurements; (2) quantum yield of fluorescence; (3) lifetime measurements; (4) fluorescence quenching; (5) anisotropy; and (6) interfacial (or evanescent) fluorescence.

2.1. Spectral Measurements

2.1.1. INTRODUCTION

No matter which fluorescence technique is employed in a given set of experiments, it is advisable to perform a spectral analysis of the fluorophore before embarking on more complicated experiments. If nothing else, this will reveal the optimal wavelengths for excitation and emission and will often provide a wealth of additional information. Three types of spectra—absorption, excitation, and emission—are typically used for planning fluorescence studies and interpreting the results. They will be discussed in Sections 2.1.2, 2.1.3, and 2.1.4, respectively.

2.1.2. ABSORPTION SPECTRA

Absorption spectra are often used to determine the composition and concentration of samples containing proteins and other biomolecules such as substrates, ligands, and drugs. An absorption spectrum is obtained by measuring absorbance as a function of wavelength. The absorbance (also known as "optical density") at a given wavelength (A) is defined as:

$$A = \log\left(\frac{I_0}{I}\right) \tag{1}$$

where I_0 and I are the intensities of the incident and transmitted light, respectively. The relationship between absorbance and concentration (c) is given by Beer's law:

$$A = \varepsilon c l \tag{2}$$

where ε and l are the extinction coefficient and path length (of the cuvette), respectively. Because of the logarithmic relationship between absorbance and transmitted light, concentrations are most accurately determined from absorbance values between 0.05 and 2.0. For absorbance values of less than 0.05, there is less than a 10% difference between the incident (I_0) and transmitted (I) light, and depending on instrument stability, such absorbance measurements may be relatively error prone. At the other extreme, less than 1% of the incident light is transmitted by the chromophore for absorbance values greater than 2. In this case, the signal-to-noise ratio is poor, and once again absorbance measurements are relatively error prone.

A typical absorption spectrum of a visible fluorophore (fluorescein) is shown in Fig. 2. Notice that there are several weak absorption bands in the ultraviolet region and then a major transition in the visible region at 490 nm. Such a spectrum is very useful for selecting the optimum excitation wavelength. This is especially important in anisotropy studies where the maximum possible anisotropy (r_0) varies with excitation wavelength (see Section 2.5). In the above example, excitation of the fluorophore at 490 nm would insure both maximum fluorescence intensity and maximum anisotropy. A further consideration in any type of fluorescence study is the optical density of the longest wavelength absorption peak. This peak is immediately adjacent to the fluorophore's emission peak and depending on the Stokes' shift, the two can overlap to some degree. Thus, at high concentrations

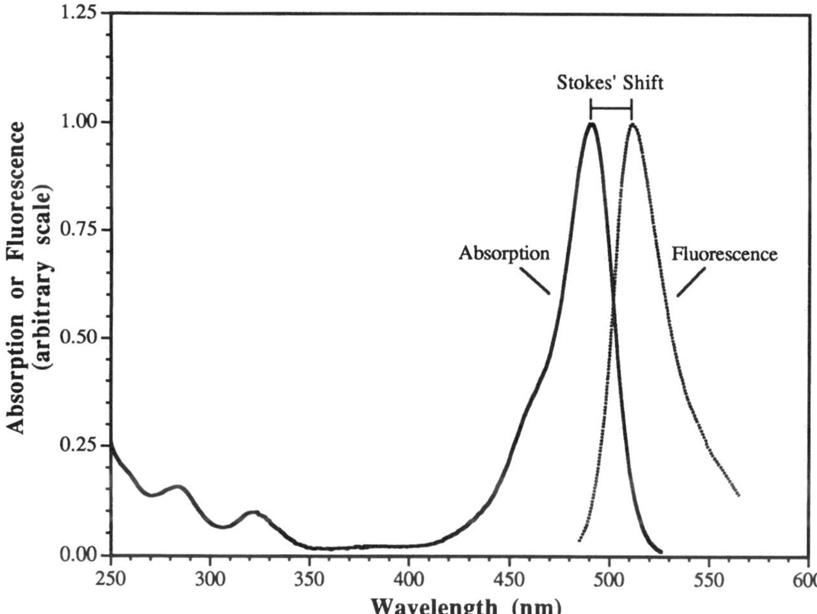

Figure 2. Absorption and fluorescence emission spectra (normalized) of fluorescein, a fluorescent dye often used to label both proteins and cells. Although the absorption spectrum contains several peaks in the ultraviolet region, the emission spectrum is the "mirror image" of the longest wavelength (lowest energy) absorption transition. The emission maximum is displaced by 25 nm to the red from the absorption maximum. This displacement is known as the Stokes' shift and is due to internal conversion and solvent relaxation effects.

fluorophores with a small Stokes' shift can reabsorb a significant amount of their own fluorescence. This phenomenon is called *trivial reabsorption*. A good rule of thumb is that the optical density of the longest wavelength absorption peak should be less than 0.1 or so. At this level, even in the worst case, less than 20% of the fluorophore's emission will be reabsorbed.

2.1.3. EXCITATION SPECTRA

An excitation spectrum is obtained by measuring the fluorescence emission of a fluorophore at a fixed wavelength, while varying the excitation wavelength. The number of photons absorbed will be proportional to the extinction coefficient of the fluorophore at each wavelength. Because of internal conversion, however, these photons will be reemitted at the fixed emission wavelength rather than at the excitation wavelength. For a given fluorophore, a plot of fluorescence intensity (measured at the fixed emission wavelength) versus excitation wavelength will have the same shape or profile as the fluorophore's absorption spectrum, except that the y axis will have an arbitrary scale. The principal advantage of excitation spectra is that they can be obtained for samples of much lower (100- to 1000-fold) concentration than required for absorption measurements. However, due to the arbitrary scale of the y axis, excitation spectra are not really suitable for concentration determinations.

Some care must be taken in obtaining an accurate excitation spectrum because the output of the xenon lamp used in most spectrofluorometers is not constant with wavelength. Thus an uncorrected excitation spectrum (referred to as a *technical excitation spectrum*) will be a convolution of the actual excitation spectrum and the spectral response curve of the lamp. On modern instruments this problem can usually be avoiding by employing an accessory called a quantum counter. The reader is referred to Lakowicz (1983) for a detailed description of quantum counters and their use.

2.1.4. EMISSION SPECTRA

An emission spectrum is obtained by plotting fluorescence intensity versus wavelength. The intensity axis is typically plotted on an arbitrary scale because the absolute value of the intensity can vary with several different factors including the concentration of the fluorophore, its quantum yield (see Section 2.2), the optical geometry of the spectrofluorometer, and the type of photoelectric device used to detect the fluorescence. As shown in Fig. 2, the emission spectrum is the mirror image of the longest wavelength (lowest energy) band of the absorption spectrum and usually occurs at a longer wavelength (Stokes' shift). Also, the position and shape of an emission spectrum will not change with excitation wavelength. This is due to internal conversion: no matter which electronic transition or vibrational level is excited, the electron will quickly relax (10^{-12} sec) down to the lowest vibrational level of the lowest lying S^1 state before emitting a photon.

As mentioned in Section 1, the magnitude of the Stokes' shift is determined by three factors: the degree of internal conversion in the S^0 and S^1 states and the extent of solvent relaxation effects. Although the internal conversion terms will be constant for a given fluorophore at constant excitation wavelength and temperature, the stabilization of the excited state due to solvent relaxation can significantly alter the Stokes' shift of the fluorophore. An extreme example of this is the fluorescence probe 6-propionyl-2-(dimethylamino)naphthalene. The emission maximum of this probe shifts from 392 nm in cyclohexane to 523 nm in water (Weber and Farris, 1979; MacGregor and Weber, 1981). Solvent relaxation effects also have major implications for intrinsic protein fluorescence, as will be seen in Section 3.1.2.

Because water is the solvent of choice for most biochemical reactions, we will limit our discussion to relaxation effects that occur in polar (high dielectric constant) solvents; the reader is referred to MacGregor and Weber (1981) or Lakowicz (1983) for a more complete discussion of solvent effects. The mechanism of dipolar solvent relaxation is shown in Fig. 3. Aromatic molecules, including most fluorophores, often exhibit a greater dipole moment in the excited state than in the ground state (MacGregor and Weber, 1981; Lakowicz, 1983). Due to the Franck-Condon principle, the dipole moment of the fluorophore changes almost instantaneously (10^{-15} sec) upon excitation, but the atomic nuclei of both the fluorophore and the solvent will remain frozen. This results in unfavorable dipole–dipole interactions with surrounding solvent molecules. Because solvent relaxation occurs much faster (10^{-11} sec) than fluorescence emission (10^{-9} to 10^{-8} sec), there is ample opportunity while the fluorophore is in the excited state for

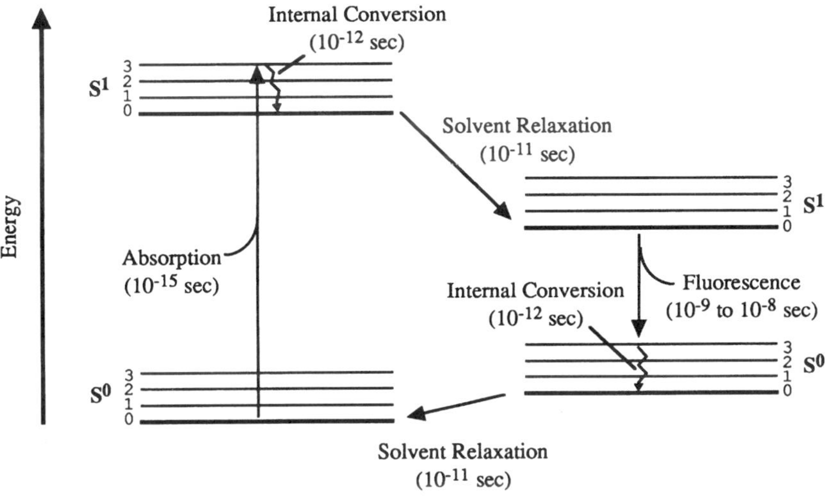

Figure 3. Jablonski diagram of solvent relaxation effects. Fluorophores often exhibit larger dipole moments in the excited state than in the ground state. In such cases dipolar solvent molecules (e.g., water) can find themselves in an energetically unfavorable configuration immediately after absorption. This is because excitation is significantly faster than solvent relaxation (10^{-15} sec vs. 10^{-11} sec) and solvent molecules cannot instantaneously realign themselves to the excited state dipole moment. However, the relaxation rate for nonviscous solvents is usually faster than fluorescence emission, so solvent molecules can reorient themselves into a more favorable configuration before emission occurs. Immediately after the fluorophore returns to the ground state, solvent molecules again find themselves in an unfavorable configuration (this time with respect to the ground state dipole moment) and a second solvent relaxation step ensues. The two relaxation steps have the effect of reducing the emission energy gap between the S^1 and S^0 states, which produces a red-shift in the emission maximum of the fluorophore relative to the same molecule in a nonpolar (or viscous) solvent (see Fig. 4).

solvent molecules to reorient themselves into a more energetically favorable configuration. This produces a red shift in the emission maximum relative to the same fluorophore in a nonpolar solvent, where dipolar relaxation effects do not occur (see Fig. 4).

Most emission spectra reported in the literature are uncorrected or "technical" spectra. As such they are the convolution of the actual emission spectrum and the spectral response curves of the emission monochromator and the photodetector (generally a photomultiplier tube). Although these response curves do not vary significantly over the limited wavelength range employed in most emission spectra, exacting applications such as determination of spectra shifts or relative quantum yields (see Section 2.2) require corrected emission spectra. A technical spectrum can be corrected by dividing the observed fluorescence intensity (at each wavelength) by a correction factor that takes both spectral response curves into account. Such correction factors are usually supplied with each spectrofluorometer by the manufacturer. The reader is referred to Lakowicz (1983) for more information about corrected emission spectra and correction factors.

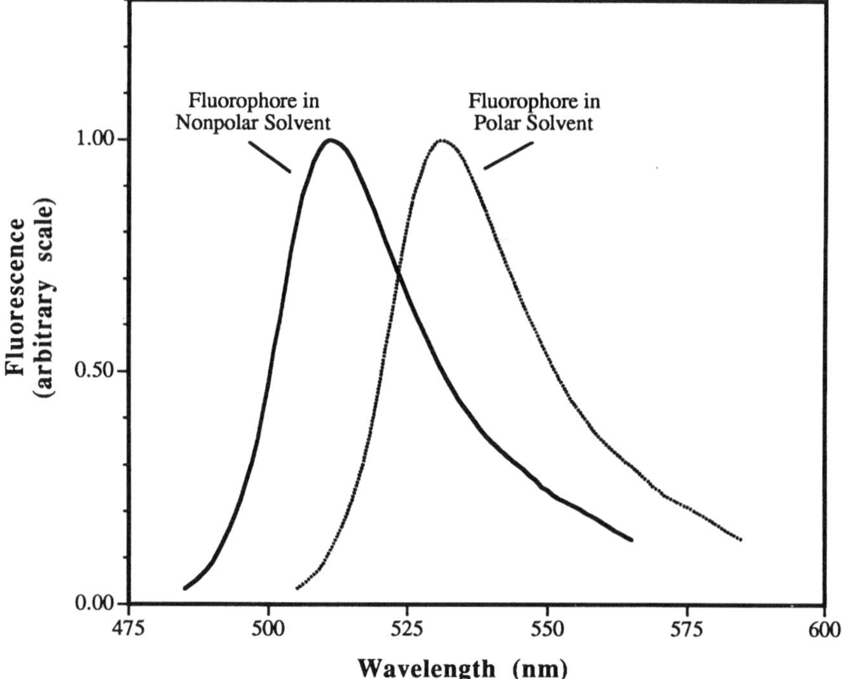

Figure 4. Effects of solvent relaxation on the emission spectrum of a fluorophore. See Fig. 3 for explanation.

2.2. Quantum Yield of Fluorescence

The quantum yield of fluorescence (Q) of a fluorophore is defined as the ratio of its rate of fluorescence emission (Γ) to the total rate that its S^1 state is depopulated (including both radiative and nonradiative processes):

$$Q = \frac{\Gamma}{\Gamma + \sum_i k_i} \tag{3}$$

where $\sum k_i$ is the sum of the different nonradiative decay rates. The quantum yield can also be viewed as the ratio of photons emitted by the fluorophore to those absorbed. Quantum yields can vary from zero for nonfluorescent compounds to close to unity for highly fluorescent compounds such as fluorescein or rhodamine. Proteins typically exhibit quantum yields between 0.05 and 0.3, although in some cases their fluorescence can be almost totally quenched by a prosthetic group (e.g., myoglobin and hemoglobin).

The absolute quantum yield is nontrivial to measure and the reader should refer to the literature for quantum yield values of their particular fluorophore (Weber and Teale, 1957; Seybold *et al.*, 1969). In contrast, it is relatively easy to measure relative changes in quantum yield. This can be done by recording the emission spectra (be sure to use corrected rather than technical emission spectra) of the fluorophore under two different conditions and then taking the ratio of the integrals (area under the curve) of the two spectra. Relative quantum yield values are useful for characterizing reactions or interactions involving proteins (e.g., protein folding, ligand binding, and enzyme kinetics) in which the fluorescence intensity is affected by the interaction (see Section 4).

2.3. Fluorescence Lifetimes

2.3.1. INTRODUCTION

The fluorescence lifetime (τ) is defined as the average time that a fluorophore remains in the excited state, and like quantum yield is a function of the rate of fluorescence emission (Γ) and the sum of nonradiative decays $\sum k_i$:

$$\tau = \frac{1}{\Gamma + \sum_i k_i} \tag{4}$$

Fluorescence lifetimes are typically in the nanosecond time scale, but can range from a few picoseconds to hundreds of nanoseconds. The fluorescence lifetime provides a time window through which dynamic processes in the microenvironment of individual fluorophores within a protein can be observed. Time-resolved measurements thus reveal rate processes that occur during the lifetime of the excited state. Examples of such excited-state reactions are collisional quenching (Section 2.4), fluctuations in the tryptophan environment of proteins (Section 3.1.3), rotations of proteins and segments thereof (Sections 2.5 and 4.2), and reorientation of solvent or other polar compounds around the dipole moment of the excited state.

In case of a single fluorescence lifetime the decay function is given by the following equation:

$$I(t) = I_0 e^{-t/\tau} \tag{5}$$

where $I(t)$ is the fluorescence intensity at time t, I_0 is the initial fluorescence intensity, and τ is the fluorescence lifetime. In many cases, however, the fluorescence decay is more complex. Essentially all fluorescence decay profiles can be fitted by nonlinear least-squares methods with a sum of single exponential decays according to:

$$I(t) = I_0 \sum_i \alpha_i e^{-t/\tau_i} \tag{6}$$

where α_i is the amplitude (preexponential) of each τ_i. However, this does not prove that the observed lifetimes and preexponentials have a physical meaning. In some cases nonexponential decays or lifetime distributions are more likely to describe the underlying physical processes (Lakowicz, 1983; Beechem and Brand, 1985; Alcala *et al.*, 1987a–c; Eftink, 1991). Distributions of decay times are usually assumed to have Gaussian or Lorentzian shapes (Eftink, 1991). If excited-state reactions occur, data fitting with Eq. (6) may yield negative preexponentials (Beechem and Brand, 1985; Alcala *et al.*, 1987b; Lakowicz, *et al.*, 1987a). In many cases, however, the experimental data, even when they are very accurate and precise, cannot distinguish between different decay models (Ludescher *et al.*, 1985; Alcala *et al.*, 1987a,c; Lakowicz *et al.*, 1987a; Fernando and Royer, 1992; Mei *et al.*, 1992). In general, any mathematical analysis of fluorescence decay data only yields an empirical description and does not prove the validity of a model. In some cases the simultaneous analysis of different data sets, called *global analysis*, may aid in determining the most likely decay parameters (Knutson *et al.*, 1983; Beechem *et al.*, 1985; Beechem and Gratton, 1988; Beechem, 1992).

2.3.2. INSTRUMENTATION

Fluorescence lifetimes can be measured with the pulse method (time-domain measurements) or with the harmonic response or phase-modulation method (frequency-domain measurements), the principles of which will be briefly discussed below. Detailed information about the principles and technical aspects can be found in the literature for the pulse (Demas, 1983; Lakowicz, 1983; O'Connor and Phillips, 1984; Canonica and Wild, 1985; Phillips *et al.*, 1985) and the phase-modulation method (Spencer and Weber, 1979; Gratton and Limkeman, 1983; Lakowicz, 1983, 1986; Gratton *et al.*, 1984a; Jameson *et al.*, 1984a; Lakowicz and Maliwal, 1985; Lakowicz *et al.*, 1985).

In the pulse method the sample is excited by a brief pulse of light and the emission decay is directly obtained after correction for the length of the light pulse. Examples of monoexponential decays are shown in Fig. 5A. The pulse method consists of several different techniques of which time-correlated single photon counting is most widely used (Demas, 1983; Lakowicz, 1983; O'Connor and Phillips, 1984). In this technique the time between the excitation pulse and the first emitted photon is measured; multiple repetition of this procedure yields the fluorescence decay curve.

In the multifrequency phase-modulation method the sample is excited with a sinusoidally modulated light source. A xenon arc lamp is often used for excitation of intrinsic protein fluorescence and an argon-ion laser for midultraviolet and visible excitation. The light is typically modulated over a range of 1 to 250 MHz, although higher frequencies can be obtained with special equipment. The photodetector is also modulated at the same frequency, but the fluorescence lifetime will cause both a delay in the phase and a decrease in the modulation of the fluorescence emission. The magnitude of these changes depends on both the lifetime of the fluorophore and the modulation frequency. The light source is modulated either by an electrooptical modulator driven by a radio frequency amplifier or by using the harmonic content of mode-locked laser pulses. Figure 5B shows typical phase and modulation data as a function of modulation frequency and fluorescence lifetime.

The time resolution and the precision of the pulse and harmonic methods are comparable and have been considerably improved during the last decade through the introduction of modern high-repetition laser light sources and microchannel plate detectors; the measurable lifetime range has been extended to the picosecond region (Alcala *et al.*, 1985; Canonica and Wild, 1985; Phillips *et al.*, 1985; Lakowicz *et al.*, 1986; Laczko and Lakowicz, 1989).

2.3.3. TIME-RESOLVED SPECTRAL MEASUREMENTS

Measurements of time-resolved emission spectra can be useful when the emission maximum is red-shifted during the decay process, due to excited-state reactions such as dipolar relaxation. In such cases the lifetime is emission wavelength-dependent. Alternatively, the emission spectra of individual lifetime components in a mixture may be different, the longer lifetime components generally being red-shifted, which can be resolved by variable-wavelength lifetime

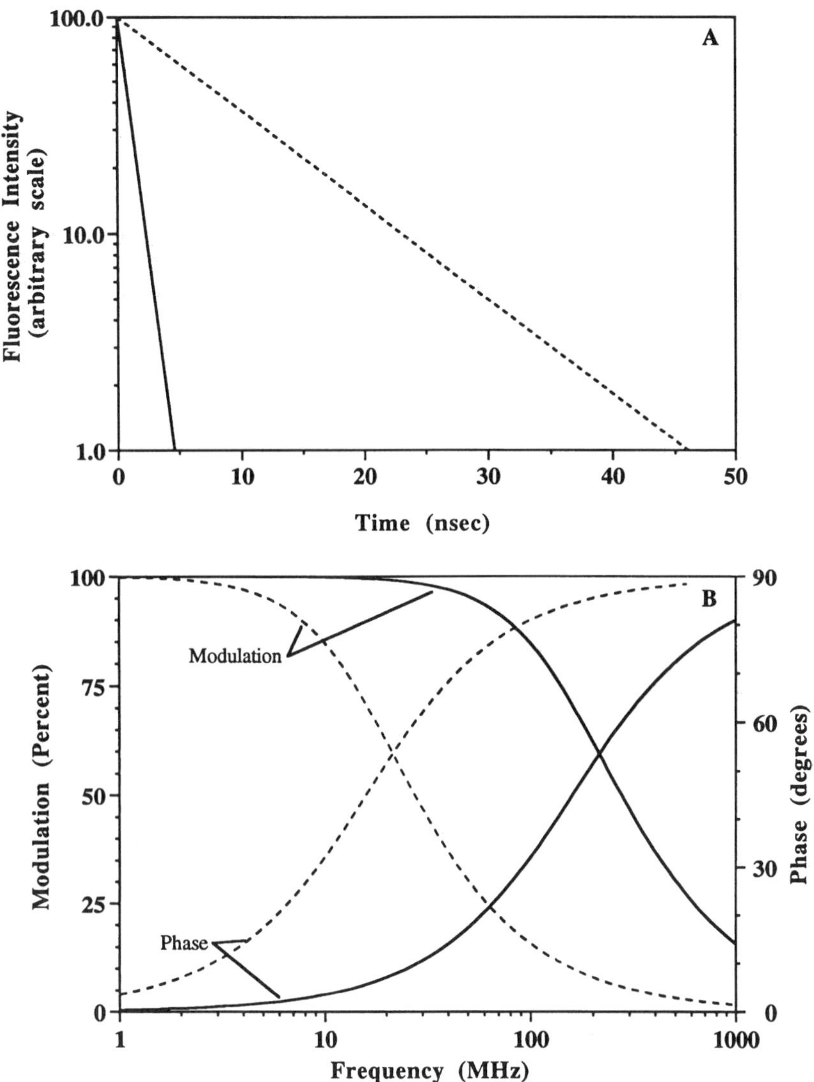

Figure 5. Theoretical curves for monoexponential fluorescence intensity decay. (A) Measurements in the time domain. (B) Measurements in the frequency domain. In both cases solid lines represent a fluorescence lifetime of 1 nsec and dashed lines represent a lifetime of 10 nsec.

measurements. For a description of time-resolved emission methodology, the reader is referred to the literature (Lakowicz, 1983, 1986; Gratton *et al.*, 1984a; Jameson *et al.*, 1984a; Gratton and Jameson, 1985; Eftink, 1991; Jameson and Hazlett, 1991).

2.4. Quenching of Fluorescence

Fluorescence quenching can be due to any number of processes that reduce the intensity of emission. Such processes include resonance energy transfer (see Section 2.6), excited-state reactions, complex formation, and collisions with quenchers. However, only the last two of these are generally included in the classical definition of fluorescence quenching. Complex formation occurs in the ground state and typically the complex is nonfluorescent. Such quenching is referred to as "static quenching" because the lifetime of the fluorophore is unaffected although its fluorescence intensity is reduced. In contrast, collisional quenching is an excited-state phenomenon in which both the lifetime and intensity are reduced in equal proportion. This type of quenching is referred to as "dynamic quenching" and is due to the collision of quencher and fluorophore during the excited state lifetime of the latter. As a consequence the quencher concentration must be relatively high (ca. 0.1 M) for dynamic quenching to occur in bulk solution.

Dynamic quenching of fluorescence was first described more than 75 years ago by Stern and Volmer (1919). The relationship between fluorescence intensity (F) and quencher concentration (Q) is known as the Stern–Volmer equation:

$$\frac{F_0}{F} = 1 + k_q\tau_0[Q] = 1 + K_Q[Q] \tag{7}$$

where F_0 and τ_0 are the fluorescence intensity and lifetime in the absence of quencher, respectively, and k_q is the bimolecular quenching rate constant. Further, it is customary to define the product of k_q and τ_0 as K_Q, the Stern–Volmer quenching constant, as shown in the right side of Eq. (7). As mentioned above, both fluorescence intensity and lifetime decrease in equal proportion in dynamic quenching; so τ_0/τ may be substituted for F_0/F in the above equation, yielding the following expression for lifetime as a function of quencher concentration:

$$\frac{\tau_0}{\tau} = 1 + k_q\tau_0[Q] = 1 + K_Q[Q] \tag{8}$$

For static quenching, the dependence of fluorescence intensity on quencher concentration is given by:

$$\frac{F_0}{F} = 1 + K_S[Q] \tag{9}$$

where K_S is the equilibrium constant (in units of M^{-1}) for formation of the static complex. In this case, however, the fluorescence lifetime is unaffected by formation of such complexes, so the τ_0/τ ratio does not vary with quencher concentration:

$$\frac{\tau_0}{\tau} = 1 \tag{10}$$

In many cases both dynamic and static quenching occur simultaneously, and intensity quenching can be described by the following quadratic equation:

$$\frac{F_0}{F} = (1 + K_Q[Q])(1 + K_S[Q]) \tag{11}$$

Interestingly, because the fluorescence lifetime is unaffected by static quenching, the τ_0/τ ratio will still be a linear function of quencher concentration [as described by Eq. (8)] even though both dynamic and static components are present. In fact, this provides a convenient means to distinguish between the two types of quenching. If the intensity quenching plot (F_0/F vs. $[Q]$) curves upward, while the lifetime quenching plot (τ_0/τ vs. $[Q]$) is linear, then both dynamic and static components are probably present.

In some cases two (or more) classes of fluorophores may exist—one accessible to collisional quenchers and one inaccessible. Intensity quenching plots in such cases will curve downward, instead of upward as observed for static quenching. This situation often occurs in multitryptophan proteins, where some tryptophan residues are buried in the interior of the protein and thus inaccessible to collisional quenchers, while others are located near the surface of the protein and will be accessible. Lehrer (1971) proposed the following modification of the Stern–Volmer equation for analyzing cases in which one class of fluorophores is inaccessible:

$$\frac{F_0}{\Delta F} = \frac{1}{f_a K_Q[Q]} + \frac{1}{f_a} \tag{12}$$

where $\Delta F = F_0 - F$ and f_a is the fraction of accessible fluorophores. The Stern–Volmer quenching constant (K_Q) in this equation is the average value of the individual quenching constants of all accessible fluorophores. If these fluorophores have either identical quenching constants or a unimodal distribution of them, then Eq. (12) is an appropriate model. For bi- or multimodal distributions, however, a more complex analysis is required (see reviews by Eftink and Ghiron, 1981; Laws and Contino, 1992).

Lakowicz (1983) lists a number of different compounds that are effective collisional quenchers. However, only a subset of these are typically used in protein studies. Eftink and Ghiron (1981) divided this subset into the following three categories based on charge: anionic quenchers (I^-, Br^-), cationic quenchers (pyridinium, imidazolium, Eu^{3+}, Ag^+, Cs^+), and neutral quenchers (acrylamide, dichloroacetamide, hydrogen peroxide, methionine, oxygen, pyridine, succinimide, trichloroethanol). The ionic quenchers are thought to only interact with fluorophores at or near the surface of the protein and thus are useful for evaluating the solvent accessibility of tryptophan and tyrosine residues (Eftink and Ghiron, 1981). Iodide is the most common of these and was employed in Lehrer's original accessibility studies (Lehrer, 1971). In the last decade, however, acrylamide has gained favor in such studies because being a neutral quencher, its quenching rate constant (k_q) is not affected by the surface charge of the protein (Eftink and Ghiron, 1981; Blatt *et al.*, 1986; Calhoun *et al.*, 1986). Finally, of all the quenchers only oxygen is able to diffuse into the interior of a protein and quench-buried fluorophores (Lakowicz and Weber, 1973a,b; Gratton *et al.*, 1984b; Jameson *et al.*, 1984b). Thus, it is ideal for studying protein dynamics, as will be discussed in Section 4.2. On the other hand, oxygen-quenching studies require special equipment that is not readily available to the average researcher.

2.5. Anisotropy

2.5.1. INTRODUCTION

Fluorescence anisotropy measurements have numerous applications in studies of protein structure, function, and dynamics. In such measurements a fluorophore is excited with plane polarized light and the polarization of its fluorescence emission is determined. Typically, a decrease in polarization is observed, which is usually due to rotational motion of the fluorophore but can also result from resonance energy transfer (see Section 2.6). A more rigorous physical description is given in Section 2.5.2. It should be mentioned that the terms *anisotropy* and *polarization* are often used interchangeably in the literature. In fact, the two are actually interrelated parameters as can be seen in Eqs. (13) and (14), below. Experimental determinations of anisotropy fall into two general categories—steady-state and dynamic—depending on the instrumentation employed. In steady-state measurements one determines the time-averaged polarization of a fluorophore, while in dynamic measurements the change in polarization is monitored over the duration of the fluorescence lifetime. Steady-state measurements are the easiest to perform and provide information about the average rotational diffusion of a fluorophore (or fluorophore–protein conjugate). Dynamic anisotropy measurements require a fluorescence lifetime instrument, but can provide information about both the global and local motions of the fluorophore. Both the theoretical and technical aspects of fluorescence anisotropy have been extensively reviewed (Lakowicz, 1983; Bentley *et al.*, 1985; Bucci and Steiner, 1988; Eftink, 1991; Jameson and Hazlett, 1991). Here we will briefly describe the basic principles of this technique.

2.5.2. STEADY-STATE ANISOTROPY

When a sample is excited with plane polarized light, only those fluorophores that have absorption dipoles oriented parallel to the electric field vector of the incident radiation are excited. This phenomenon is commonly known as *photoselection*. Fluorescence intensity is collected at right angles to the incident beam through a polarizer oriented either parallel ($I_{//}$) or perpendicular ($I_{\perp}$) to the electric field vector of the exciting light. Polarization (p) and anisotropy (r) are defined as:

$$p = \frac{I_{//} - I_{\perp}}{I_{//} + I_{\perp}} \tag{13}$$

$$r = \frac{I_{//} - I_{\perp}}{I_{//} + 2I_{\perp}} \tag{14}$$

The use of anisotropy is preferred because such values are additive, which considerably simplifies theoretical expressions for time-resolved anisotropy and ligand binding.

The limiting or intrinsic anisotropy (r_0) is obtained in the absence of molecular motion and depends on (1) the probability that a fluorophore is excited by a polarized light beam (photoselection) and (2) the angle between the absorption and emission dipoles:

$$r_0 = \tfrac{2}{5}\left(\frac{3\cos^2(\alpha) - 1}{2}\right) \tag{15}$$

The term $\frac{2}{5}$ results from the probability of photoselection and α is the angle between the excitation and the emission dipole. Hence, r_0 can range from 0.4 (α = 0°) to −0.2 (α = 90°). In general, the intrinsic anisotropy r_0 depends on the excitation wavelength. For instance, r_0 of tryptophan shows a strong wavelength dependence in the excitation range of 280 to 300 nm (Lakowicz, 1983).

The observed steady-state anisotropy (r) is a function of r_0, the rotational correlation time (ϕ), and the fluorescence lifetime (τ) of the fluorophore according to the Perrin equation:

$$\frac{r_0}{r} = \left(1 + \frac{\tau}{\phi}\right) \tag{16}$$

It can be seen that r is most sensitive to changes in rotational motion if the values of τ and ϕ are in the same order of magnitude. If $\phi \gg \tau$, no substantial depolarization will be observed because the fluorescence intensity will have decayed before any measurable motion occurs and r will approach r_0; if $\phi \ll \tau$, the orientation of the fluorophore will be random before decay of fluorescence occurs, and r approaches zero.

It should be noted that rotational diffusion is sometimes expressed as the rotational relaxation time (ρ) or as the rotational diffusion coefficient (D_{rot}), which are related to the rotational correlation time as $\rho = 0.5D_{rot}^{-1} = 3\phi$. From the Stokes-Einstein relation of diffusion, the rotational correlation time of a sphere may be described as:

$$\phi = \frac{V\eta}{RT} \tag{17}$$

where V is the hydrated molecular volume of the rotating species, η the viscosity of the solution, R the gas constant, and T the absolute temperature. From Eqs. (16) and (17), it follows that a plot of $1/r$ versus T/η should be linear with the y intercept equal to $1/r_0$; a nonlinear plot indicates the presence of more than one rotational mode.

2.5.3. DYNAMIC ANISOTROPY

Average rotational correlation times can only be calculated from steady-state anisotropy measurements with Eq. (16) if the lifetime is known. Moreover, Eq. (17) is valid only for the isotropic rotation of a spherical particle, but many proteins cannot be considered spherical. The time-dependent or dynamic anisotropy is given by:

$$r(t) = r_0 \sum_i \beta_i e^{-t/\phi_i} \tag{18}$$

where β_i is the amplitude of each ϕ_i. The anisotropic decay of an ellipsoid is described by three rotational correlation times, while as many as five different rotational correlation times may be required to define the anisotropic decay of an asymmetric particle in general (Belford *et al.*, 1972; Lakowicz *et al.*, 1983; Bucci and Steiner, 1988). The amplitudes (β_i) depend on the geometry of the protein and the average orientation of the probe with respect to the protein.

Internal motion is often superimposed on the overall rotation of proteins and other biopolymers. Examples of internal motions are rotation of the (intrinsic or extrinsic) fluorophore about its bond to the main chain, wobbling of the fluorophore within a cone, and segmental motion of a protein domain. These types of internal flexibility, each of which can contribute to the overall depolarization

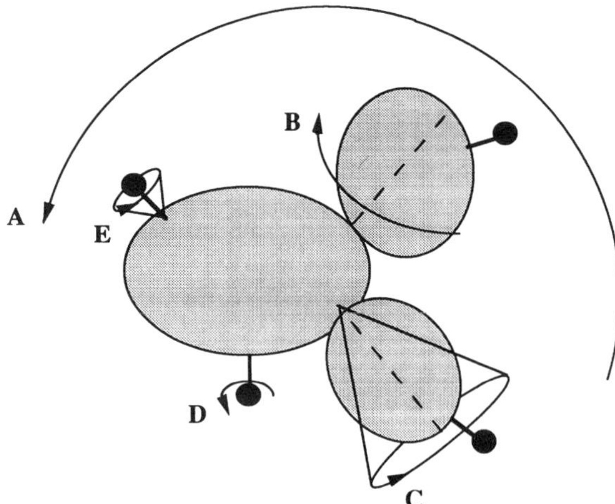

Figure 6. Schematic illustration of possible rotational motions in a hypothetical three-domain protein with four fluorophores (which may be intrinsic or extrinsic): (A) overall rotation of the entire protein; (B) rotation of a domain around a flexible joint; (C) wobble of a domain within a cone; (D) rotation of a fluorophore around its axis; and (E) wobble of a fluorophore within a cone.

of the fluorescence, are schematically shown in Fig. 6. When segmental flexibility is superimposed on the overall rotation, the anisotropy decay can be approximated by:

$$r(t) = r_0[\beta e^{-t/\phi_F} + (1 - \beta)]e^{-t/\phi_P} \tag{19}$$

where β is the amplitude belonging to the rotational correlation time ϕ_F of the segmental motion and ϕ_P is the overall rotational correlation time.

With currently available instrumentation, resolution of two or three rotational modes is considered as the practical limit (Lakowicz *et al.*, 1983; Bucci and Steiner, 1988). Although rotational correlation time is independent of fluorescence lifetime, the two should be in the same order of magnitude for accurate determination of ϕ. Thus, it is often desirable to label large proteins with a long lifetime fluorophore such as dimethylaminonaphthalenesulfonate (DANS) or pyrene (see Section 3.2) in order to examine their rotational motion. On the other hand, intrinsic tryptophan fluorescence can be used to probe the internal motions of proteins (see Section 4.2). The measurement of picosecond motions can be improved by reducing the tryptophan lifetime with collisional quenchers (Lakowicz *et al.*, 1983, 1987b). Most proteins show internal motion of tryptophan residues (for reviews, see Beechem and Brand, 1985; Bucci and Steiner, 1988; Eftink, 1991).

Measurements of dynamic anisotropy can be performed with time-domain or frequency-domain techniques. In the time domain fluorescence intensities of the parallel and perpendicular components are measured as a function of time; while in the frequency domain the phase difference and modulation ratio of the perpendicular and parallel components of the emission are measured at variable frequency from which the anisotropy decay function can be calculated (Lakowicz, 1983, 1986; Gratton *et al.*, 1984a; Jameson and Hazlett, 1991). Examples of theoretical curves for monoexponential anisotropy decays in the time and the frequency domain are shown in Fig. 7.

2.6. Energy Transfer

2.6.1. INTRODUCTION

Fluorescence energy transfer is an excited-state reaction in which the excited-state energy of a fluorophore (referred to as the "donor") is transferred to another molecule (acceptor), which may be either another fluorophore or a chromophore. Energy transfer takes place without the appearance of a photon via a dipole–dipole interaction between the donor and acceptor molecules.

2.6.2. FÖRSTER'S THEORY OF RESONANCE ENERGY TRANSFER

As first shown by Förster (1948), the rate of energy transfer (k_T) is determined by several factors, including: (1) the fluorescence lifetime of the donor (τ_d); (2) the extent of overlap between the emission spectrum of the donor with the absorption spectrum of the acceptor [defined as J, the overlap integral; see Eq. (22), below]; (3) the distance between donor and acceptor (r); and (4) the relative orientation (κ) of this pair. Based on a quantum mechanical treatment of resonance transfer between two well-separated molecules, Förster derived the following equation:

$$k_T = \frac{1}{\tau_d}\left(\frac{R_0}{r}\right)^6 \tag{20}$$

where R_0 (the characteristic distance) is defined as the distance at which the energy transfer efficiency is 50%. Further, Förster showed that R_0 was a function of both the orientation factor and the overlap integral mentioned above:

$$R_0 = 979\left(\frac{J\kappa^2\phi_d}{n^4}\right)^{\frac{1}{6}} \text{ (in nm)} \tag{21}$$

where n and ϕ_d are the index of refraction of the solvent and the quantum yield of the donor, respectively. The orientation factor (κ^2) can vary from 0 to 4, with the

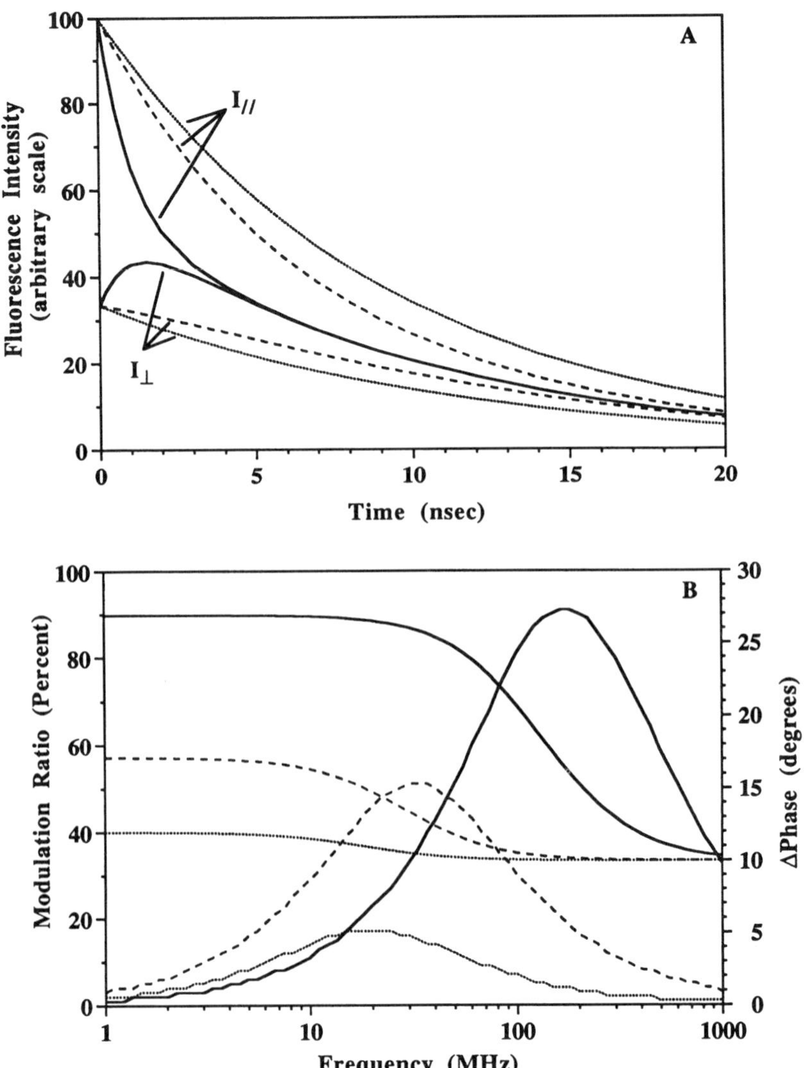

Figure 7. Monoexponential fluorescence anisotropy decay curves for a fluorophore with a lifetime of 10 nsec and an intrinsic anisotropy of 0.4. Simulated data are presented for three different rotational correlation times [ϕ = 1 nsec (solid lines), ϕ = 10 nsec (dashed lines), and ϕ = 50 nsec (dotted lines)]. (A) Time-domain curves, where $I_{//}$ and $I_{\perp}$ denote the parallel and perpendicular components of the fluorescence intensity, respectively. (B) Frequency-domain curves of the phase difference ($\Delta Phase = Phase_{//} - Phase_{\perp}$; bell-shaped curves) and the modulation ratio [modulation ratio = $100(m_{\perp}/m_{//})$].

former value being observed when the transition dipoles of the donor and acceptor are perpendicular to each other and the latter when they are aligned and parallel. If the transition dipoles are antiparallel (i.e., with an angle of 180° between them), then a value of 1 is expected. In practice, the experimentalist works with an ensemble of donor and acceptor molecules with random orientation. In this case κ^2 is equal to $\frac{2}{3}$. The overlap integral is defined as:

$$J = \int_0^\infty F_d(\lambda)\varepsilon_a(\lambda)\lambda^4\, d\lambda \tag{22}$$

where $F_d(\lambda)$ and $\varepsilon_a(\lambda)$ are the emission spectrum of the donor and the extinction coefficient of the acceptor as functions of wavelength, respectively. Finally, the energy transfer efficiency (E) is a function of both the distance between donor and acceptor and the characteristic distance:

$$E = \frac{R_0^6}{R_0^6 + r^6} \tag{23}$$

Förster's theory was confirmed in a careful set of studies by Stryer and Haugland (1967) and Gabor (1968) in which a series of short homopolymers of L-proline were synthesized (varying in length from 1 to 12 amino acids) and a naphthyl donor and a DANS acceptor were conjugated at either end of each peptide. Because poly-L-proline forms a rigid helix, it was possible to compute precisely the distance between the two fluorophores. Analysis of the distance dependency of energy transfer efficiency showed that the exponent in Eq. (23) was indeed close to 6 (5.9 ± 0.3).

2.6.3. ENERGY TRANSFER AS A SPECTROSCOPIC RULER

As suggested by Stryer and Haugland (1967), the distance dependency of energy transfer efficiency can be used as a "spectroscopic ruler" to measure the distance between two sites in a macromolecule. Such measurements have been performed with several different biologically active peptides including adrenocorticotropin (ACTH), angiotensin II, bradykinin, oxytocin, and several different enkephalins (see review by Schiller, 1985). In these studies the distances between naturally occurring tyrosine and tryptophan residues within the oligopeptide chain were determined and used to predict the conformation of the peptides in solution. In cases where the peptides contained only one fluorescent residue, or none at all, extrinsic fluorophores were conjugated to specific sites within the peptide.

Determination of intramolecular distances using energy transfer has been relatively successful in the case of oligopeptides where it is possible to determine the characteristic distance (R_0) with a high degree of certainty. In the case of proteins, however, such studies are far more difficult because of uncertainties in several of the parameters used to calculate the characteristic distance. In particular it is difficult to obtain accurate values for the quantum yield of the donor (ϕ_d) and the orientation factor (κ^2). This is because the values of these parameters vary depending on whether the fluorophore is located in bulk solution or exposed to the protein. An additional problem results from the inverse sixth power dependency of energy transfer efficiency (E) on distance. For instance, an uncertainty of 10 to 20% in E can translate into a 0.5- to 1-nm error in distance. Nevertheless, energy transfer has been used successfully in a few cases to characterize the size and shape of proteins. An excellent example of this is the classical study by Wu and Stryer (1972) in which the approximate molecular dimensions of the photoreceptor protein, rhodopsin, were determined. Finally, a more common application of energy transfer in proteins is for detection of ligand-binding events. In this case the exact distance between donor and acceptor is immaterial because energy transfer is only used phenomenologically to determine whether or not binding has occurred. Applications in ligand binding are discussed in greater length in Section 4.3.

2.7. Interfacial Fluorescence Spectroscopy

2.7.1. INTRODUCTION

If one attempts to observe proteins near or at interfaces using conventional fluorescence spectroscopy (e.g., front-face illumination), their contribution to the total fluorescence would be too small when compared with the overwhelming contribution from solution proteins that are not associated with the interface. Therefore, the study of structure and dynamics of molecules near or at interfaces require a new fluorescence spectroscopic technique that can preferentially elicit response from interfacial species. The basis for the selective excitation of interface-bound protein molecules relies on the optical effect of total internal reflection and enhancement of the electromagnetic field at the dielectric interface. The surface counterpart of fluorescence spectroscopy is called total internal reflection fluorescence (TIRF) or evanescent wave spectroscopy (EWS). A brief description of the optical effect of total internal reflection is presented here for better understanding of this and other spectroscopic techniques that utilize total internal reflection. A more detailed description of excitation and emission of fluorescence at dielectric interfaces can be found in the literature (Reichert, 1989; Axelrod *et al.*, 1992).

2.7.2. THE EVANESCENT FIELD

A collimated light beam arriving at the interface formed between the transparent solid and aqueous solution undergoes total internal reflection if its incident angle, θ, is larger than the critical angle (Fig. 8). The incident and the reflected light constructively interfere and create a standing wave in the optically denser transparent solid. In the case of the interface between two dielectrics, the electric field amplitude right at the interface is not zero but has a finitc value. The electric field decreases with the distance from the interface in the optically rarer aqueous solution. This decaying electromagnetic field is called an evanescent surface wave. The interfacial amplitude of the electric field of perpendicularly polarized light, E^0 (expressed relative to the electric field of the incident wave), is given by:

$$E^0 = \frac{2\cos\theta}{\sqrt{1 - \frac{n_2}{n_1}}} \tag{24}$$

where n_2 and n_1 are the refractive indices of the rarer and denser medium, respectively. The decay of the electric field amplitude of the evanescent surface wave with the distance from the interface is exponential:

$$E(z) = E^0 e^{-z/d_p} \tag{25}$$

where the depth of penetration, d_p, is defined by the optical properties of the system:

$$d_p = \frac{\lambda}{2\pi n_1 \sqrt{\sin^2\theta - \left(\frac{n_2}{n_1}\right)^2}} \tag{26}$$

and λ is the wavelength of the totally reflected light.

2.7.3. EVANESCENT EXCITATION

A fluorescent molecule transported by diffusion and/or convection toward the interface will enter the volume illuminated with the evanescent surface wave. Once inside the electric field of the evanescent surface wave, it will interact with evanescent photons by absorbing a small fraction of the radiation after which an emission of a fluorescence photon may take place. In the first approximation the intensity of the emitted fluorescence is proportional to the strength of the electric field of the evanescent surface wave and to the surface density of the molecules. Due to the exponential nature of the evanescent surface wave the molecules

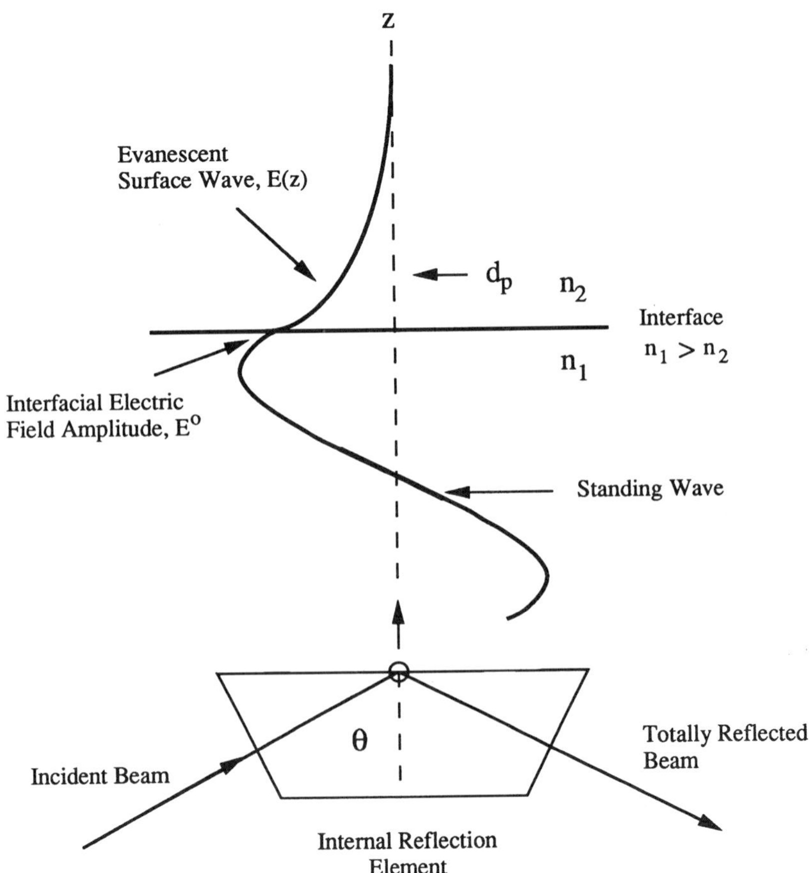

Figure 8. Schematic illustration of total internal reflection (lower panel). The upper panel shows the interface between the two dielectrics with the respective refractive indices, n_1 (denser medium) and n_2 (rarer medium), standing wave pattern in the optically denser medium and the evanescent surface wave in the optically rarer medium. The interfacial electric field amplitude E^0 and the decay of the electric field with distance, $E(z)$, are indicated.

situated right at the interface will fluoresce with the strongest intensity. Since the depth of penetration is on the order of one third of the wavelength of the excitation light, the TIRF technique predominantly excites the fluorescence of interfacially bound molecules. Hence, it is a surface-sensitive technique.

The fluorescence emitted by a molecule near or at a dielectric interface is anisotropic and its spatial distribution depends on the nature of the dielectric interface. Generally, a large part of total fluorescence is emitted into the optically denser phase, especially around angles that are close to the angles of total internal reflection for this interface. Hence, one can observe the fluorescence using the same optical element that is used to create the effect of total internal reflection.

2.7.4. INSTRUMENTATION

Total internal reflection fluorescence experiments require an optically flat, transparent solid surface. Such surfaces can be chemically modified or coated with a thin film of polymer in order to achieve a desired chemical surface composition. In the case of thin films the total internal reflection effect will not be affected provided that the film is transparent to both the excitation and emitted light.

Typically, a TIRF instrument consists of an excitation light source (laser or xenon lamp), an excitation monochromator with light delivering optics, an internal reflection element, fluorescence collection optics with a high numerical aperture collection lens, an emission monochromator, and a photon counting system (Hlady *et al.*, 1985). The internal reflection element can be a 70° dovetail quartz prism or a hemicylindrical prism coupled to a silica plate by an index-matching fluid. The other surface of the plate is exposed to a flow cell so that the protein solution can be conveniently introduced and exchanged with buffer. A version of a TIRF cell constructed to study fluorophores at liquid–liquid interface has been described by Morrison and Weber (1987). In the last five years, TIRF spectroscopy has benefited from the availability of two-dimensional photon detectors, like charge-coupled devices (CCD). Such detectors enable spectroscopic imaging of adsorbed protein layers and thus ally TIRF spectroscopy with the already existing TIRF microscopy (Axelrod, 1989).

3. PROTEIN FLUORESCENCE

Proteins are naturally fluorescent. Three different aromatic amino acid residues (phenylalanine, tryptophan, and tyrosine) fluoresce to some degree, but the extinction coefficient and quantum yield of phenylalanine are so low that it does not contribute significantly to protein fluorescence (Lakowicz, 1983). Intrinsic protein fluorescence is ideal for probing many structural aspects of proteins including folding, conformation, dynamics, and ligand binding. The phenomenon of intrinsic fluorescence will be discussed in Section 3.1 and applications in Section 4.

Although intrinsic protein fluorescence is the method of choice in many cases, there are several compelling reasons for labeling proteins with extrinsic fluorophores. First, protein fluorescence is relatively insubstantial compared to the fluorescence of an extrinsic fluorophore such as fluorescein. This is due to the low quantum yield of tryptophan and tyrosine and the weak intensity of the xenon arc lamp below 300 nm. Second, unless you are working with a single tryptophan protein, the presence of several different emitters within one protein may complicate the interpretation of time-resolved data. Third, the fluorescence lifetimes of tryptophan and tyrosine are often too short to allow examination of the global motion and segmental flexibility of proteins by anisotropy measurements. Finally, monitoring of ligand binding or enzyme kinetics often requires that the ligand or substrate be fluorescent, rather than the protein itself. These and other issues involving extrinsic fluorophores will be examined in Section 3.2 and applications will be discussed in Section 4.

3.1. Intrinsic Fluorescence

3.1.1. ABSORPTION AND EXCITATION

The absorption spectra of proteins contain a large peak at or below 220 nm and a much smaller peak at about 280 nm. The former is due to a number of chemical moieties including the polypeptide linkage, the aromatic amino acids, and the charged amino acids, while the latter is due primarily to tryptophan and tyrosine. Although tryptophan possesses a higher extinction coefficient and quantum yield than tyrosine, proteins typically contain far more tyrosine residues than tryptophan. Thus, tyrosine can in some cases contribute significantly to intrinsic protein fluorescence. Probably more common, however, is fluorescence energy transfer from tyrosine to tryptophan (see Section 2.6 for a discussion of energy transfer). This can complicate interpretation of intrinsic fluorescence studies because the conformation and dynamics of both tyrosine and tryptophan enter in. Thus, it is often desirable to selectively excite tryptophan in order to eliminate any contribution from tyrosine. This is accomplished by choosing an excitation wavelength of 295 nm (or higher) where the extinction coefficient of tyrosine is negligible compared to that of tryptophan.

3.1.2. EMISSION

As mentioned above, phenylalanine is only weakly fluorescent and does not contribute significantly to the intrinsic fluorescence of proteins. For completeness, however, we have included it in our ensuing discussion of fluorescent amino acids. Tryptophan, tyrosine, and phenylalanine all absorb light in the UV region of the spectrum and emit fluorescence between 250 and 400 nm (Demchenko, 1986). In water at neutral pH the emission spectrum of phenylalanine is centered around 280 nm and exhibits fine structure. In contrast, the emission spectra of both tyrosine and tryptophan are broad and structureless. Phenylalanine exhibits the weakest fluorescence (quantum yield of 0.02) of the three and the lowest extinction coefficient, which explains why there have been very few observations of phenylalanine fluorescence in proteins (Permyakov *et al.*, 1981). Tyrosine is more fluorescent than phenylalanine, but still emits relatively weakly compared to tryptophan. Tyrosine fluorescence is best observed in proteins that lack tryptophan residues altogether. In water tyrosine fluoresces with a quantum yield of 0.14. Its spectrum is centered around 303–305 nm (Teale and Weber, 1957), with a full-width half-maximum (FWHM) bandwidth of 34 nm (Longworth, 1971).

For most proteins, tryptophan is the major contributor to intrinsic fluorescence, especially when fluorescence is excited at 295 nm, well past the absorption maxima of both phenylalanine and tyrosine. In water tryptophan's emission spectrum is typically centered between 348 to 350 nm with a FWHM of 60 nm and a quantum yield of 0.23 (Teale and Weber, 1957; Longworth, 1971). Tryptophan residues are often used as local reporter groups in proteins because of their relatively large, solvent-dependent Stokes' shift, which amounts to 60 nm in water (6000 cm^{-1}). In comparison, the fluorescence emission maximum of indole (tryptophan's chromophoric moiety) in nonpolar solvents is located at about 310 nm. This maximum shifts toward longer wavelengths as the polarity of the solvent increases. For this reason, tryptophan fluorescence in proteins is strongly dependent on the local environment of the fluorophore. In addition, almost all protein polar groups can quench tryptophan fluorescence by a variety of mechanisms. Thus, tryptophan's quantum yield also depends on its position and interactions with neighboring residues. Permyakov (1993) gives an extensive inventory of possible tryptophan fluorescence-quenching mechanisms.

Several different spectral forms of fluorescence emission have been identified in proteins; each is thought to originate from a particular environment inside the protein molecule (Permyakov, 1993). These are illustrated in Fig. 9 and a brief description of each is given below.

Spectral form A is similar to the emission of the unperturbed indole moiety in a neutral hydrophobic environment. It contains a few discrete spectral peaks within the emission spectrum. Only two proteins are known to actually display such a spectrum: azurin, a small globular bacterial protein that contains a single tryptophan residue (Burstein *et al.*, 1977), and bacteriorhodopsin, a membrane protein from *Halobacterium halobium* (Permyakov and Shnyrov, 1983).

Spectral form S corresponds to the emission of an unperturbed indole fluorophore, inside the protein's hydrophobic core, that forms a 1:1 excited-state complex (called exciplex) with a neighboring polar group. This type of emission

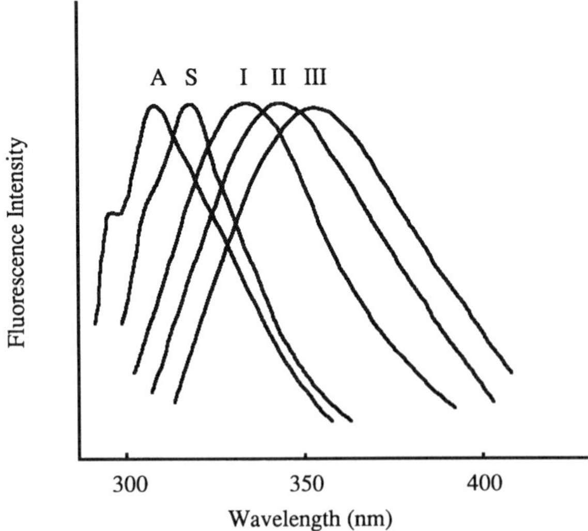

Figure 9. Intrinsic fluorescence spectra (normalized) of the six tryptophan spectral forms in a protein molecule (redrawn after Permyakov, 1993). Spectral form A is similar to the emission of the unperturbed indole moiety in a neutral hydrophobic environment; spectral form S corresponds to the emission of an unperturbed indole fluorophore, inside the protein's hydrophobic core, that forms a 1:1 excited state complex (called exciplex) with a neighboring polar group; spectral form I is similar to spectral form S, except that a 2:1 exciplex is formed with neighboring polar groups within the protein; spectral form II results when the indole ring is exposed to water molecules that have much lower mobility than that of bulk water; and spectral form III is observed once the indole ring comes into contact with bulk water.

spectrum (λ_{max} = 316–317 nm with two shoulders at 305–307 and 320–330 nm, respectively) is observed in a few proteins, such as L-asparaginase and a family of ribonucleases from *Aspergillus* (Permyakov, 1993).

Spectral form I is similar to spectral form S, except that a 2:1 exciplex is formed with neighboring polar groups within the protein. The spectrum is without structure, with a maximum at 330–332 nm and a width of 48–50 nm. Proteins like actin and chymotrypsin exhibit this spectral form.

Spectral form II results when the indole ring is exposed to water molecules that have much lower mobility than that of bulk water. Many proteins have tryptophan residues exposed to bound water and display this type of fluorescence (λ_{max} = 340–342 nm; FWHM 53–55 nm).

Spectral form III is observed once the indole ring comes into contact with bulk water. Its fluorescence emission spectrum nearly coincides with that of free tryptophan in water (λ_{max} = 350–353 nm; width 59–61 nm). The process of protein unfolding is usually accompanied with the appearance of this red-shifted tryptophan fluorescence.

Intrinsic fluorescence can yield a significant amount of information about protein molecules if one combines a number of steady-state and time-resolved fluorescence techniques. A lucid strategy for the analysis of protein fluorescence spectra in terms of structural and dynamic changes in the environment of tryptophan residues is presented by Demchenko (1986). Spectral shifts in intrinsic protein fluorescence have often been interpreted as changes in the environment of tryptophan residues, which in turn are often associated (sometimes incorrectly) with more global conformational changes. However, one cannot emphasize enough the danger of overinterpretation of such spectral changes. A single tryptophan residue in a protein reports only about its local environment and not about the whole protein molecule. The situation is even more complex for proteins with several tryptophan residues where each "samples" only its local environment, or for proteins containing both tyrosine and tryptophan residues. In the latter, tyrosine fluorescence can be observed only as the difference between the spectrum excited at 275 nm, where the tyrosine contribution is maximal, and the spectrum excited at 295 nm, where only tryptophan fluoresces. Finally, changes in the polarity of a fluorophore's environment do not necessarily result in concomitant spectral changes. As a rule, environmental polarity can only affect the emission wavelength if dipolar relaxation is rapid compared to fluorescence emission (Demchenko, 1986). For example, both spectral forms I and II exhibit a blue-shifted emission spectrum (relative to tryptophan in water), which is often indicative of a nonpolar environment. In both cases, however, polar groups are present in the local environment of tryptophan, but they are too immobile to relax during its excited-state lifetime.

3.1.3. FLUORESCENCE LIFETIMES OF PROTEINS

As mentioned above, the emission maximum of tryptophan groups depends on their local environment. Likewise, the fluorescence lifetime of protein tryptophan groups is very sensitive to environmental factors. For instance, intramolecular quenching groups include ionized groups (e.g., histidyl, carboxyl, arginyl), disulfide bonds, methionine sulfur groups, and carbonyl groups of peptide bonds (Longworth, 1971; Permyakov, 1993). The extent of quenching will be proportional to the likelihood of collision and therefore reflects the proximity to the fluorophore and the motional freedom of such groups. Complexation with metal ions can also change the lifetime of proteins considerably (Chen, 1976; Eftink, 1991).

Although the photophysics of tryptophan and other indole compounds is quite complicated (Steiner and Kirby, 1969; Szabo and Rayner, 1980; Lakowicz, 1983;

Creed, 1984; Beechem and Brand, 1985), the tryptophan derivative *N*-acetyltryptophanamide exhibits a monoexponential decay in aqueous solutions. This indicates that multiexponential decays of tryptophan protein fluorescence (see Section 4.1) are a direct consequence of heterogeneity and/or structural fluctuations of the local environment of the tryptophan residues. Tryptophan fluorescence lifetimes vary more than 100-fold in different proteins (Beechem and Brand, 1985). Proteins usually exhibit multiple lifetimes, even if they contain only one tryptophan residue (Beechem and Brand, 1985; Alcala *et al.*, 1987c; Eftink, 1991). This indicates that variations in the microenvironment of residues occur during their excited state. If this is the case, it may be reasonable to assume that tryptophan residues in proteins exhibit quasicontinuous lifetime distributions rather than a summation of discrete lifetimes, as has been extensively discussed by Alcala and co-workers (Alcala *et al.*, 1987a–c). For different proteins it has been shown that distribution analysis fits the tryptophan fluorescence decay profiles at least as well as a summation of discrete lifetimes (Rosato *et al.*, 1987; Bismuto *et al.*, 1988, 1989; Harris and Hudson, 1990; Mei *et al.*, 1992; Royer, 1992). Moreover, the broadening of the apparent lifetime distributions on protein denaturation has been explained by a larger number of possible interactions of tryptophan residues with other groups in the unfolded state (Bismuto *et al.*, 1988; Fernando and Royer, 1992; Mei *et al.*, 1992).

The interpretation of the fluorescence decay of multitryptophan proteins is difficult, although the apparent decay pattern of such proteins is not necessarily more complex than that of single-tryptophan proteins. Approaches to elucidate the origin of the observed fluorescence lifetimes in multitryptophan proteins involve the investigation of single-tryptophan mutants (Harris and Hudson, 1990; Atkins *et al.*, 1991; Royer, 1992).

3.2. Extrinsic Fluorophores

Although certain protein cofactors, coenzymes, and prosthetic groups are fluorescent and thus can be classified as extrinsic fluorophores, the term is more traditionally applied to fluorescent dyes that are attached to the protein via a conjugation reaction. Several such fluorescent dyes are listed in Table I. Their excitation and emission wavelengths range from the ultraviolet-absorbing coumarins to the long-wavelength red emission of the naphthofluoresceins and cyanine dyes. The selection of a dye generally depends on the application. In cases where sensitivity is critical, such as ligand binding and fluorescence microscopy, the visible-absorbing fluorescein dyes (fluorescein, rhodamine-110, tetramethylrhodamine, rhodamine B, rhodamine X, and naphthofluorescein) with high extinction coefficients and quantum yields are frequently used. In contrast, ultraviolet-absorbing dyes such as pyrene and isomers of DANS are often employed in

Table I. Physical Properties of Common Fluorophores

Fluorescent dye	Absorption maximum (nm)	Extinction coefficient (M^{-1} cm^{-1})	Emission maximum (nm)	pK_a
5-Dimethylaminonaphthalene-1-sulfonate (DANS or DANSYL)	335[a]	4,600	515[a,b]	—
Pyrene	341	42,000	377, 395[b]	—
6-Aminoquinoline	350	3,700	540	—
β-Methylumbelliferone	360	17,000	450	8.2
6-Propionyl-2-dimethylamino-naphthalene (PRODAN)	360	18,000	497[a]	—
4-Chloromethylumbelliferone	372	16,000	470	8.5
1-Anilinonaphthalene-8-sulfonate (ANS)	372[a]	8,000	480[a]	—
2-Dimethylaminonaphthalene-5-sulfonate (2,5-DANS)	374[a]	2,900	480[a,b]	—
2′-Hydroxyphenyl-4-(3H)-quinoxazolinone	375	19,000	490	8.5[c]
5′-Methoxy-2′-hydroxyphenyl-4-(3H)-quinoxazolinone	375	19,000	550	8.5[c]
5′-Chloro-2′-hydroxyphenyl-6-chloro-4-(3H)-quinoxazolinone	375	19,000	510	8.5[c]
4′-Methoxy-2′-hydroxyphenyl-4-(3H)-quinoxazolinone	375	19,000	450	8.5[c]
β-Trifluoromethylumbelliferone	385	16,000	501	7.3
3-Carboxy-7-hydroxycoumarin	388	32,000	445	7.4
6-Acryloyl-2-dimethylamino-naphthalene (acrylodan)	391	20,000	500	
Fluorescein	490	90,000	514	6.4
5-Chloromethylfluorescein	490	85,000	515	6.5
5,6-Carboxyfluorescein	492	82,000	514	6.4
5-N-Dodecanoylaminofluorescein	495	85,000	518	6.5[d]
2′,7′-Dichlorofluorescein	503	90,000	529	5.1
Rhodamine-110	505	98,000	534	>12
Erythrosin	528	86,000	553, 690[e]	<7
Tetramethylrhodamine	547	90,000	572	—
5,5′-Disulfo-3,3,3′,3′-tetramethyl-1,1′-dihexanoylcarbo-cyanine (CY3)	550	130,000	565	—
Lissamine Rodamine	556	93,000	576	—
Rhodamine-X	570	93,000	596	—
Resorufin	571	55,000	585	<7
3-Dodecylresorufin	571	56,000	584	6.5[d]
Texas Red	589	85,000	615	—
Naphthofluorescein	594	55,000	663	7.6
5,5′-Disulfo-3,3,3′,3′-tetramethyl-1,1′-dihexanoyldicarbo-cyanine (CY5)	652	>200,000	667	—

[a]Excitation and emission wavelengths are environment dependent.
[b]Relatively long fluorescence lifetime (>10 nsec) when coupled to proteins.
[c]Precipitating fluorophores are maximally fluorescent below the pK_a value.
[d]pK_a values for membrane soluble fluorophores are environment dependent.
[e]Erythrosin is phosphorescent. The first value (553 nm) is its emission maximum for fluorescence, while the second (690 nm) is for phosphorescence.

fluorescence anisotropy studies because of their long fluorescence lifetimes. In cases where a fixed-wavelength light source and emission filters are employed (e.g., lifetime measurements, flow cytometry, laser scanning microscopes), excitation and emission wavelength are typically the deciding factors in dye selection, although, virtually any fluorophore may be employed when working with a spectrofluorometer that contains monochromators.

Fluorescent dyes can be conjugated to several different types of nascent chemical groups within proteins. The ϵ-amino group of lysine and the α-amino group of the N-terminus are the most common of these, but the thiol group of cysteine can also be labeled as well. Several different amine-reactive derivatives of common fluorophores are available, including N-hydroxysuccinimidyl esters, sulfonyl halides, and isothiocyanates. These derivatives form stable amide, sulfonamide, or thiourea linkages, respectively, with chemical stability in roughly that order. In a typical conjugation reaction, the protein is dissolved in a buffer solution of moderately high pH (e.g., 0.1 M $NaHCO_3$, pH 8–9) and allowed to react for 24 hr at room temperature with a 10–100 molar excess of the reactive fluorophore dissolved in an organic solvent (acetonitrile, dimethylformamide, dimethylsulfoxide, etc.). The protein–dye conjugate is usually purified by gel filtration chromatography. Its degree of labeling (DOL or DL) is determined by taking the ratio of dye to protein absorbance, at their respective absorption maxima. The reader is referred to Wei *et al.* (1992) for a comprehensive discussion of this topic. Overlabeling is undesirable, as quenching of spatially interacting fluorophores can occur as well as changes in the protein's isoelectric point or other structural perturbations. As a rule of thumb, labeling ratios of about 1–3 moles of dye per protein are often optimum.

As mentioned above, the thiol functionality is found in cysteine residues, but can also be formed by reduction of disulfide bridges using dithiothreitol (DTT), for example. Although less prevalent than primary amines, thiol groups can easily be modified in a highly selective fashion by reaction with either the iodoacetamide or maleimide derivatives of common fluorophores. Although these derivatives can cross-react with amines, conjugation reactions are typically performed at neutral pH (or lower) where most amines are charged and therefore unreactive. Otherwise, the protocol for derivatization of thiol groups is essentially the same as that used for amine conjugation, as described above. The reader is referred to recent reviews and texts regarding these and other techniques (Haugland, 1991; Kasten, 1989; Waggoner, 1990).

There is no universal fluorescent dye that can be used to interrogate all aspects of protein structure–function relationships. Often, several different fluorescent conjugates of the same protein (preferably with a different dye in each case) need to be studied in order to remove biases imparted by each individual dye and conjugation site. In such cases global analysis (Knutson *et al.*, 1983; Beechem *et al.*, 1985; Beechem and Gratton, 1988; Beechem, 1992) can be used to determine the physical properties of the system (e.g., rotational correlation times, activation energies, quenching constants). Multiple fluorescent labels attached to a single protein molecule are rarely useful in conformational or protein dynamics studies, but can provide valuable information in fluorescence microscopy experiments.

4. APPLICATIONS

4.1. Protein Folding

The most straightforward application of fluorometry in monitoring protein folding is the study of denaturation. Unfolding of proteins is usually accompanied by a red-shifted emission maximum, depolarization of tryptophan fluorescence, enhanced accessibility to quenchers such as iodide, and changes in fluorescence lifetimes. Here we will give recent examples of more subtle conformational changes as probed by fluorescence methodology.

Three independent studies of acid-induced conformational changes in polyclonal (Lin *et al.*, 1989) and monoclonal (Buchner *et al.*, 1991; Jiskoot *et al.*, 1991a) antibodies showed that a dramatic increase in intrinsic fluorescence intensity occurred when these proteins were exposed to acidic conditions ($pH \leqslant 3$). Interestingly, this increase was not accompanied by a red-shift in the emission maximum. In addition, Buchner *et al.* (1991) reported a concomitant increase in hydrophobicity using the hydrophobic fluorescent probe Nile Red, and Jiskoot *et al.* (1991a) observed an increased accessibility to collisional quenchers. These observations could be due to the swelling of the antibody molecules at low pH, which would increase the distance between tryptophan and intramolecular quencher groups without changing the average microenvironment of tryptophan. Supported by other techniques, it was concluded that immunoglobulin G does not completely unfold at low pH but undergoes a structural transition to a molten globulelike state that has been proposed for several proteins at low pH (Goto *et al.*, 1990).

A similar increase in intrinsic fluorescence intensity without an emission red-shift has been observed for bovine growth hormone (bGH) at low pH or at intermediate concentrations of denaturants such as urea and guanidinium hydrochloride (Havel *et al.*, 1988; Holzman *et al.*, 1990). Furthermore, Brems and Havel (1989) showed that under either set of conditions, bGH formed stable monomeric and multimeric intermediates that resembled the molten globule state. Although bGH is a single-tryptophan protein, lifetime measurements revealed multiexponential decays with longer average lifetimes for the intermediate and unfolded states. These observations suggested that bGH folds into a closely packed native structure in which tryptophan fluorescence is effectively quenched by intramolecular groups. This hypothesis was supported by the observation that energy transfer from tyrosine to tryptophan occurred with high efficiency in the native state but not in the (partially) unfolded states. Collisional quenching by iodide, acrylamide, and trichloroethanol was least effective in the multimeric intermediate state, indicating steric and/or chemical restrictions. Tryptophan was most readily quenched by these quenchers at high denaturant concentrations and its emission spectrum was red-shifted, as expected for a completely unfolded protein (Havel *et al.*, 1988). In comparison, the intrinsic fluorescence of human growth hormone (also a single-tryptophan protein) is quenched upon unfolding, but stable folding intermediates are not formed (Brems *et al.*, 1990).

Three lifetime components were observed by Willis and Szabo (1992) for the single tryptophan residue at position 23 of human parathyroid hormone (hPTH) (84 amino acids, partly α-helical structure). The decay parameters changed in 6 M guanidinium hydrochloride but not in helix-promoting solvents, which indicated that the tryptophan residue is located in the α-helical region. A similar decay profile was observed for a fragment of hPTH that contained only the first 34 residues. This suggested that the secondary structure of this fragment was not affected by the deleted region (35-84). The three apparent lifetimes were attributed to three rotamers of the Trp-23 side chain. Comparison with NMR data indicated that the individual fractional contributions of the lifetime components reflect preferential orientations of the tryptophan side chain induced by the secondary structure of hPTH.

Three lifetime components were also observed by Atkins *et al.* (1991) for glutamine synthetase, an ATP-binding protein that contains two tryptophan residues located at positions 57 and 158, respectively. Interestingly, single-tryptophan mutants exhibited the same three lifetime values but with different fractional contributions. In both cases, a three-component model fit the data better than distributions. Binding of ATP increased the fraction of the long (5 nsec) component of Trp-57 but did not alter the lifetime values. Further, iodide quenching of Trp-57 was reduced upon binding of ATP. The iodide accessibility and lifetime parameters of Trp-158 were unaffected by ATP binding. From these results it was concluded that ATP "tightens" the conformation of the Trp-57 loop into a form that is less solvent accessible.

4.2. Protein Dynamics

The traditional view of protein structure from X-ray diffraction and modeling studies was that the interior of a protein exhibited nearly the same packing density and degree of order as small organic molecules in the crystalline state. As a corollary, many protein chemists thought that proteins were uninteresting from a dynamic perspective, spending most of their lives in the conformation observed in the crystal structure. Although in most cases the crystal probably does reflect the average conformation of the molecule, a number of studies—both fluorescence and NMR—suggest that proteins continuously fluctuate on the nanosecond time scale (see reviews by Careri *et al.*, 1979; Beechem and Brand, 1985; Lakowicz, 1986).

The seminal work in this area was an article by Lakowicz and Weber (1973b) in which oxygen was used to quench the intrinsic fluorescence of more than ten different model proteins. Because oxygen is a small, nonpolar molecule, it can readily dissolve in nonpolar solvents. Further, it does not form static complexes with fluorophores and its bimolecular rate constant (k_q) for the quenching of indole and tryptophan is diffusion controlled in both aqueous buffers and nonpolar solvents such as dodecane (Lakowicz and Weber, 1973a). Thus, oxygen is a good probe for examining the physical state of the interior of a protein. If the interior exists in a fluid state, similar to a nonpolar solvent, then the k_q value obtained for the quenching of intrinsic fluorescence should be close to the value obtained in a nonpolar solvent. On the other hand, if the interior exists in a near crystalline state, then oxygen may not be able to penetrate the interior of the protein at all and little or no quenching would be observed (as is the case in iodide quenching studies). Lakowicz and Weber found that the bimolecular quenching rate constants (k_q) of the model proteins were two- to fivefold lower than the value observed for free tryptophan. This suggested that the interiors of the model proteins more closely resembled a fluid than a solid, even though small diffusional barriers to oxygen were probably present.

At the time that Lakowicz and Weber performed the above experiments, sophisticated time-resolved fluorescence instrumentation was not available, so they had to rely on an indirect method (oxygen quenching) to identify fluctuations in the nanosecond time scale. In this case, the fluorescence lifetime of tryptophan itself acted as a "molecular stop watch" to time events on a scale that could not be observed directly. However, recent advances in instrumentation have enabled us to investigate firsthand the behavior of tryptophan on time scales of nanoseconds and below. Two time-resolved techniques in particular—lifetime distributions and dynamic anisotropy—have been used to examine protein fluctuations. Lifetime distributions are based on the theory that if the microenvironment of a tryptophan residue fluctuates on the same time scale as its fluorescence lifetime, then a distribution of lifetimes is expected within a population of protein molecules, rather than a single exponential decay. This theory was convincingly demonstrated by Alcala *et al.* (1987a–c), and since then, lifetime distributions have become a standard tool of the trade.

Although distributed lifetimes are indicative of conformational fluctuations, they do not provide any information about the nature of these fluctuations. Dynamic anisotropy, on the other hand, can be used to examine the rotational motions of both individual tryptophan residues and the protein in general. Many such studies have been performed in the last few years. Typically two rotational components are observed: one correlated to the local motion of the fluorophore (either tryptophan or a fluorescence probe) and the other to the global motion of the protein (Hazlett *et al.*, 1989; Jullian *et al.*, 1989; Ferreira, 1989; Elofsson *et al.*, 1991; Wang *et al.*, 1992).

Finally, dynamic phosphorescence anisotropy has been used in a couple of cases to study the interactions of proteins in bilayers. Myers *et al.* (1992) labeled immunoglobulin E (IgE) with erythrosin and allowed it to complex with Fc receptors located either on the surface of rat basophilic leukemia (RBL) cells or on membrane vesicles derived from RBL cells. Dynamic anisotropy was measured both with and without addition of a cross-linker that caused the IgE/receptor complex to dimerize. The authors found that in the monomeric state the IgE/receptor complex could rotate freely on the surface of both RBL cells and vesicles. However, upon addition of the cross-linker, the complex on the surfaces of the cells became immobile, while that on the surface of the vesicles could still rotate but with twice the rotational correlation time of the monomer. In another study a glycoprotein (HLA-A2) from the class I major histocompatibility complex was labeled with erythrosin and incorporated into liposomes (Chakrabarti *et al.*, 1992). Dynamic anisotropy was used to measure the rotational diffusion of HLA-A2 in the bilayer. The results suggested that HLA-A2 self-associates to form aggregates with radii ca. four to seven times larger than the monomer. Further, this aggregation process could be partially inhibited by β2-microglobulin, which is known to block HLA-A2 association on the surfaces of B-lymphoblast cells.

4.3. Ligand Binding

Ligand binding is an essential step in a number of different interactions involving proteins such as drug-receptor interactions, drug binding to plasma proteins, and antigen–antibody interactions. It can be detected by monitoring changes in either protein fluorescence or ligand fluorescence. The former will be discussed in Section 4.3.1 and the latter in Section 4.3.2.

4.3.1. USING PROTEIN FLUORESCENCE TO DETECT LIGAND BINDING

Ligand binding can affect intrinsic protein fluorescence in a number of different ways, including spectral shifts and changes in quantum yield, lifetime, and anisotropy. The first three of these are often correlated, especially in cases where ligand binding affects the microenvironment of tryptophan. For example,

Ferreira (1989) observed both a blue shift and an increase in the quantum yield of the intrinsic fluorescence of parvalbumin upon binding of cationic calcium (Ca^{2+}). Although the fluorescence lifetime did not change significantly upon Ca^{2+} binding, the width of the lifetime distribution decreased by a factor of two. In another study, Chabbert *et al.* (1991) investigated the binding of two different synthetic peptides to five different tryptophan-containing mutants of calmodulin. Changes were observed in both fluorescence intensity and lifetime upon binding, although the extent of such changes differed among the five mutants. These observations were attributed to changes in the microenvironments of the tryptophan residues brought about by ligand-induced changes in calmodulin's conformation. The different behavior of the five mutants illustrates that conformational changes are not necessarily uniform throughout a protein and care must be taken in extrapolating the results of intrinsic fluorescence studies to the global level.

In some cases, ligand binding results in quenching of the protein's intrinsic fluorescence due to resonance energy transfer. In order for this to occur, however, the absorption spectrum of the ligand must overlap with the emission spectrum of tryptophan and both moieties have to be located relatively close to each other (≤ 5 nm, or so). Heme proteins such as hemoglobin and myoglobin are the best-characterized examples of this phenomenon. Both exhibit extremely low intrinsic fluorescence due to energy transfer from tryptophan to the heme moiety. This mechanism was first studied in hemoglobin by Weber and Teale (1959), but there have been numerous studies since then (review by Beechem and Brand, 1985; Bismuto *et al.*, 1989; Willis *et al.*, 1990).

In cases where a significant level ($> 50\%$) of fluorescence quenching results upon binding, the affinity of the protein–ligand interaction can easily be calculated. An early example of this was the binding of the dinitrophenyl (DNP) hapten to anti-DNP antibodies, in which up to 80% quenching of intrinsic fluorescence was observed (Velick *et al.*, 1960). A more modern example is the binding of the antibiotic chloramphenicol to a bacterial acetyltransferase that is implicated in chloramphenicol resistance (Ellis *et al.*, 1991). In another recent example, Dughi *et al.* (1993) used fluorescence quenching to examine the binding of the hormone thyroxin (T4) to a mutant form of human serum albumin found in patients with familial dysalbuminemic hyperthyroxinemia (FDH). Readers wishing to use intrinsic fluorescence quenching to determine affinities of protein–ligand interactions are referred to Day (1990), in which the methods are clearly outlined.

4.3.2. USING LIGAND FLUORESCENCE TO DETECT BINDING

In some cases the ligand of interest will be naturally fluorescent and its binding can be monitored directly; alternately, it can be labeled with a functional derivative of one of the fluorescent probes listed in Table I. In either case, once the fluorescent ligand is in hand, the next step is to determine whether any observable changes in its fluorescent properties occur upon binding. If the ligand is substantially smaller in size than the protein, then a change in anisotropy is almost certain. Such changes are frequently accompanied by a change in the fluorescence intensity of the ligand: either enhancement or quenching can be observed. Binding studies based on intensity changes are often preferable to those based on anisotropy for several reasons. First, not all spectrofluorometers are equipped with polarizers, so anisotropy measurements may not be available to the researcher. Second, anisotropy measurements in general are not as sensitive as intensity measurements. This is because the polarizer rejects a significant portion of the incident light. Third, in cases where ligand binding is accompanied by a change in fluorescence intensity, anisotropy data must be corrected for the change in quantum yield before an accurate binding isotherm can be constructed (Herron, 1984; Wei and Herron, 1993). The three different methods for detecting the binding of fluorescent ligands (enhancement, quenching, and anisotropy) will be discussed in greater detail in Sections 4.3.2a, 4.3.2b, and 4.3.2c, respectively.

4.3.2a. Fluorescence Enhancement

A classical example of fluorescence enhancement is the binding of 1-anilino-8-naphthalene sulfonate (ANS) to bovine serum albumin (BSA). Daniel and Weber (1966) observed that ANS had a very low quantum yield in solution, but became strongly fluorescent upon binding to BSA. Later, Kolb and Weber (1975) showed that serum albumin possesses two classes of binding sites for small molecules such as ligands or drugs: a low-affinity class and a high-affinity class. This finding had major impact on the field of pharmacokinetics because drug binding to human serum albumin can significantly affect the unbound drug concentration in plasma.

In another study, Blumenthal *et al.* (1985) observed that the fluorescence of tryptophan-containing peptides derived from myosin light-chain kinase (MLCK) was enhanced by 40% upon binding to calmodulin, a protein that is responsible for the calcium dependence of a number of intracellular processes. Interestingly, calmodulin does not contain any tryptophan residues, an observation that greatly simplifies the interpretation of binding studies with the above peptides. The above fluorescence enhancement assay has found great use in the study of intracellular regulation, because tryptophan-containing peptides derived from a number of different calmodulin-dependent enzymes bind to calmodulin and exhibit similar enhancement effects (see reviews by Blumenthal and Krebs, 1988; Anderson and Malencik, 1989). Recently, peptides derived from MLCK have been labeled with 6-acryloyl-2-dimethylaminonaphthalene (acrylodan), a fluorescent label that fluoresces in the visible region (see Table I). The fluorescence of the peptide–

acrylodan conjugate is enhanced by four- to fivefold upon binding and the visible emission offers the advantage that dissociation constants (K_d) in the low nanomolar range can be detected (see review by Blumenthal, 1993).

Additional examples of fluorescence enhancement upon ligand binding include: (1) the binding of a fluorescent analogue of guanidine diphosphate (GDP) to elongation factor Tu (Eccleston *et al.*, 1987); (2) drug binding to α_1-acid glycoprotein (Johansen *et al.*, 1992); (3) a similar study with human serum albumin (Chatelain *et al.*, 1994); and (4) the binding of a peptide mimic of human chorionic gonadotrophin (hCG) to an anti-hCG antibody (Wei *et al.*, 1994). The reader is referred to Daniel and Weber (1966) or Wei *et al.* (1994) for a detailed description of fluorescence enhancement methodology and techniques for analyzing ligand binding.

4.3.2b. Fluorescence Quenching

Although the literature is not replete with examples of protein–ligand interactions in which ligand fluorescence is quenched, there is one well-known hapten–antibody system (antifluorescein) where the fluorescence of the hapten (fluorescein) decreases significantly upon binding. Antifluorescein antibodies were first described by Lopatin and Voss (1971) and typically quench fluorescein fluorescence by more than 90% (Watt and Voss, 1977). Considering fluorescein's high quantum yield in aqueous solutions ($Q \geqslant 0.9$, at pH $\geqslant 8$) (Seybold *et al.*, 1969), it is not surprising that fluorescein is quenched upon binding. However, the degree to which it is quenched is somewhat unusual. Immunoglobulins do not contain any intrinsic chromophoric groups that exhibit spectral overlap with fluorescein, so the mechanism of quenching does not involve resonance energy transfer.

Two different groups (Watt and Voss, 1977; Templeton and Ware, 1985) have investigated the quenching mechanism and concluded that it involves formation of a charge transfer complex between fluorescein, in the excited state, and a neighboring tryptophan residue. This mechanism is consistent with the three-dimensional structure of the antigen-binding fragment of an antifluorescein antibody in complex with fluorescein (Herron *et al.*, 1989, 1994). This structure revealed that there were two tryptophan residues in intimate contact with the fluorescein moiety, either of which could participate in formation of a charge transfer complex. Nevertheless, a recent study by Omelyanenko *et al.* (1993) showed that other factors are also involved in the quenching of fluorescein, including the local pH of the antigen-combining site and the formation of a salt link with a buried arginine residue. Several other physicochemical parameters of the antifluorescein system have been investigated using fluorescence quenching. These include binding isotherms (Herron, 1984), ligand-binding kinetics (Kranz *et al.*, 1981, 1982), and thermodynamics (Herron *et al.*, 1986). The reader is referred to Herron (1984) for a review of methods used to characterize antifluorescein antibodies.

4.3.2c. Fluorescence Anisotropy

As mentioned previously, fluorescence anisotropy can often be used to measure ligand binding when other methods have failed. The difference in size (or more correctly hydrodynamic volume) between ligand and protein is probably the most critical factor in anisotropy studies. For instance, if a small ligand (ca. 500 mol. wt.) with a rotational correlation time (ϕ) of 1 nsec and a fluorescence lifetime (τ) of 5 nsec were to bind to a protein (50,000 mol. wt.) with a rotational correlation time of 50 nsec, then according to Eq. (16), the anisotropy of the ligand would increase fivefold upon binding (from 0.07 to 0.36). Compare this to a case in which two protein monomers, each with a rotational correlation time of 50 nsec, associate to form a dimer. Assuming that a fluorescent probe, again with a 5-nsec lifetime, is conjugated to one of the monomers, its anisotropy would only increase from 0.36 to 0.38 upon dimerization. Thus, better resolution is achieved when there is a large difference in the rotational correlation time of the two species. Another related factor is the lifetime of the fluorophore. We saw in the first example that the lifetime of the fluorophore was greater than the rotational correlation time of the ligand, while in the second example, it was much smaller. This observation leads to another generalization about anisotropy studies: the lifetime of the fluorophore should be roughly equivalent, if not greater than, the rotational correlation time of the unbound ligand.

Anisotropy (either steady-state or dynamic) has been used in several recent studies to monitor ligand binding. These included: (1) the binding of NADH to malate dehydrogenase (Jameson *et al.*, 1989); (2) the binding of a phenytoin–fluorescein conjugate to an antiphenytoin antibody (Bright, 1989); (3) the binding of an epitope peptide to an antibody that was elicited to an outer membrane protein from *Neisseria meningitidis* (Jiskoot *et al.*, 1991b); and (4) the aforementioned example of the binding of a peptide that mimics hCG to an anti-hCG antibody (Wei and Herron, 1993). Readers who wish to use fluorescence anisotropy as a tool to measure ligand binding are referred to Weber and Daniel (1966) and Wei and Herron (1993) for a discussion of methods and analysis.

4.4. Enzyme Kinetics

Proteins are sometimes employed therapeutically because of their enzymatic activities (e.g., streptokinase, tissue plasminogen activator, and urokinase). Fluorescence spectroscopy can be an effective tool in such cases either for assaying small quantities of an enzyme or for measuring its activity. Over the last two decades a large number of fluorescent probes have been developed specifically for enzyme kinetics and represent a flourishing area of interest in cell biology, enzymology, histology, cytometry, and molecular biology. Techniques that employ fluorogenic substrates (i.e., substrates that *become* fluorescent after enzyme action) are typically several orders of magnitude more sensitive than chromogenic substrates and rival radiochemical and chemiluminescent methods of enzyme assay. In addition, a number of different fluorescent substrates have been developed that change their emission wavelength upon enzyme action. Both types of substrates are available for a wide variety of enzymes, including glycosidases (see Sections 4.4.2 and 4.4.3), proteases (Livingston *et al.*, 1981; Leytus *et al.*, 1983), phosphatases (Rotman *et al.*, 1963; Livingston *et al.*, 1981), esterases (Parvari *et al.*, 1983; Berman and Leonard, 1990), cytochrome P450 (Burke *et al.*, 1985), peroxidases (Brandt and Keston, 1965; Cathcart *et al.*, 1983), and kinases (Housey *et al.*, 1988). Fluorescent probes are also available for enzyme-linked immunoassays (Kato *et al.*, 1975; Baruch *et al.*, 1991) and DNA hybridization assays (Lichter and Ward, 1990), and are becoming available to probe enzyme activity inside individual cells *in vivo* (see Sections 4.4.2 and 4.4.3).

General strategies for using fluorescent probes will be discussed in Section 4.4.1. In addition, several other topics will be examined including: (1) selection of fluorescent probes; (2) advantages of certain fluorophores over others; (3) fluorescence properties and sensitivity to environmental conditions (pH, polarity of the medium, quenching, etc.); and (4) stability and relative activity parameters. Specific examples of fluorescent or fluorogenic substrates are almost too numerous to mention, so our discussion will focus on substrates developed for two enzymes in particular: β-galactosidase and β-glucuronidase. The reader may be familiar with these enzymes because they are often employed both in enzyme-linked assays and as markers in molecular biology studies. In many respects the substrates developed for these enzymes are archetypal of this class of fluorescent probes as a whole, and much of the information presented can easily be extrapolated to other substrate–enzyme systems. Fluorescent probes for β-galactosidase and β-glucuronidase will be discussed in Section 4.4.2 and 4.4.3, respectively.

4.4.1. FLUORESCENT ENZYME SUBSTRATES—GENERAL CONSIDERATIONS

Development of new fluorescent assays for enzymes usually requires optimization of assay conditions. One of the most important factors is the pH of the buffer system. Not only does the pH need to be in the optimum range for the enzyme, but it also needs to be high enough to ensure optimum fluorescence. This is because many common dyes fluoresce maximally at pH values well above their pK_a values (Table I), where they exist as charged species. In cases where these requirements are inconsistent, the reaction can be run at physiological pH and a stop buffer (usually high pH) can be employed to both terminate the reaction and provide maximum fluorescence. For *in vivo* experiments, however, a fluorescent probe should be chosen that is either insensitive to pH or exhibits a pK_a value less than or equal to the intracellular pH. In some cases, intracellular pH information can be obtained by measuring pH-dependent fluorescence spectra *in vivo*. Another factor is that the charged and uncharged fluorophore species often exhibit different absorption spectra. In this case, care must be taken in selecting the optimum excitation wavelength.

The fluorescence intensity and photostability of the fluorescent substrate are also considerations in kinetic assays. The molar fluorescence intensity of a fluorophore is related to the product of its quantum yield (Q) and molar extinction coefficient (ϵ). Since many common fluorophores exhibit similar quantum yield values (0.5–0.9), a rough estimate of fluorescence intensity can be obtained by comparing extinction coefficient values. It is important to note that the quantum yield is often lowered by conjugation to a protein or other environmental factors. Prolonged exposure to incident radiation can cause the fluorescence of the sample to diminish with time through photobleaching, which is essentially a photochemical reaction of the absorbing compound (often with the aid of molecular oxygen). It is estimated, for example, that fluorescein can survive only about 10^5 excitations before decomposing (Mathies and Stryer, 1986). In general, longer-wavelength-absorbing fluorophores (rhodamine, resorufin, etc.) are more photostable than shorter wavelength (i.e., fluorescein-based) dyes. Several antifade reagents [e.g., diazabicyclooctane, 1,2,3-trihydroxybenzene (pyrogallol), dithiothreitol, etc.] are commercially available for most *in vitro* and some *in vivo* applications. These reagents act as free-radical scavengers and can extend the useful life of an experiment. The new precipitating fluorophores (quinoxazolinones; see Table I) exhibit significant photostability in most applications. Their fluorescence, however, is only appreciable in the solid (crystalline) state, and they are not suitable for covalent labeling to proteins.

The chemical stability of fluorogenic substrates or fluorescent derivatives is another important consideration. For substrates, relative stability can be estimated from the pK_a value of the phenolic form of the fluorophore. In general, the lower the pK_a value, the more susceptible the substrate will be to chemical hydrolysis (acid–base decomposition). Three types of linkages (glycosides, esters, and ethers) are commonly employed in fluorogenic substrates. Glycosides contain an acetal linkage and are the least chemically stable of the three, while ethers are the most stable. In addition, fluorogenic substrates prepared from fluorescein, napththofluorescein, or similar compounds exist in the lactone form which is more susceptible to chemical hydrolysis. Long-wavelength rhodamine peptidase substrates and resorufin derivatives have been shown to have the greatest long-term stability, particularly for *in vivo* applications.

Fluorescent substrates have been used to measure enzyme activity in solutions and suspensions, in single cells, and on solid surfaces. Limiting factors include the enzymatic turnover rate and background absorbance or fluorescence, either generated by nonspecific hydrolysis or contaminating species. In order to measure enzyme kinetics, the substrate must usually be present in concentrations above the K_m value of the enzyme. For *in vivo* experiments, maintaining this concentration may become diffusion dependent, and care must be taken to insure that proper kinetic measurements are being observed. In enzyme immunoassays, the measured rate of enzyme activity (which is proportional to the concentration of enzyme) can be directly related to the concentration of a second biomolecule. Such a technique also takes advantage of the amplification of signal generated by continuous turnover of substrate ("enzyme amplification").

4.4.2. β-GALACTOSIDASE

β-Galactosidase has been employed both in enzyme-linked immunoassays (Kato *et al.*, 1975) and as a marker for *lacZ* gene fusions (Fiering *et al.*, 1991). The enzyme used in these applications is usually derived from *Escherichia coli* and has a broad specificity for a variety of β-galactopyranosides and a high turnover rate and good stability. Substrates for this enzyme include a wide variety of galactosides derived from fluorescent phenols. Currently the most sensitive substrate for this enzyme is fluorescein di-β-D-galactopyranoside (FDG), which is enzymatically converted to fluorescein via fluorescein monogalactoside (Huang, 1991). Enzymatic hydrolysis of FDG can be followed by the increase in either absorbance or fluorescence. Although the enzymatic turnover rate for FDG is considerably lower than for the common galactosidase substrate *o*-nitrophenyl galactoside (ONPG) (about 0.3 on a molar basis), the absorbance of fluorescein alone is about fivefold greater than that of *o*-nitrophenol, and fluorescence measurement results in several orders of magnitude higher sensitivity. Because of this sensitivity, the FDG–β-galactosidase system has been used in a variety of coupled-enzyme assay techniques.

Resorufin galactoside (Hofmann and Sernetz, 1984), and napththofluorescein di-β-D-galactopyranoside have the longest wavelength spectral properties of existing fluorescent substrates. Resorufin fluorescence can be monitored at neutral pH, using excitation at 570 nm where the substrates have minimal absorbance. Resorufin substrates are generally stable except in the presence of thiols such as DTT or mercaptoethanol, and have thus been used in long-duration assays. Like fluorescein digalactoside, napththofluorescein digalactoside is colorless and nonfluorescent until hydrolyzed to the red fluorescent naphthofluorescein. Measurement of naphthofluorescein fluorescence is highest above its pK_a (pH 7.6) (see Table I).

Substrates derived from 7-hydroxy-4-methylcoumarin (β-methylumbelliferone) are some of the oldest and most commonly used fluorescent substrates. The bright blue fluorescence of the phenolic hydrolysis product of methylumbelliferyl β-D-galactopyranoside (MUG) is well resolved spectrally from that of the substrate. For maximum fluorescence, the pH should be adjusted above the pK_a of the dye (pK_a = 8.2). In contrast, 3-carboxyumbelliferyl β-D-galactopyranoside (CUG) exhibits both higher absorbance (ϵ = 32,000 vs. 17,000 for MUG) and a lower pK_a (pK_a = 7.4) than MUG, which permits continuous measurement of enzymatic hydrolysis at a lower pH where the turnover rate is usually higher. In addition, the carboxylic CUG is more water soluble. This property facilitates its use at high concentrations ($\gg K_m$), and has thus been incorporated into automated (microtiterplate reader) assays. Trifluoromethylumbelliferyl β-D-galactopyranoside (TUG) is an additional galactosidase substrate with a lower pK_a (7.2) and longer wavelength excitation and emission spectra than MUG. These properties give the substrate higher relative fluorescence at physiological pH and improved detectability from endogenous fluorescent species.

The *E. coli lacZ* gene is an important genetic marker for the expression of recombinant genes in mammalian cells. Chimeric fusions of the *lacZ* gene in single cells have been detected by flow cytometry using FDG (Nolan *et al.*, 1988) and intrinsic β-galactosidase activity in single cells during a cell cycle (Yashphe and Halvorson, 1976). Cells expressing *lacZ* can be detected and sorted shortly after infection (Nicolas *et al.*, 1988). The purity of the fluorescent substrate in these techniques is very important, since an extremely low fluorescence background of the reagent is necessary to eliminate false-positive nonexpressing cells.

In order to detect β-galactosidase activity *in vivo*, several new substrates have recently been developed (Zhang *et al.*, 1991). Lipophilic versions of the fluorescein, coumarin, and resorufin galactosides described above have been demonstrated to specifically label *lacZ*-positive cells in culture at low concentration. A 12-carbon alkyl (lipophilic) modification of these substrates provides the best combination of cell permeability and product retention (Jasin and Zalamea, 1992; Westerfield *et al.*, 1992). The chloromethylfluorescein derivative of FDG and the coumarin derivative of MUG have also been used to improve retention of the fluorescent product in *lacZ* positive cells *in vivo*. These compounds can also act as substrates for the ubiquitous glutathione S-transferase system, which conjugates them *in vivo* to the tripeptide glutathione (GSH). Finally, several new substrates prepared from the 2′-hydroxyphenyl-4-(3H)quinoxazolinones ("ELF") precipitating fluorophores have also shown utility in labeling for β-galactosidase *in vivo* (Naleway *et al.*, 1994). Since the product of these substrates is difficult to quantitate, their use has been limited to qualitative analyses at present.

4.4.3. β-GLUCURONIDASE

The most widely used fluorescent substrate for detection of β-glucuronidase activity is 4-methylumbelliferyl β-D-glucuronide (MUGlcU). This compound is virtually nonfluorescent until cleaved by β-glucuronidase to release 4-methylumbelliferone (MU, also known as 7-hydroxy-4-methyl coumarin). Using excitation at 363 nm and measuring emission at 447 nm, background fluorescence from the substrate is negligible. It is important to note, however, that the substrate does exhibit slight intrinsic fluorescence (emission maximum 375 nm) if excited at 316 nm. The excitation source and/or appropriate filter sets should be chosen, therefore, with this in mind. An example would be a DAPI/Hoechst filter set with excitation bandwidth at 365 ± 30 nm and emission at wavelengths greater than 420 nm.

In general, measurements of β-glucuronidase activity using MUGlcU are two to three orders of magnitude more sensitive than other common glucuronidase substrates (5-bromo-4-chloro-3-indolyl-β-D-glucuronide or *p*-nitrophenyl-β-D-glucuronide). The product itself is a pH-sensitive fluorescent probe (pK_a 8.2) that exhibits maximal fluorescence at pH values above its pK_a. At physiological pH values, the fluorescence is relatively low. This is due to the equilibrium between the phenolic form of the product and its phenoxide form at high pH values. It is the phenoxide form of the dye that gives maximal fluorescence at 447 nm. The phenolic form of the dye actually has different fluorescence properties (excitation maximum 323 nm, emission maximum 386 nm). Treatment of samples with 0.2 M sodium carbonate buffer (pH 9.5) after incubation serves the dual purpose of stopping the enzyme reaction and maximizing the fluorescence by increasing the pH of the system above the pK_a of MU.

Other fluorescent glucuronidase substrates have been synthesized from 4-trifluoromethylumbelliferone, resorufin, and fluorescein (Naleway, 1992). The first of these, 4-trifluoromethylumbelliferyl β-D-glucuronide (TUGlcU), exhibits nearly the same enzymatic turnover rate as MUGlcU. In addition, it offers the following two advantages over MUGlcU: (1) it is not normally necessary to terminate the reaction with a stop buffer because the pK_a of the cleavage product (trifluoromethylumbelliferone, TU) is lower than that of 4-methylumbelliferone (pK_a values of 7.3 and 8.2 for TU and MU, respectively); and (2) the excitation and emission maxima of trifluoromethylumbelliferone (393 nm and 502 nm, respectively) are red-shifted relative to MU, which improves the sensitivity of detection.

Resorufin β-D-glucuronic acid (Res-GlcU) is cleaved by β-glucuronidase to produce resorufin, which has a high extinction coefficient (56,000), a high quantum yield, and desirable wavelength properties (excitation maximum 571 nm; emission maximum 584 nm) that are in a range where most cells do not absorb or fluoresce. Further, resorufin fluoresces maximally at neutral pH values and is relatively stable, except in the presence of reducing agents (e.g., DTT) in which it readily undergoes reduction to a nonfluorescent product. The di-glucuronide derivative of fluorescein (FDGlcU) has also been used in β-glucuronidase activity measurements, but with limited success, mainly due to its poor membrane permeability and autohydrolysis properties.

Fluorogenic β-glucuronidase substrates have been used to investigate the following pharmaceutical and diagnostic applications: (1) phagocytosis by polymorphonuclear leukocytes (Suzuki *et al.*, 1988); (2) immune lysis of liposomes (Thompson and Gaber, 1985); and (3) bacterial contamination in food. An important nonpharmaceutical application involves fusion of the β-glucuronidase (GUS) gene from *E. coli* into plant cells as a marker of plant gene fusions (Jefferson, 1987). Detection of β-glucuronidase activity using fluorescent substrates has proven to be orders of magnitude more sensitive than absorption measurements. However, attempts to detect β-glucuronidase activity *in vivo* using fluorescent derivatives have met with only limited success, mainly because of the impermeable nature of these substrates to the cell membrane. Potential substrates for such analysis are the acetoxymethyl esters of glucuronides that have been used to help load certain ion indicators for intracellular measurements. Finally, intrinsic fluorescence in cellular extracts, usually from large amounts of chlorophyll, or absorption from endogenous chromophores, which results in quenching, are serious problems in fluorescent analysis of β-glucuronidase activity in plant tissues.

4.5. Interfacial Protein Studies

Experimental observations have shown that man-made surfaces can affect the biological function of surface-bound proteins. In particular, protein adsorption often leads to surface-induced denaturation and loss of biological function (Andrade, 1985), although the exact nature and extent of the denaturation process are somewhat elusive. Thus, the rationale for a majority of interfacial protein studies is to ascertain how surfaces affect the structure, function, and orientation of adsorbed proteins. As in solution studies, fluorescence spectroscopy is well-suited to address these issues. Almost every fluorescence technique employed in solution can also be applied to interfacial assemblies of proteins using TIRF spectroscopy or conventional fluorescence techniques provided that the majority of observed fluorescence originates from the surface-bound species. The reader is referred to Hlady (1994) for a review of these topics.

The first decision to be made in designing an interfacial fluorescence experiment is the selection of the fluorophore. Extrinsic fluorophores have been used extensively in studying adsorption kinetics and equilibria (Cheng *et al.*, 1985). In addition, small fluorescent ligands (e.g., ANS) that bind to the adsorbed protein have been used as probes to report on the microenvironment of the adsorbed layer (Hlady and Andrade, 1989). It should be mentioned, however, that the introduction of extrinsic fluorophores may lead to changes in the interfacial behavior of a given protein, thus influencing the same processes about which the probe is supposed to report. Also, it is important to have knowledge of the number and position of extrinsic fluorophores in the protein molecule in order to properly interpret the results of an experiment.

An alternative is to use intrinsic (tryptophan and/or tyrosine) fluorescence to monitor protein adsorption and other interfacial processes. Since the fluorescence of a tryptophan residue is sensitive to dipolar relaxation and polarity changes that occur in its local environment, this approach is useful in conformational studies of proteins that contain a limited number of tryptophan residues (Hlady and Andrade, 1988). A recent example is a study of recombinant human growth hormone (rhGH) in which fluorescence emission and circular dichroism spectra were used to examine the conformation of rhGH adsorbed to silica particles (Clark *et al.*, 1993). A negative aspect of utilizing intrinsic protein fluorescence is its susceptibility to photodegradation under UV illumination. The reader is referred to a review by Hlady *et al.* (1985) for a more complete discussion of the use of intrinsic fluorescence in protein adsorption studies.

A shortcoming of all interfacial fluorescence studies is the lack of a firm quantitative relationship between the intensity of emitted fluorescence and the absolute surface concentration of bound protein. The fluorescence quantum yield of either intrinsic or extrinsic fluorophores may change due to: (1) the influence of the dielectric interface on fluorescence emission; (2) lateral protein–protein interactions; and (3) the close proximity of the fluorophores in the surface protein layer. If absolute surface concentrations are required, interfacial fluorescence measurements need to be calibrated and/or supplemented by another more quantitative technique. Several TIRF calibration schemes have been proposed for this purpose based on the use of either radiolabeled proteins (Hlady *et al.*, 1986) or a strong photobleaching laser pulse (Zimmermann *et al.*, 1990).

The accessibility of tryptophan residues in multitryptophan proteins and the average conformation of such proteins in the adsorbed state can be inferred from fluorescence-quenching studies. For example, Horsley *et al.* (1991) used iodide quenching to probe the conformation of adsorbed hen and human lysozyme molecules. Table II lists the Stern–Volmer quenching parameters for both types of lysozyme adsorbed to different silica surfaces. When adsorbed to hydrophobic silica surfaces, both lysozymes showed almost no change in fractional accessibility (f_a), while the effective quenching constant (K_Q) increased to a different extent in each case. In contrast, adsorption of either type of lysozyme to unmodified silica caused a twofold increase in f_a and a decrease in K_Q. The change in fractional accessibility suggested that the conformation of lysozyme was perturbed upon adsorbing to unmodified silica surfaces, probably resulting in a more

Table II. Stern-Volmer Quenching Parameters for Human and Hen Lysozymes Adsorbed at Different Silica Surfaces

Sample	Fraction of accessible fluorophores (f_a)	Quenching constant (K_Q) (M^{-1})
Buffer solution[a]		
Hen lysozyme	0.37 ± 0.02	2.8 ± 0.3
Human lysozyme	0.18 ± 0.03	2.6 ± 0.6
Hydrophobicized silica[b]		
Hen lysozyme	0.34 ± 0.04	3.3 ± 0.7
Human lysozyme	0.20 ± 0.04	4.2 ± 1.6
Unmodified silica surface[c]		
Hen lysozyme	0.56 ± 0.10	1.7 ± 0.4
Human lysozyme	0.53 ± 0.23	0.9 ± 0.4
APS-silica surface[d]		
Hen lysozyme	0.36 ± 0.03	5.7 ± 1.1
Human lysozyme	0.68 ± 0.10	5.0 ± 1.6

[a]Phosphate-buffered saline, pH 7.4 (PBS).
[b]Silica surfaces were silanized with dichlorodimethylsilane.
[c]Unmodified silica surfaces are negatively charged in PBS.
[d]Silica surfaces were silanized with aminopropyltriethoxysilane (APS) and are positively charged in PBS.

open conformation in which the previously shielded tryptophan residues became exposed to solvent and hence more accessible to iodide ions. The decrease in K_Q can be attributed to electrostatic repulsion between iodide ions and the negatively charged silica surface. When the silica surface was modified to carry positive charges (APS–silica), K_Q increased due to electrostatic attraction. The fractional accessibility of adsorbed human lysozyme increased as well, indicating that this molecule was less conformationally stable at positively charged surfaces than hen lysozyme.

Protein prosthetic groups such as the heme moiety are often useful in interfacial protein fluorescence studies. The average orientation of adsorbed cytochrome *c* has been deduced from linear dichroism of its surface fluorescence (Fraaje *et al.*, 1990). In addition, protein–protein interactions can be studied by immobilizing one type of protein to the interface and monitoring how it interacts with other types of proteins in bulk solution. The formation of an antigen–antibody complex is one example of such interactions and TIRF spectroscopy has become a favorite optical technique for development of fluorescence immunosensors and fluoroimmunoassays (Andrade *et al.*, 1990; Herron *et al.*, 1993).

Resonance energy transfer can be used to examine conformational changes in adsorbed proteins by measuring the average distance between donors and acceptors in double labeled proteins (Burghardt and Axelrod, 1983; Brynda *et al.*, 1990). The interpretation of such experiments is not always straightforward; the rate of energy transfer in an adsorbed layer may differ from that in a three-dimensional system because of the two-dimensional geometry and/or the presence of the dielectric interface (Kellerer and Blumen, 1984). Interestingly, protein aggregation at the interface can also cause a decrease in the average distance between donors and acceptors. The changes in energy transfer efficiency can then be interpreted in terms of the organization of the adsorbed layer, rather than as a conformational change within a (single) adsorbed molecule (Zheng, 1994).

The dynamics of proteins in solution can be successfully probed by time-resolved fluorescence measurements (see Sections 2.3, 2.5, and 4.2). However, application of such techniques to surface-bound proteins is complicated by the high surface density of protein in the adsorbed state. At high concentrations of adsorbed protein, intrinsic fluorophores are brought into close proximity to each other and can interact through a number of mechanisms, including homoenergy transfer (i.e., energy transfer in which both donor and acceptor are the same type of fluorophore). Such effects will dramatically influence the observed decay rates of emitting species by making them nonexponential (Rumbles *et al.*, 1991). As a rule, lifetimes of extrinsically labeled proteins at interfaces also become shorter. Such measurements can be interpreted in terms of protein supramolecular organization in the adsorbed layer rather than in terms of protein structure and dynamics (Suci and Hlady, 1990; Crystall *et al.*, 1993).

5. CONCLUSIONS

Fluorescence spectroscopy possesses several unique qualities that makes it suitable for examining the structure and function of proteins. First, it is one of the most sensitive techniques discussed in this volume, being able to detect picogram quantities (or less) of protein. This exquisite sensitivity is exploited in both fluorescent immunoassays and fluorescence microscopy to detect minute amounts of material in biological specimens and in cells. Both of these applications are relevant to the pharmaceutical scientist because they are routinely employed during the purification and formulation of therapeutic proteins, for identification of new potentially therapeutic proteins, and for evaluation of the efficacy of targeted drug delivery systems.

Second, the fluorescence emission spectrum is very sensitive to the environment of the fluorophore and can be used to examine the physical state of the protein. For example, the emission maximum of a protein will often shift 20–30 nm to the red upon denaturation. Thus, emission spectra can be employed during formulation of protein pharmaceuticals as a monitor of protein stability. Although an invariant emission maximum does not guarantee that the protein is active, a red shift in the emission spectrum after a particular formulation step strongly suggests that a major (and probably destabilizing) conformational change has occurred. Such observations should be confirmed initially by other techniques. However, once the exact meaning of a spectral shift has been established, it is often convenient to use fluorescence measurements alone as a measure of quality control.

Third, the fluorescence lifetime of the fluorophore can be utilized as sort of a "molecular stopwatch" to examine processes that occur on a nanosecond time scale. Several important reactions or interactions of proteins can be investigated using techniques such as fluorescence lifetime, quenching, anisotropy, and resonance energy transfer. These include: (1) protein folding; (2) conformational fluctuations of proteins; (3) the hydrodynamics and segmental flexibility of proteins; (4) ligand-binding events such as drug binding to plasma proteins and drug-receptor interactions; and (5) determination of intra- and intermolecular distances. Almost any of these may be employed during formulation to monitor the structure and/or activity of the protein.

Fourth, in cases where proteins are employed therapeutically because of their enzymatic activities, fluorogenic substrates provide a convenient means to monitor activity during formulation. With the wide variety of such probes presently available, this approach is applicable to most common enzymatic activities. For reasons mentioned above, fluorogenic substrates are several orders of magnitude more sensitive than their chromogenic counterparts, which offers unique possibilities for examining the efficacy of enzyme therapy *in vivo*. Although this *in vivo* approach has not been employed with the present generation of protein pharmaceuticals, the availability of a number of new fluorescent probes for *in vivo* applications may facilitate such studies in the near future.

Finally, total internal reflection fluorescence provides a convenient means to investigate the interactions of proteins at interfaces. Although often overlooked by pharmaceutical scientists, protein adsorption to glass and plastic is a major factor in both the quantitation and activity of protein products. This is especially true when such products are formulated at relatively low concentrations (< 0.1 mg/ml). In general, any fluorescent technique that can be used in bulk solution can also be employed at solid–liquid interfaces, although time-resolved TIRF measurements are still very difficult to perform and not readily available to the general practitioner.

ACKNOWLEDGMENTS. This work was supported in part by a NATO Science Fellowship awarded to W. J. by NWO, the Dutch Organization for Scientific Research; and U. S. Public Health Service Grant No. AI 22898 awarded to J. N. H.

REFERENCES

Alcala, J. R., Gratton, E., and Jameson, D. J., 1985, A multifrequency phase fluorometer using the harmonic content of a mode-locked laser, *Anal. Instrum.* **14:**225–250.

Alcala, J. R., Gratton, E., and Prendergast, F. G., 1987a, Resolvability of fluorescence lifetime distributions using phase fluorometry, *Biophys. J.* **51:**587–596.

Alcala, J. R., Gratton, E., and Prendergast, F. G., 1987b, Fluorescence lifetime distributions in proteins, *Biophys. J.* **51:**597–604.

Alcala, J. R., Gratton, E., and Prendergast, F. G., 1987c, Interpretation of fluorescence decays in proteins using continuous lifetime distributions, *Biophys. J.* **51:**925–936.

Anderson, S. R., and Malencik, D. A., 1989, Fluorescence studies of the calcium-dependent functions of calmodulin, in: *Fluorescent Biomolecules, Methodologies and Applications* (D. M. Jameson and G. D. Reinhart, eds.), Plenum Press, New York, pp. 217–245.

Andrade, J. D. (ed.), 1985, *Surface and Interfacial Aspects of Biomedical Polymers*, Vol. 2, *Protein Adsorption*, Plenum Press, New York.

Andrade, J. D., Lin, J.-N., Hlady, V., Herron, J. N., Christensen, D., and Kopecek, J., 1990, Immunosensors: Remaining problems in the development of remote, continuous, multichannel devices, in: *Biosensor Technology, Fundamentals and Applications* (R. P. Buck, W. E. Hatfield, M. Umana, and E. F. Bowden, eds.), Marcel Dekker, New York, pp. 219–249.

Atkins, W. M., Stayton, P. S., and Villafranca, J. J., 1991, Time-resolved fluorescence studies of genetically engineered *Escherichia coli* glutamine synthetase. Effects of ATP on the tryptophan-57 loop, *Biochemistry* **30:**3406–3416.

Axelrod, D., 1989, Total internal reflection fluorescence microscopy, in: *Fluorescence Microscopy of Living Cells in Culture B* (D. L. Taylor and Y.-L. Wang, eds.), Vol. 30, *Methods in Cell Biology*, Academic Press, San Diego, pp. 245–270.

Axelrod, D, Hellen, E. H., and Fulbright, R. M., 1992, Total internal reflection fluorescence, in: *Topics in Fluorescence Spectroscopy*, Vol. 3, *Biochemical Applications* (J. R. Lakowicz, ed.), Plenum Press, New York, pp. 289–344.

Baruch, D., Glickstein, H., and Cabantchik, Z. I., 1991, *Plasmodium falciparum*: Modulation of surface antigenic expression of infected erythrocytes as revealed by cell fluorescence ELISA, *Exp. Parasitol.* **73:**440–450.

Beechem, J. M., 1992, Global analysis of biochemical and biophysical data, *Methods Enzymol.* **210:**37–54.

Beechem, J. M., and Brand, L., 1985, Time-resolved fluorescence of proteins, *Annu. Rev. Biochem.* **54:**43–71.

Beechem, J. M., and Gratton, E., 1988, Fluorescence spectroscopy data analysis environment: A second generation global analysis program, *Proc. S.P.I.E.* **909:**70–81.

Beechem, J. M., Ameloot, M., and Brand, L., 1985, Global and target analysis of complex decay phenomena, *Anal. Instrum.* **14:**379–402.

Belford, G. G., Belford, R. L., and Weber, G., 1972, Dynamics of fluorescence polarization in macromolecules, *Proc. Natl. Acad. Sci. USA* **69:**1392–1393.

Bentley, K. L., Thompson, L. K., Klebe, R. J., and Horowitz, P. M., 1985, Fluorescence

polarization: A general method for measuring ligand binding and membrane viscosity, *BioTechniques* **3:**356–366.
Berman, H. A., and Leonard, K., 1990, Ligand exclusion on acetylcholinesterase, *Biochemistry* **29:**10640–10649.
Bismuto, E., Gratton, E., and Irace, G., 1988, Effect of unfolding on the tryptophanyl fluorescence lifetime distribution in apomyoglobin, *Biochemistry* **27:**2132–2136.
Bismuto, E., Irace, G., and Gratton, E., 1989, Multiple conformational states in myoglobin revealed by frequency domain fluorometry, *Biochemistry* **28:**1508–1512.
Blatt, E., Husain, A., and Sawyer, W. H., 1986, The association of acrylamide with proteins. The interpretation of fluorescence quenching experiments, *Biochim. Biophys. Acta* **871:**6–13.
Blumenthal, D. K., 1993, Development and characterization of fluorescently-labeled myosin light chain kinase calmodulin-binding domain peptides, *Mol. and Cell. Biochem.* **127/128:**45–50.
Blumenthal, D. K., and Krebs, E. G., 1988, Calmodulin-binding domains on target proteins, in: *Molecular Aspects of Cellular Regulation*, Vol. 5, *Calmodulin* (P. Cohen and C. B. Klee, eds.), Elsevier, Amsterdam, pp. 341–355.
Blumenthal, D. K, Takio, K., Edelman, A. M., Charbonneau, H., Titani, K., Walsh, K. A., and Krebs, E. G., 1985, Identification of the calmodulin-binding domain of skeletal muscle myosin light chain kinase, *Proc. Natl. Acad. Sci. USA* **82:**3187–3191.
Brandt, R., and Keston, A. S., 1965, Synthesis of diacetyldichlorofluorescin: A stable reagent for fluorometric analysis, *Anal. Biochem.* **11:**6–9.
Brems, D. N., and Havel, H. A., 1989, Folding of bovine growth hormone is consistent with the molten globule hypothesis, *Proteins* **5:**93–95.
Brems, D. N., Brown, P. L., and Becker, G. W., 1990, Equilibrium denaturation of human growth hormone and its cysteine-modified forms, *J. Biol. Chem.* **265:**5504–5511.
Bright, F. V., 1989, Multifrequency phase fluorescence study of hapten–antibody complexation, *Anal. Chem.* **61:**309–313.
Brynda, E., Hlady, V., and Andrade, J. D., 1990, Protein packing in adsorbed layers studied using excitation energy transfer, *J. Colloid Interface Sci.* **139:**374–380.
Bucci, E., and Steiner, R. F., 1988, Anisotropy decay of fluorescence as an experimental approach to protein dynamics, *Biophys. Chem.* **30:**199–224.
Buchner, J., Renner, M., Lilie, H., Hinz, H.-J., Jaenicke, R., Kiefhaber, T., and Rudolph, R., 1991, Alternatively folded states of an immunoglobulin, *Biochemistry* **30:**6922–6929.
Burghardt, T. P., and Axelrod, D., 1983, Total internal reflection fluorescence study of energy transfer in surface adsorbed and dissolved bovine serum albumin, *Biochemistry* **22:**979–985.
Burke, M. D., Thompson, S., Elcombe, C. R., Halpert, J., Haaparanta, T., and Mayer, R. T., 1985, Ethoxy-, pentoxy-, and benzyloxyphenoxazones and homologues: A series of substrates to distinguish between different induced cytochromes P-450, *Biochem. Pharmacol.* **34:**3337–3345.
Burstein, E. A., Permyakov, E. A., Yashon, V. A., Burkhanov, S. A., and Finnazzi-Agrò, A., 1977, The fine structure of luminescence spectra of azurin, *Biochem. Biophys. Acta* **491:**155–159.
Calhoun, D. B., Vanderkooi, J. M., Holtom, G. R., and Englander, S. W., 1986, Protein

fluorescence quenching by small molecules: Protein penetration versus solvent exposure, *Proteins* **1**:109–115.

Canonica, S., and Wild, U. P., 1985, Single photon counting with synchronously pumped dye laser excitation, *Anal. Instrum.* **14**:331–357.

Careri, G., Fasella, P., and Gratton, E., 1979, Enzyme dynamics: The statistical physics approach, *Annu. Rev. Biophys. Bioeng.* **8**:69–97.

Cathcart, R., Schwiers, E., and Ames, B. N., 1983, Detection of picomole levels of hydroperoxides using a fluorescent dichlorofluorescein assay, *Anal. Biochem.* **134**: 111–116.

Chabbert, M., Lukas, T. J., Watterson, D. M., Axelsen, P. H., and Prendergast, F. G., 1991, Fluorescence analysis of calmodulin mutants containing tryptophan: Conformational changes induced by calmodulin-binding peptides from myosin light chain kinase and protein kinase II, *Biochemistry* **30**:7615–7630.

Chakrabarti, A., Matko, J., Rahman, N. A., Barisas, B. G., and Edidin, M., 1992, Self-association of class I major histocompatibility complex molecules in liposome and cell surface membranes, *Biochemistry* **31**:7182–7189.

Chatelain, P., Matteazzi, J.-R., and Laruel, R., 1994, Binding of fantofarone, a novel Ca^{2+} antagonist, to serum albumin: A fluorescence study, *J. Pharm. Sci.* **83**:674–676.

Chen, R. F., 1976, The effect of metal cations on intrinsic protein fluorescence, in: *Biochemical Fluorescence*, Vol. 2 (R. F. Chen and H. Hedelhoch, eds.), Marcel Dekker, New York, pp. 573–606.

Cheng, Y., Lok, B. K., and Robertson, C. R., 1985, Interactions of macromolecules with surfaces in shear fields using visible wavelength total internal reflection fluorescence, in: *Surface and Interfacial Aspects of Biomedical Polymers*. Vol. 2, *Protein Adsorption* (J. D. Andrade, ed.), Plenum Press, New York, pp. 121–160.

Clark, S. R., Billstein, P., Mandenius, C. F., and Elwing, H., 1993, Fluorometric investigation of recombinant human growth hormone adsorbed to silica nanoparticles, *Anal. Chim. Acta* **290**:21–26.

Creed, D., 1984, The photophysics and photochemistry of the near-UV absorbing amino acids-I. Tryptophan and its simple derivatives, *Photochem. Photobiol.* **39**:537–562.

Crystall, B., Rumbles, G., Smith, T. A., and Phillips, D., 1993, Time-resolved evanescent wave induced fluorescence measurements of surface adsorbed bovine serum albumin, *J. Colloid Interface Sci.* **155**:247–250.

Daniel, E., and Weber, G., 1966, Cooperative effects in binding by bovine serum albumin. I. The binding of 1-anilino-8-naphthalenesulfonate. Fluorimetric titrations, *Biochemistry* **5**:1893–1900.

Day, E. D., 1990, *Advanced Immunochemistry* 2nd ed., Wiley-Liss, New York.

Demas, J. N., 1983, *Excited State Lifetime Measurements*, Academic Press, New York.

Demchenko, A. P., 1986, *Ultraviolet Spectroscopy of Proteins*, Springer-Verlag, Berlin.

Dughi, C., Bhagavan, N. V., and Jameson, D. M., 1993, Fluorescence investigations of albumin from patients with familial dysalbuminemic hyperthyroxinemia, *Photochem. Photobiol.* **57**:416–419.

Eccleston, J. F., Gratton, E., and Jameson, D. M., 1987, Interaction of a fluorescent analogue of GDP with elongation factor Tu: Steady-state and time-resolved fluorescent studies, *Biochemistry* **26**:3902–3907.

Eftink, M. R., 1991, Fluorescence techniques for studying protein structure, in: *Methods of Biochemical Analysis*, Vol. 35, *Protein Structure Determination* (C. H. Suelter, ed.), John Wiley & Sons, New York, pp. 127–205.

Eftink, M. R., and Ghiron, C. A., 1981, Fluorescence quenching studies with proteins, *Anal. Biochem.* **114:**199–227.

Ellis, J., Murray, I. A., and Shaw, W. V., 1991, Intrinsic fluorescence of chloramphenicol acetyltransferase: Responses to ligand binding and assignment of the contributions of tryptophan residues by site-directed mutagenesis, *Biochemistry* **30:**10799–10805.

Elofsson, A., Rigler, R., Nilsson, L., Roslund, J., Krause, G., and Holmgren, A., 1991, Motion of aromatic side chains, picosecond fluorescence, and internal energy transfer in *Escherichia coli* thioredoxin studied by site-directed mutagenesis, time-resolved fluorescence spectroscopy, and molecular dynamics simulations, *Biochemistry* **30:** 9648–9656.

Fernando, T., and Royer, C. A., 1992, Unfolding of the *trp* repressor studied using fluorescence spectroscopic techniques, *Biochemistry* **31:**6683–6691.

Ferreira, S. T., 1989, Fluorescence studies of the conformational dynamics of parvalbumin in solution: Lifetime and rotational motions of the single tryptophan residue, *Biochemistry* **28:**10066–10072.

Fiering, S. N., Roederer, M., Nolan, G. P., Micklem, D. R., Parks, D. R., and Herzenberg, L. A., 1991, Improved FACS-Gal: Flow cytometric analysis and sorting of viable eukaryotic cells expressing reporter gene constructs, *Cytometry* **12:**291–301.

Förster, T., 1948, Intermolecular energy migration and fluorescence, *Ann. Phys.* (Leipzig) **2:**55–57 (trans. R. S. Knox).

Fraaje, J. G. E. M., Kleijn, J. M., van der Graff, M., and Dijt, J. C., 1990, Orientation of adsorbed cytochrome *c* as a function of the electrical potential of the interface studied by total internal reflection fluorescence, *Biophys. J.* **52:**367–379.

Gabor, G., 1968, Radiationless energy transfer through a polypeptide chain, *Biopolymers* **6:** 809–816.

Goto, Y., Calciano, L. J., and Fink, A. L., 1990, Acid-induced folding of proteins, *Proc. Natl. Acad. Sci. USA* **87:**573–577.

Gratton, E., and Jameson, D. M., 1985, New approach to phase and modulation resolved spectra, *Anal. Chem.* **57:**1694–1697.

Gratton, E., and Limkeman, M., 1983, A continuously variable frequency cross-correlation phase fluorometer with picosecond resolution, *Biophys. J.* **44:**315–324.

Gratton, E., Jameson, D. M., and Hall, R. D, 1984a, Multifrequency phase and modulation fluorometry, *Annu. Rev. Biophys. Bioeng.* **13:**105–124.

Gratton, E., Jameson, D. M., Weber, G., and Albert, B., 1984b, A model of dynamic quenching of fluorescence in globular proteins, *Biophys. J.* **45:**789–794.

Harris, D. L., and Hudson, B. S., 1990, Photophysics of tryptophan in bacteriophage T4 lysozymes, *Biochemistry* **29:**5276–5285.

Haugland, R. P., 1991, Fluorescent labels, in: *Biosensors with Fiberoptics* (D. L. Wise and L. B. Wingard, eds.), Humana Press, Clifton, NJ, pp. 85–109.

Havel, H. A., Kauffman, E. W., and Elzinga, P. A., 1988, Fluorescence quenching studies of bovine growth hormone in several conformational states, *Biochim. Biophys. Acta* **955:** 154–163.

Hazlett, T. L., Johnson, A. E., and Jameson, D. M., 1989, Time-resolved fluorescence studies on the ternary complex formed between bacterial elongation factor Tu, guanosine 5′-triphosphate, and phenylalanyl-tRNA^{Phe}, *Biochemistry* **28:**4109–4117.

Herron, J. N., 1984, Equilibrium and kinetic methodology for the measurement of binding properties in monoclonal and polyclonal populations of antifluorescyl-IgG antibodies, in: *Fluorescein Hapten: An Immunological Probe* (E. W. Voss, Jr., ed.), CRC Press, Boca Raton, FL, pp. 49–76.

Herron, J. N., Kranz, D. M., Jameson, D. M., and Voss, E. W., Jr., 1986, Thermodynamic properties of ligand binding by monoclonal antifluorescyl antibodies, *Biochemistry* **25:**4602–4609.

Herron, J. N., He, X.-M., Mason, M. L., Voss, E. W., Jr., and Edmundson, A. B., 1989, Three-dimensional structure of a fluorescein-Fab complex crystallized in 2-methyl-2, 4-pentanediol, *Proteins* **5:**271–280.

Herron, J. N., Christensen, D. A., Hlady, V., Janatova, V., Wang, H.-K., and Wei, A.-P., 1993, Fluorescent immunosensors using planar waveguides, in: *Advances in Fluorescence Sensing Technology* (J. R. Lakowicz and R. B. Thompson, eds.), SPIE, Bellingham, WA, pp. 28–39.

Herron, J. N., Terry, A. H., Johnston, S., He, X.-M., Guddat, L. W., Voss, E. W., Jr., and Edmundson, A. B., 1994, High resolution structures of the 4-4-20 Fab-fluorescein complex in two solvent systems: effects of solvent on structure and antigen-binding affinity, *Biophys. J.* **67:**2167–2183.

Hlady, V., 1994, Spectroscopic and other techniques for studying adsorption of bioproducts at interfaces, in: *Interfacial Behavior of Bioproducts* (J. L. Brash and P. Wojciechovski, eds.), M. Dekker, New York, in press.

Hlady, V., and Andrade, J. D., 1988, Fluorescence emission from adsorbed bovine serum albumin and albumin bound 1-anilinonaphthalene-8-sulfonate studied by TIRF, *Colloids Surfaces* **32:**359–368.

Hlady, V., and Andrade, J. D., 1989, A TIRF titration study of 1-anilinonaphthalene-8-sulfonate binding to silica-adsorbed bovine serum albumin, *Colloids Surfaces* **42:** 85–96.

Hlady, V., Van Wagenen, R. A., and Andrade, J. D., 1985, Total internal reflection intrinsic fluorescence (TIRIF) spectroscopy applied to protein adsorption, in: *Surface and Interfacial Aspects of Biomedical Polymers*, Vol. 2, *Protein Adsorption* (J. D. Andrade, ed.), Plenum Press, New York, pp. 81–119.

Hlady, V., Reinecke, D. R., and Andrade, J. D., 1986, Fluorescence of adsorbed protein layers: Quantitation of total internal reflection fluorescence, *J. Colloid Interface Sci.* **111:**555–569.

Hofmann, J., and Sernetz, M., 1984, Immobilized enzyme kinetics analysed by flow-through microfluorimetry: Resorufin β-*D*-galactopyranoside as a new fluorogenic substrate for β-galactosidase, *Anal. Chim. Acta* **163:**67–72.

Holzman, T. F., Dougherty, J. J., Brems, D. N., and MacKenzie, N. E., 1990, pH-induced conformational states of bovine growth hormone, *Biochemistry* **29:**1255–1261.

Horsley, D., Herron, J., Hlady, V., and Andrade, J. D., 1991, Fluorescence quenching of adsorbed hen and human lysozymes, *Langmuir* **7:**218–222.

Housey, G. M., Johnson, M. D., Hsiao, W. L. W., O'Brian, C. A., Murphy, J. P.,

Kirschmeier, P., and Weinstein, I. B., 1988, Overproduction of protein kinase c causes disordered growth control in rat fibroblasts, *Cell* **52:**343–354.

Huang, Z., 1991, Kinetic fluorescence measurement of fluorescein di-β-*D*-galactoside hydrolysis by β-galactosidase: Intermediate channeling in stepwise catalysis by a free single enzyme, *Biochemistry* **30:**8535–8540.

Jameson, D. M., and Hazlett, T. L., 1991, Time-resolved fluorescence in biology and biochemistry, in: *Biophysical and Biochemical Aspects of Fluorescence Spectroscopy* (T. G. Dewey, ed.), Plenum Press, New York, pp. 105–133.

Jameson, D. M., Gratton, E., and Hall, R. D., 1984a, The measurement and analysis of heterogeneous emissions by multifrequency phase and modulation fluorometry, *Appl. Spectrosc. Rev.* **20:**55–106.

Jameson, D. M., Gratton, E., Weber, G., and Albert, B., 1984b, Oxygen distribution and migration within $Mb^{DES\ Fe}$ and $Hb^{DES\ Fe}$. Multifrequency phase and modulation fluorometry study, *Biophys. J.* **45:**795–803.

Jameson, D. M., Thomas, V., and Zhou, D.-M., 1989, Time-resolved fluorescence studies on NADH bound to mitochondrial malate dehydrogenase, *Biochim. Biophys. Acta* **994:** 187–190.

Jasin, M., and Zalamea, P., 1992, Analysis of *Escherichia coli* β-galactosidase expression in transgenic mice by flow cytometry of sperm, *Proc. Natl. Acad. Sci. USA* **89:**10681–10685.

Jefferson, R. A., 1987, Assaying chimeric genes in plants: The GUS gene fusion system, *Plant Mol. Biol. Rep.* **5:**387–405.

Jiskoot, W., Bloemendal, M., Van Haeringen, B., Van Grondelle, R., Beuvery, E. C., Herron, J. N., and Crommelin, D. J. A., 1991a, Non-random conformation of a mouse IgG2a monoclonal antibody at low pH, *Eur. J. Biochem.* **201:**223–232.

Jiskoot, W., Hoogerhout, P., Beuvery, E. C., Herron, J. N., and Crommelin, D. J. A., 1991b, Preparation and application of a fluorescein-labeled peptide for determining the affinity constant of a monoclonal antibody-hapten complex by fluorescence polarization, *Anal. Biochem.* **196:**421–426.

Johansen, A.-K., Willassen, N.-P., and Sager, G., 1992, Fluorescent studies of β-adrenergic ligand binding to α_1-acid glycoprotein with 1-anilino-8-naphthalene sulfonate, isoprenaline, adrenaline and propranolol, *Biochem. Pharmacol.* **43:**725–729.

Jullian, C., Brunet, J. E., Thomas, V., and Jameson, D. M., 1989, Time-resolved fluorescence studies on protoporphyrin IX-apohorseradish peroxidase, *Biochim. Biophys. Acta* **997:**206–210.

Kasten, F. H., 1989, The origins of modern fluorescence microscopy and fluorescent probes, in: *Cell Structure and Function by Microspectrofluorometry* (E. Kohen and J. S. Hirschberg, eds.), Academic Press, New York, pp. 3–50.

Kato, K., Hamaguchi, Y., Fukui, H., and Ishikawa, E., 1975, Enzyme-linked immunoassay I. Novel method for synthesis of the insulin-beta-galactosidase conjugate and its applicability for insulin assay, *J. Biochem.* **78:**235–237.

Kellerer, H., and Blumen, A., 1984, Anisotropic excitation transfer to acceptors randomly distributed on surfaces, *Biophys. J.* **46:**1–8.

Knutson, J. R., Beechem, J. R., and Brand, L., 1983, Simultaneous analysis of multiple fluorescence decay curves: A global approach, *Chem. Phys. Lett.* **102:**501–507.

Kolb, D. A., and Weber, G., 1975, Cooperativity of binding of anilinonaphthalene-sulfonate to serum albumin induced by a second ligand, *Biochemistry* **14:**4476–4481.

Kranz, D. M., Herron, J. N., Giannis, D. E., and Voss, E. W., Jr., 1981, Kinetics and mechanism of deuterium oxide-induced fluorescence enhancement of fluorescyl ligand bound to specific heterogeneous and homogeneous antibodies, *J. Biol. Chem.* **256:**4433–4438.

Kranz, D. M., Herron, J. N., and Voss, E. W., Jr., 1982, Mechanisms of ligand binding by monoclonal anti-fluorescyl antibodies, *J. Biol. Chem.* **257:**6987–6995.

Laczko, G., and Lakowicz, J. R., 1989, A 6 GHz frequency-domain fluorometer, *Biophys. J.* **55:**190a.

Lakowicz, J. R., 1983, *Principles of Fluorescence Spectroscopy*, Plenum Press, New York.

Lakowicz, J. R., 1986, Fluorescence studies of structural fluctuations in macromolecules as observed by fluorescence spectroscopy, *Methods Enzymol.* **131:**518–567.

Lakowicz, J. R., and Maliwal, B. P., 1985, Construction and performance of a variable-frequency phase-modulation fluorometer, *Biophys. Chem.* **21:**61–78.

Lakowicz, J. R., and Weber, G., 1973a, Quenching of fluorescence by oxygen. A probe for structural fluctuations in macromolecules, *Biochemistry* **21:**4161–4170.

Lakowicz, J. R., and Weber, G., 1973b, Quenching of protein fluorescence by oxygen. Detection of structural fluctuations in proteins on the nanosecond time scale, *Biochemistry* **21:**4171–4179.

Lakowicz, J. R., Maliwal, B. P., Cherek, H., and Balter, A., 1983, Rotational freedom of tryptophan residues in proteins and peptides, *Biochemistry* **22:**1741–1752.

Lakowicz, J. R., Maliwal, B. P., and Gratton, E., 1985, Recent developments in frequency-domain fluorometry, *Anal. Instrum.* **14:**193–223.

Lakowicz, J. R., Laczko, G., and Gryczynski, I., 1986, 2-GHz frequency-domain fluorometer, *Rev. Sci. Instrum.* **57:**2499–2506.

Lakowicz, J. R., Cherek, H., Gryczynski, I., Joshi, N., and Johnson, M. L., 1987a, Analysis of fluorescence decay kinetics measured in the frequency domain using distributions of decay times, *Biophys. Chem.* **28:**35–50.

Lakowicz, J. R., Cherek, H., Gryczynski, I., Joshi, N., and Johnson, M. L., 1987b, Enhanced resolution of fluorescence anisotropy decays by simultaneous analysis of progressively quenched samples, *Biophys. J.* **51:**755–768.

Laws, W. R., and Contino, P. B., 1992, Fluorescence quenching studies: Analysis of nonlinear Stern-Volmer data, *Methods Enzymol.* **210:**448–463.

Lehrer, S. S., 1971, Solute perturbation of protein fluorescence. The quenching of the tryptophan fluorescence of model compounds and lysozyme by iodide ion, *Biochemistry* **10:**3254–3263.

Leytus, S. P., Patterson, W. L., and Mangel, W. F., 1983, New class of sensitive fluorogenic substrates for serine proteases, *Biochem. J.* **215:**253–260.

Lichter, P., and Ward, D. C., 1990, Is non-isotopic *in situ* hybridization finally coming of age? *Nature* **345:**93–94.

Lin, J. N., Andrade, J. D., and Chang, I.-N., 1989, The influence of adsorption of native and modified antibodies on their activity, *J. Immunol. Methods* **125:**67–77.

Livingston, D. C., Brocklehurst, J. R., Cannon, J. F., Leytus, S. P., Wehrly, J. A., Peltz, S. W., Peltz, G. A., and Mangel, W. F., 1981, Synthesis and characterization of a new fluorogenic active-site titrant of serine proteases, *Biochemistry* **20:**4298–4306.

Longworth, J. W., 1971, Luminescence of polypeptides and proteins, in: *Excited States of Proteins and Nucleic Acids* (R. F. Steiner and I. Weinryb, eds.), Plenum Press, New York, pp. 319–484.

Lopatin, D. E., and Voss, E. W., Jr., 1971, Fluorescein. Hapten and antibody active-site probe, *Biochemistry* **10:**208–213.

Ludescher, R. D., Volwerk, J. J., De Haas, G. H., and Hudson, B. S., 1985, Complex photophysics of the single tryptophan of porcine pancreatic phospholipase A_2, its zymogen, and an enzyme/micelle complex, *Biochemistry* **24:**7240–7249.

MacGregor, R. B., and Weber, G., 1981, Fluorophores in polar media. Spectral effects of the Langevin distribution of electrostatic interactions, *Proc. N.Y. Acad. Sci.* **366:** 140–154.

Mathies, R. A., and Stryer, L., 1986, Single-molecule fluorescence detection: A feasibility study using phycoerythrin, in: *Applications of Fluorescence in the Biomedical Sciences* (D. L. Taylor, A. S. Waggoner, R. F. Murphy, F. Lanni, and R. Birge, eds.), Alan R. Liss, New York, pp. 129–140.

Mei, G., Rosato, N., Silva, N., Jr., Rusch, R., Gratton, E., Savini, I., and Finazzi-Agrò, A., 1992, Denaturation of human Cu/Zn superoxide dismutase by guanidine hydrochloride: A dynamic fluorescence study, *Biochemistry* **31:**7224–7230.

Morrison, L. E., and Weber, G., 1987, Biological membrane modeling with a liquid/liquid interface. Probing mobility and environment with total internal reflection fluorescence, *Biophys. J.* **52:**367–379.

Myers, J. N., Holowka, D., and Baird, B., 1992, Rotational motion of monomeric and dimeric immunoglobulin E-receptor complexes, *Biochemistry* **31:**567–575.

Naleway, J. J., 1992, Histochemical, spectrophotometric, and fluorometric GUS substrates, in: *GUS Protocols: Using the GUS Gene as a Reporter of Gene Expression* (S. R. Gallagher, ed.), Academic Press, New York, pp. 61–87.

Naleway, J. J., Fox, C. M. J., Robinhold, D., Terpetschnig, E., Olson, N. A., and Haugland, R. P., 1994, Synthesis and use of new fluorogenic precipitating substrates, *Tet. Lett.* **35:**8569–8572.

Nicolas, J. F., Jouin, H., Bennerot, C. C., and Rocancourt, D., 1988, Stable and unstable expression of *lacZ* recombinant retrovirus in multipotential murine embryonal stem cells, *Cytometry* (Suppl.) **2:**33 (Abstr. 215).

Nolan, G. P., Fiering, S., Nicolas, J. F., and Herzenberg, L. A., 1988, Fluorescence-activated cell analysis and sorting of viable mammalian cells based on β-*D*-galactosidase activity after transduction of *Escherichia coli lacZ*, *Proc. Natl. Acad. Sci. USA* **85:** 2603–2607.

O'Connor, D. V., and Phillips, D., 1984, *Time-correlated Single Photon Counting*, Academic Press, New York.

Omelyanenko, V. G., Jiskoot, W., and Herron, J. N., 1993, Role of electrostatic interactions in the binding of fluorescein by anti-fluorescein antibody 4-4-20, *Biochemistry* **32:**10423–10429.

Parvari, R., Pecht, I., and Soreq, H., 1983, A microfluorometric assay for cholinesterases, suitable for multiple kinetic determinations of picomoles of released thiocholine, *Anal. Chem.* **133:**450–456.

Permyakov, E. A., 1993, *Luminescent Spectroscopy of Proteins*, CRC Press, Boca Raton, 1993.

Permyakov, E. A., and Shnyrov, V. L., 1983, A spectrofluorometric study of the environment of tryptophans in bacteriorhodopsin, *Biophys. Chem.* **18:**145–152.

Permyakov, E. A., Yarmolenko, V. V., Kalinchenko, L. P., Morozova, L. A., and Burstein, E. A., 1981, Calcium binding to α-lactalbumin: Structural rearrangement and association constant evaluation by means of intrinsic protein fluorescence changes, *Biochem. Biophys. Res. Commun.* **100:**191–197.

Phillips, D., Drake, R. C., O'Connor, D. V., and Christensen, R. L., 1985, Time correlated single-photon counting (TCSPC) using laser excitation, *Anal. Instrum.* **14:**267–292.

Reichert, W. M., 1989, Evanescent detection of adsorbed films: Assessment of optical considerations for absorbance and fluorescence spectroscopy at the crystal/solution and polymer/solution interfaces, *CRC Rev. Biocomp.* **5:**173–205.

Rosato, N., Finazzi-Agrò, A., Gratton, E., Stefanini, S., and Chiancone, E., 1987, Time-resolved fluorescence of apoferritin and its subunits, *J. Biol. Chem.* **262:**14487–14491.

Rotman, B., Zderic, J. A., and Edelstein, M., 1963, Fluorogenic substrates for β-galactosidases and phosphatases derived from fluorescein (3,6-dihydroxyfluoran) and its monomethyl ether, *Proc. Natl. Acad. Sci. USA* **50:**1–6.

Royer, C. A., 1992, Investigation of the structural determinants of the intrinsic fluorescence emission of the *trp* repressor using single tryptophan mutants, *Biophys. J.* **63:**741–750.

Rumbles, G., Brown, A. J., and Phillips, D., 1991, Time-resolved evanescent wave induced fluorescence spectroscopy, *J. Chem Soc. Faraday Trans.* **87:**825–830.

Schiller, P. W., 1985, Application of fluorescence techniques in studies of peptide conformations and interactions, in: *The Peptides*, Vol. 7, (V. J. Hruby, ed.) Academic Press, New York, pp. 115–164.

Seybold, P. G., Gouterman, M., and Callis, J., 1969, Calorimetric, photometric and lifetime determinations of fluorescence yields of fluorescein dyes, *Photochem. Photobiol.* **9:**229–242.

Spencer, R. D., and Weber, G., 1979, Measurements of subnanosecond fluorescence lifetimes with a cross-correlation phase fluorometer, *Ann. NY Acad. Sci.* **158:** 361–376.

Steiner, R. F., and Kirby, E. P., 1969, The interaction of the ground and excited states of indole derivatives with electron scavengers, *J. Phys. Chem.* **73:**4130–4135.

Stern, O., and Volmer, M., 1919, Über die abklingungszeit der fluoreszenz, *Physik. Zeitschr.* **20:**183–188.

Stokes, G. G., 1852, On the change of refrangibility of light, *Phil. Trans. R. Soc. Lond.* **142:**463–562.

Stryer, L., and Haugland, R. P., 1967, Energy transfer: A spectroscopic ruler, *Proc. Natl. Acad. Sci. USA* **58:**719–726.

Suci, P., and Hlady, V., 1990, Fluorescence lifetime components of Texas Red-labeled bovine serum albumin: Comparison of bulk and adsorbed states, *Colloids Surfaces* **51:** 89–104.

Suzuki, K., Uchida, T., Sakatani, T., Sasagawa, S., and Hosaka, S., 1988, Measurement of active phagocytosis by polymorphonuclear leukocytes by fluorescence liberation from phagocytized microspheres, *J. Leukocyte Biol.* **39:**475–488.

Szabo, A. G., and Rayner, D. M., 1980, Fluorescence decay of tryptophan conformers in aqueous solution, *J. Am. Chem. Soc.* **102:**554–563.

Teale, F. W. J., and Weber, G., 1957, Ultraviolet fluorescence of the aromatic amino acids, *Biochem. J.* **65:**476–482.

Templeton, E. F. G., and Ware, W. R., 1985, Charge transfer between fluorescein and tryptophan as a possible interaction in the binding of fluorescein to anti-fluorescein antibody. *Molec. Immunol.* **22:**45–55.

Thompson, R. B., and Gaber, B. P., 1985, Improved fluorescence assay of liposome lysis, *Anal. Lett.* **18**(B15)**:**1847–1863.

Velick, S. F., Parker, C. W., and Eisen, H. N., 1960, Excitation energy transfer and the quantitative study of the antibody hapten reaction, *Proc. Natl. Acad. Sci. USA* **46:** 1470–1482.

Waggoner, A. S., 1990, Fluorescent Probes for Cytometry, in: *Flow Cytometry and Sorting*, 2nd ed. (M. R. Melamed, T. Lindmo, M. L. Mendelsohn, eds.), Wiley-Liss, New York, pp. 209–225.

Wang, C.-K., Mani, R. S., Kay, C. M., and Cheung, H. C., 1992, Conformation and dynamics of bovine brain S-100a protein determined by fluorescence spectroscopy, *Biochemistry* **31:**4289–4295.

Watt, R. M., and Voss, E. W., Jr., 1977, Mechanism of quenching of fluorescein by antifluorescein IgG antibodies, *Immunochemistry* **14:**533–541.

Weber, G., and Daniel, E., 1966, Cooperative effects in binding by bovine serum albumin. II. The binding of 1-anilino-8-naphthalenesulfonate. Polarization of the ligand fluorescence and quenching of the protein fluorescence, *Biochemistry* **5:**1900–1907.

Weber, G., and Farris, F. J., 1979, Synthesis and spectral properties of a hydrophobic fluorescence probe: 6-propionyl-2-(dimethylamino)naphthalene, *Biochemistry* **18:** 3075–3078.

Weber, G., and Teale, F. J. W., 1957, Determination of the absolute quantum yield of fluorescent solutions, *Trans. Faraday Soc.* **53:**646–655.

Weber, G., and Teale, F. J. W., 1959, Electronic energy transfer in haem proteins, *Discuss. Faraday Soc.* **27:**134–141.

Wei, A.-P., and Herron, J. N., 1993, Use of synthetic peptides as tracer antigens in fluorescence polarization immunoassays of high molecular weight antigens, *Anal. Chem.* **65:**3372–3377.

Wei, A.-P., Herron, J. N., and Christensen, D. A., 1992, Characterization of fluorescent dyes for optical immunosensors based on fluorescence energy transfer, in: *Biosensor Design and Application* (P. R. Mathewson and J. W. Finley, eds.), ACS Symposium Series No. 511, American Chemical Society, Washington, DC, pp. 105–120.

Wei, A.-P., Blumenthal, D. K., and Herron, J. N., 1994, Antibody-mediated fluorescence enhancement based on shifting the intramolecular dimer—monomer equilibrium of fluorescent dyes, *Anal. Chem.* **66:**1500–1506.

Westerfield, M., Wegner, J., Jegalian, B. G., DeRobertis, E. M., and Puschel, A. W., 1992, Specific activation of mammalian *Hox* promoters in mosaic transgenic zebrafish, *Genes Dev.* **6:**591–598.

Willis, K. J., and Szabo, A. G., 1992, Conformation of parathyroid hormone: Time-resolved fluorescence studies, *Biochemistry* **31:**8924–8931.

Willis, K. J., Szabo, A. G., Zuker, M., Ridgeway, J. M., and Albert, B., 1990, Fluorescence decay kinetics of the tryptophyl residues of myoglobin: Effect of heme ligation and evidence for discrete lifetime components, *Biochemistry* **29:**5270–5275.

Wu, C. W., and Stryer, L., 1972, Proximity relationship in rhodopsin, *Proc. Natl. Acad. Sci. USA* **69:**1104–1108.

Yashphe, J., and Halvorson, H. O., 1976, β-*D*-Galactosidase activity in single yeast cells during cell cycle of *Saccharomyces lactis*, *Science* **191:**1283–1284.

Zhang, Y. Z., Naleway, J. J., Larison, K. D., Huang, Z., and Haugland, R. P., 1991, Detecting *lacZ* gene expression in living cells with new lipophilic, fluorogenic β-galactosidase substrates, *FASEB J.* **5:**3108–3113.

Zheng, T., 1994, Packing of immunoglobulin G in the adsorbed state: An energy transfer study, Master's thesis, University of Utah, Salt Lake City.

Zimmermann, R. M., Schmidt, C. F., and Gaub, H. E., 1990, Absolute quantities and equilibrium kinetics of macromolecular adsorption measured by fluorescence photobleaching in total internal reflection, *J. Colloid Interface Sci.* **139:**268–280.

2

Structural Information on Proteins from Circular Dichroism Spectroscopy Possibilities and Limitations

Michael Bloemendal and W. Curtis Johnson, Jr.

1. INTRODUCTION

In this chapter we give a general survey of the application of circular dichroism (CD) spectroscopy for structural studies of proteins. In addition to the commonly used far-UV and near-UV/visible CD, the less well-known vibrational CD will be discussed. Although magnetic CD (MCD) has a fundamentally different origin and is beyond the scope of this chapter, some theoretical and experimental considerations will be given in Section 2. For a detailed treatise, we refer the reader to Sutherland and Holmqvist (1980), Sharonov (1991), and Cheesman *et al.* (1991). Time-resolved CD is discussed briefly in Section 6.

The chapter is primarily meant as a survey for those who are interested in applying CD and who wish to enlarge their knowledge on the possibilities, limitations, and experimental details, including the extraction of information from the measurements. As such it does not pretend to give a comprehensive overview of the recent literature on CD, and neither will we indulge in extensive theoretical considerations. In order to be as transparent as possible, every technique will be

Michael Bloemendal • Department of Protein and Molecular Biology, Royal Free Hospital School of Medicine, London NW3 2PF, England. *W. Curtis Johnson, Jr.* • Department of Biochemistry and Biophysics, Oregon State University, Corvallis, Oregon 97331.

Physical Methods to Characterize Pharmaceutical Proteins, edited by James N. Herron *et al.*, Plenum Press, New York, 1995.

treated separately. After a general introduction, we detail the potential information, the experimental aspects, and the limitations of the technique. We will end each section with some recent examples to illustrate the technique.

2. WHAT IS CIRCULAR DICHROISM?

2.1. General Considerations and History

Circular dichroism (the difference in absorption between left- and right-handed circularly polarized light) and optical rotatory dispersion (ORD) (the wavelength dependence of the optical rotation) are two manifestations of optical activity that are related through the Kronig-Kramer transform. Optical activity was first described in 1811 by Agaro, who detected that plane polarized light is rotated by quartz. Eight years later, Biot found that optical rotation can also be observed in solution, and that it depends on concentration, path length, and wavelength (Pasteur 1848; Cotton, 1896; Jirgensons, 1973; Charney, 1979). Pasteur (1848) suggested that optical activity is related to the asymmetry of the molecule, which was formulated by LeBel and Van't Hoff as the impossibility to superimpose the mirror image of a molecule, or a molecular group, on its original (Jirgensons, 1973). The asymmetry may be caused by covalently bound groups or, in case of a macromolecule, by the conformation. A completely different source of optical activity was detected in 1846 by Faraday, who observed that it can be induced by the application of a magnetic field (Charney, 1979). Magnetically induced optical activity is insensitive to molecular conformation, but it depends on ligand geometry around the chromophore. Cotton (1896) found that optically active materials absorb right- and left-handed circularly polarized light to a different extent.

From the above it will be clear (1) that CD is related to optical rotation, and (2) that CD, and hence ORD, provides information on asymmetries in the molecules under study.

2.2. CD Chromophores and Their Information

Any asymmetric group has a CD. This may involve inherent asymmetry, as in the chiral α-carbon of the peptide bond. Such chromophores yield CD irrespective of their chemical surrounding, although the latter may modify the optical properties. However, in many biological molecules asymmetry is caused by the asymmetric environment, and hence CD provides information on this structure.

Table I. Some Relevant Electronic Absorption Bands in Proteins[a]

Group	Location (nm)	log ϵ_{max}
Peptide	±210	±2.0
	±190	±3.8
His	211	3.8
Cys	250	2.5
Trp	280	3.7
	219	4.7
Tyr	274	3.1
	222	3.9
	193	4.7
Phe	257	2.3
	206	4.0
	188	4.8

[a]From Campbell and Dwek (1984).

Since CD is an absorption difference, it is only observed in absorption bands. Table I shows the prevailing electronic absorptions in proteins. In principle they all yield CD. However, due to the preponderance of peptide bonds relative to other chromophores in proteins, the far-UV CD spectrum is dominated by the contribution of peptide bonds. Their CD band position and intensity depend on their conformation. Hence, this part of the spectrum can be used to determine the secondary structure content of proteins. Above 250 nm, only the aromatic residues (and to some extent cysteine) absorb light. This part of the spectrum yields information on the tertiary structure. When visible chromophores, like ligands and/or metal ions, are available, the 400- to 800-nm part of the CD spectrum can also be used to obtain information on the tertiary structure of proteins.

In principle, structural knowledge on proteins can also be obtained from vibrational absorptions in the infrared part of the spectrum. A survey of some important vibrational transitions is given in Table II. The application of infrared CD to biomolecules is still limited due to a series of experimental problems, the most important one probably being the strong water absorption in this region. Recently, vibrational CD has been used to study protein structures and ligand binding.

The only amino acid that has a strong magnetic CD signal is tryptophan. However, all metallic ligands in proteins and metal–protein complexes show MCD. Hence, this technique is very useful for the study of structure–function relationships of such proteins. Its sensitivity and high spectral resolution enables the detection of weak transitions unresolved in the absorption spectrum.

Table II. Some Relevant Vibrational Absorption Bands in Proteins

Group	Location (cm^{-1})	Remarks
NH (stretch), α-helix[a]	3300–3290	—
NH (stretch), β-sheet[a]	3300–3280	—
Amide I, α-helix[a]	1660–1650 (1695, 1659, 1641)	1695/1641 peaks weak
Amide I, β-sheet[a]	1630 (1682, 1634)	1682 peak weak
Amide I, random coil[a]	(1656)	—
Amide I, turn[a]	(1684, 1654, 1633)	1654/1633 peaks weak
Amide II, α-helix[a]	1550–1540 (1546, 1516)	1516 peak weak
Amide II, β-sheet[a]	1525–1520 (1510, 1530, 1550)	1550/1510 peaks weak[c]
Amide II, random coil[a]	1520	—
Saturated aliph. acids[b]	1725–1700	—
Saturated esters[b]	1750–1720	—
H_2O (stretch)[a]	(3900–3200)	Broad intense band
H_2O (bend)[a]	(1645)	Broad band

[a]From Campbell and Dwek (1984); in parentheses, peak positions according to Lee (1985) for amide II and water, and to Lee *et al.* (1990) for amide I.
[b]From Gans (1980).
[c]1510 Band in antiparallel, 1550 in parallel β-sheet.

Although CD is an absorption difference, the position of the CD bands may be and often is shifted with respect to that of the ordinary absorption due to exciton coupling, which in addition to vibrational coupling may also yield multiple CD peaks from a single absorption band (Bayley, 1980).

2.3. Parameters and Units

The most straightforward parameter to use in CD is $\Delta\epsilon$, which is directly related to normal absorption and follows from the Lambert-Beer law as

$$\Delta\epsilon = \Delta A/cL = (A_L - A_R)/cL \tag{1}$$

where A_L and A_R are the absorption of left- and right-handed polarized light, respectively, c is the concentration in mole/dm^3, L is the path length in centimeters, and $\Delta\epsilon$ is the molar absorption difference in liter·mole^{-1}·cm^{-1}. Note that c is the concentration of the chromophores; e.g., for secondary structure determination from far-UV CD, this is the concentration of peptide bonds. The alternative parameter is the molar ellipticity, $[\theta]_\lambda$, which originates from the fact that CD causes a linearly polarized light beam to become elliptically polarized after passing through an optically active sample. It can be shown (Campbell and Dwek, 1984) that

$$[\theta]_\lambda = 3300\ \Delta\epsilon \tag{2}$$

where $[\theta]_\lambda$ is in degrees·dl·mole^{-1}·dm^{-1}.

In MCD, a strong magnetic field, B_0, is applied to the sample in the propagation direction of the light. In first-order approximation, the MCD is proportional to the magnetic field. Hence,

$$\Delta\epsilon_B = \Delta A_B/cLB_0 \tag{3}$$

where the subscript B denotes CD caused by the magnetic field only, and $\Delta\epsilon_B$ is given in dm^3·mole^{-1}·cm^{-1}·T^{-1}.

In some studies the intensity of the CD is expressed as the anisotropy ratio, g, which is defined as

$$g = (A_L - A_R)/[0.5(A_L + A_R)] \tag{4}$$

where g is theoretically related to the electric and magnetic transition dipole moments and to the rotational and dipole strength (Bormett *et al.*, 1992).

2.4. Instrumentation

The essential parts of a conventional CD spectrophotometer are a light source (usually a xenon, hydrogen, or deuterium lamp), a monochromator to select light of the desired wavelength, a polarizer to produce plane polarized light, and a photoelastic modulator to convert the plane polarized light into alternating left- and right-handed circularly polarized light. A lock-in amplifier separates the small modulated signal detected by the photomultiplier from the noise. Recently the use of a multichannel detector to measure all wavelengths simultaneously has been described (Lewis *et al.*, 1992; Nölting *et al.*, 1992). Since CD measures absorption differences, effects are small (~10^{-4} of regular absorption), and highly sensitive detectors and bright light sources are a prerequisite for a good signal-to-noise ratio. Since photoelastic modulator frequencies are in the high micro- or low millisecond range, completely different devices have to be used for ultrafast time-resolved (pico- to microseconds) CD. These are reviewed by Lewis *et al.* (1992).

Commercial CD spectrophotometers are available from several sources. Provided that they are carefully maintained, they will perform well. However, for specific purposes (e.g., far-UV, magnetic, vibrational, or ultrafast time-resolved CD), home-designed instruments have been described by Johnson (1971), Osborne *et al.* (1973), Brahms and Brahms (1980), Eglinton *et al.* (1980), Keiderling (1981), Bokma *et al.* (1987), Sharonov (1991), and Lewis *et al.* (1992). For MCD measurements a magnet is incorporated in the spectrometer. Depending on the application, this can be a permanent, electro- or superconducting magnet (Sutherland and Holmqvist, 1980; Sharonov, 1991, Lewis *et al.*, 1992).

A discussion of possible instrumental artifacts in CD spectrometry, in particular the possible occurrence of circular birefringence, linear birefringence, and linear dichroism, is given by Shindo and Nakagawa (1985), Wu *et al.* (1990), and Lewis *et al.* (1992). For most purposes these effects can be neglected. However, proper maintenance of the equipment is crucial. A CD spectrophotometer should be regularly calibrated, since deterioration of the optics causes the relative contribution of the scattered light to become more important, resulting in errors in intensity and peak positions. Although several compounds have been suggested for calibration, (+)-10-camphor sulfonic acid (CSA) is commonly used. It is easily available and provides a two-point calibration with $\Delta\epsilon = 2.36$ at 290.5 nm and -4.9 at 192.5 nm (Chen and Yang, 1977). A 0.5-nm path length cell with 1 mg/ml CSA will suffice. CSA is hygroscopic, so its concentration must be determined spectroscopically. The extinction coefficient of the anhydrous material at 285 nm is 34.5 (Lowry and French, 1932).

2.5. CD versus ORD

Since CD is an absorption difference, it is observed only in an absorption band. Optical rotation, although related to CD, is a refractive index phenomenon (for ORD, see, e.g., Charney, 1979; Cantor and Schimmel, 1980; Bayley, 1980). As such, it is measurable at wavelengths remote from the absorption band. However, it involves the weighted sum of the effects of all CD bands, so in theory the whole spectrum has to be measured for accurate interpretation. Moreover, since CD detection is technically superior to ORD detection and hence more sensitive, ORD is rarely used anymore for the study of protein structure. However, recently Hvidt and co-workers have used ORD for the study of temperature-induced secondary structure changes in frog αβ-tropomyosin at high salt concentration (Hvidt and Lehrer, 1992). As will be discussed in Section 3.3, secondary structure determination by CD is hampered by high NaCl concentrations.

3. FAR-UV CIRCULAR DICHROISM

3.1. Introduction

The most utilized form of CD spectroscopy is far-UV CD for determining the secondary structure content of proteins. Commercially available CD spectrophotometers for the far-UV are widely used. However, without a thorough understanding of the pitfalls of such equipment and the methods to extract secondary structure information from the measurements, interpretation of results may be misleading. Extensive reviews on far-UV CD have been written by Johnson (1988, 1990) and Manning (1989). In this section we shall summarize the most important conclusions of these and related papers. In particular we wish to alert workers to overinterpretation of results.

3.2. What Does It Measure?

As noted in Table I and discussed in Section 2.2, far-UV CD spectra are dominated by the contribution of the peptide bonds, although some side chains may also contribute. Since the CD spectrum of a protein is highly sensitive to the conformation of peptide bonds, it can be used to monitor the secondary structure of proteins. However, it should be stressed that the far-UV CD of the peptide bond essentially reflects the dihedral angles (see Fig. 1). This means that individual bonds with dihedral angles close to those occurring in a certain type of secondary structure will show the spectrum of that particular structure, irrespective of whether an identifiable length of this structure is present. Table III summarizes the dihedral angles of the most common secondary structures. A single residue with ϕ,ψ angles of -57, -47 will show an α-helical CD spectrum, whereas four or five residues are the minimum to recognize an α-helical loop. Conversely, when helices have strongly distorted angles, like those of transferrins, CD will underestimate the helical content (Hadden *et al.*, 1994).

3.3. Experimental Details

Low-noise far-UV CD experiments require an intense light source. Such sources easily convert oxygen into ozone, which rapidly damages the optics. Either a vacuum chamber or very good nitrogen flushing prevents this. Deteriora-

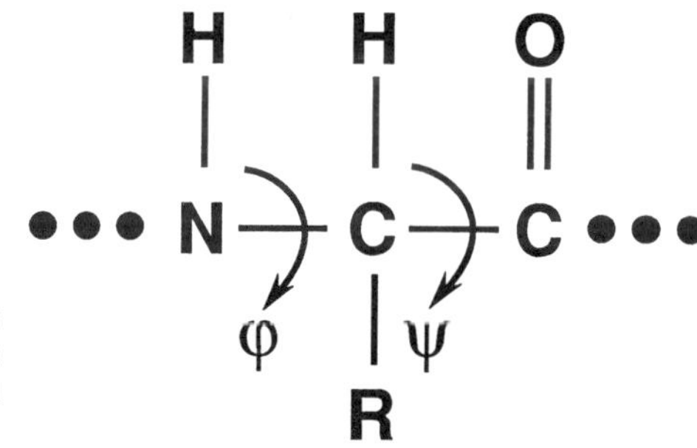

Figure 1. Dihedral angles, according to IUPAC recommendations (*Biochemistry* **9:** 3471–3478, 1970), shown for ϕ=0° and ψ=0°.

Table III. Dihedral Angles for Peptide Bonds in Different Secondary Structures[a]

Structure	ϕ_i	ψ_i	ϕ_{i+1}	ψ_{i+1}
α-helix	−57	−47		
Parallel β-sheet	−119	+113		
Antiparallel β-sheet	−139	+135		
Collagen	−76	+160		
	−68	+156		
	−67	+168		
β-Turn I	−60	−30	−90	0
β-Turn I′	+60	+30	+90	0
β-Turn II	−60	+120	+80	0
β-Turn II′	+60	−120	−80	0
β-Turn III	−60	−30	−60	−30
β-Turn III′	+60	+30	+60	+30
β-Turn V*	−80	+80	−80	+80

[a]All angles transformed to IUPAC-IUB recommendations (*Biochemistry*, 1970, **9**:3471–3478); *i* denotes the residue, *i*+1 the residue next to *i* (turns are defined by a sequence of residues); all turns according to Lewis *et al.* (1973); other structures according to Madison and Schellman (1972).

tion of the optics is easily monitored by the two-point CSA calibration, for which $-\Delta\epsilon(192.5)/\Delta\epsilon(290.5)$ should be at least 2.0. Ideally this should be tested at least once a week. The optics can often be renewed by replacing the first mirror after the source and cleaning the front surface of the prisms. Although spectra can be measured to 184 nm on most commercial CD instruments and to 170 nm on the newest when the equipment is well maintained, special setups have been developed for shorter wavelengths. Descriptions can be found in Johnson (1971), Gross and Schnepp (1977), and Toumadje *et al.* (1992).

The limiting factor in far-UV CD spectroscopy on proteins, however, is usually the transparency of the buffer, since almost any compound absorbs light in this spectral region. A table of solvents with their absorption as function of path length is given by Johnson (1986, 1990). Water absorbs all light below 176 nm in a 50-μm path length; a 1-mm cell with water has an absorption of 1 at 186 nm. The presence of chloride anion, which is often used in biological buffers, prevents measurements below 205 nm. Gdn/HCl and urea absorb strongly below 220 nm. The transparency of buffer additives like sodium dodecyl sulfate (SDS), ethylenediaminotetraacetate (EDTA), and so on is also discussed by Johnson (1990). In order to preclude artifacts due to buffer absorption, the total absorption of the sample should always be checked before the measurement and kept below 1 for the entire spectral range studied. In practice this means that for measurements to

184 nm a 0.2-mm path length is the upper limit, although better results will be obtained with 0.05- or 0.1-mm cells. Typical protein concentrations for such path lengths are 0.2 to 1 mg/ml.

When using very short path length cuvettes the path length has to be checked; it may deviate considerably from the data given by the manufacturer. Errors in path length change $\Delta\epsilon$, which in turn influences the secondary structure estimation. From the infrared interference fringes of an empty cell, the path length in centimeters (L) may be calculated as $L = \Delta n\lambda_1\lambda_2/2(\lambda_1-\lambda_2)$, where λ_1 and λ_2 are long- and short-wavelength absorption minima, respectively, and Δn the number of maxima between them (Bree and Lyons, 1956; Johnson, 1988).

Another point of concern is the determination of the protein concentration. As we will discuss in Section 3.4, reliable secondary structure content can only be obtained when the protein concentration of the sample (and hence $\Delta\epsilon$) is accurately known. Traditional methods like Lowry's (Lowry *et al.*, 1951; Smith *et al.*, 1985) and Bradford's (1976) are only reliable when the specific response of the studied protein is known, since this response varies with a factor of almost three for different proteins (Lowry *et al.*, 1951; Smith *et al.*, 1985). It is advisable to use specific absorptions at a certain wavelength, provided that these are accurately known. The concentration of Trp- and Tyr-containing proteins can be determined accurately using an extinction coefficient contribution of 5690 liter·mole^{-61}·cm^{-1} per Trp, 1280 liter·mole^{-1}·cm^{-1} per Tyr, and 120 liter·mole^{-1}·cm^{J-1} per Cys at 280 nm for Gdn·HCl denatured protein (Elwell and Schellman, 1977).

3.4. Estimation of the Secondary Structure

A series of possible errors both on the instrumental and analysis level and their effects on the calculations are discussed by Hennessey and Johnson (1982) and by Manning (1989). The most common ones are discussed in this section.

A first concern in extracting secondary structure from far-UV CD spectra is the information content. This has been addressed in detail using eigenvalue analysis (Hennessey and Johnson, 1981; Manavalan and Johnson, 1985; Toumadje *et al.*, 1992). They show that measurements truncated at 200 nm, 190 nm, 178 nm, and 168 nm contain 2, 4, 5, and 6 independent pieces of information within the noise level of the experiments. Effectively, this means that measurements from 200 to 250 nm provide only a reliable percentage of α-helix, which dominates the CD, whereas a description in terms of α-helix, parallel, and antiparallel β-sheet, turn and "other" requires measurements down to at least 184 nm. In general, introducing the constraints that the sum of secondary structures is unity and all fractions are positive leads to unreliable results (Johnson, 1986, 1990) and can even disturb estimation of the amount of α-helix.

Several methods have been developed to quantify the secondary structure of proteins from their far-UV CD spectra. One of the oldest uses model polypeptides with assumed pure secondary structure. The spectrum of the unknown structure is then linearly fitted in terms of these model compounds (Greenfield and Fasman, 1969; Brahms and Brahms, 1980). In more advanced methods the pure secondary structure spectra are derived from reference proteins with structure known from X-ray diffraction data on their crystals (Saxena and Wetlaufer, 1971; Chen *et al.*, 1972, 1974; Grosse *et al.*, 1974; Chang *et al.*, 1978; Siegel *et al.*, 1980; Bolotina and Lugauskas, 1986; Pribic *et al.*, 1993).

Care has to be taken in a simple linear approach, because CD spectra of the peptide bond may be affected by contributions of aromatic side chains, disulfide bridges, and other conformations. Aromatic residues, even if they do not have significant far-UV CD themselves, may produce shifts and form changes in the peptide band due to exciton coupling (Kuwajima *et al.*, 1991). Moreover, a certain structure does not necessarily yield a unique CD spectrum, as by its own virtue (see Section 2.1) the CD spectrum of a chromophore is sensitive to its environment. Hence, the secondary structure CD spectrum is influenced by the tertiary structure of a protein. Moreover, the α-helix spectrum depends on chain length (Goodman *et al.*, 1969; Chen *et al.*, 1972) and the vicinity of other helixes (Manning, 1989), whereas β-sheets and turns show a range of differently shaped spectra (Chang *et al.*, 1978).

Therefore, more advanced methods have been developed that use a large reference set of proteins with known secondary structure. If the reference set is large enough, all specific effects will be present in the set. There will be different sized α-helixes, different types of sheets and turns, and so on. However, when using a large reference set, the number of spectra will exceed the information content of the data. The reference CD spectra will be linearly dependent; since they contain experimental error, the problem is overdetermined and a large number of solutions will fit the data within that error. This implies that it is essential to know the exact protein concentration and cell path length, since not only the shape of the spectra (peak positions) but also their absolute magnitudes influence the results.

Several approaches are currently in use to solve the overdetermination problem. Provencher and Glöckner (1981) used a least-squares fit that includes a regularizer term to stabilize the solution. The effect of this regulation is that solutions emphasizing a particular reference protein are penalized unless they are really mandatory for a good fit. The computer program (Provencher, 1982a,b), which is available on request, uses two constraints, namely, all secondary structure fractions found should be positive and the sum of fractions should be 1. Although these constraints seem obvious, they have the disadvantage that any result looks reasonable. As shown by Johnson and co-workers (Johnson 1986, 1988, 1990), it is better to perform the fits without constraints and use these as a quality check. The Provencher method provides a range of solutions of which the proper one has to be selected. As Provencher (1982a) points out, the "best solution" given by the program should not automatically be selected. Manavalan and Johnson (1987) have shown that the best results are obtained choosing the solution consistent with the information content of the data (five or six degrees of freedom for spectra measured to 180 nm).

An alternative approach, namely, singular-value decomposition combined with variable selection, was chosen by Johnson and co-workers (Hennessey and Johnson, 1981; Compton and Johnson, 1986; Manavalan and Johnson, 1987). The program is available from the authors. In variable selection, sets of CD spectra of the reference proteins are removed from the reference set until certain criteria are satisfied. In this way conflicting information is eliminated. Criteria for the quality of a solution are: (1) sum of secondary structures between 0.9 and 1.1 (preferably 0.95 and 1.05); (2) all fractions larger than −0.05 (ideally positive), and when a larger set of reference proteins gives a positive fraction, it should not become negative by removing proteins from this set; (3) the fit of the reconstituted CD within the noise level of the experimental data; and (4) basis set having as many reference spectra as possible. Many subsets of the reference proteins will give similar results. Their solutions are averaged and the standard deviation of this average is given as the uncertainty in the determined fractions (Toumadje *et al.*, 1992). It is clear that this is not the statistical uncertainty of a fraction as found from the experimental CD curve but a measure of the variance in the solutions.

A method that provides real estimates of uncertainties in the secondary structure fractions was developed by Van Stokkum *et al.* (1990). A multivariate linear model with noise (Gauss-Markoff model) provides both the secondary structure fractions and the estimated errors in the fractions for a protein of unknown structure. The method uses the same set of reference proteins as Johnson and co-workers. Since there are numerous nonlinear effects in the relation between secondary structure and CD spectra (e.g., chain length and environment dependence of helixes and sheets), a locally linearized model is introduced. Whereas variable selection removes conflicting information from the reference set, locally linearizing synthesizes a minimum reference subset necessary for a proper result from the total reference set. Again a number of solutions is found. As selection criteria, fractions between −0.05 and 1.05 combined with a sum of fractions as close as possible to unity are chosen. Alternatively, a minimum in the sum of errors of the fractions can be used as the selection standard.

A comparison between the three methods is given by Van Stokkum *et al.* (1990). From a reference set of 22 proteins each protein was successively removed and fitted in terms of the remaining 21 proteins. The results for the secondary structure are compared with that deduced from the X-ray structures. Table IV summarizes the results. It is obvious that all models (even the simple linear model) give the same average quality of α-helix prediction. For the Provencher method

Table IV. Comparison of Root Mean Square Residuals × 100 of the Secondary Structure Fractions Obtained from Far-UV CD-Spectra with Different Methods[a]

	Structural classes[b]				
Method	H	A	P	T	O
Linear Model[c]	9	16	7	11	17
LLM-I[c,d]	9	12	8	7	9
LLM-II[c,d]	7	12	7	7	8
SVD+VS[e]	6	10	8	6	9
PROV-I[f]	12	21	6	11	8
PROV-II[f]	9	13	6	7	8

[a]All data are based on 22 protein spectra from 178–240 nm (Van Stokkum *et al.*, 1990), except from SVD+VS, which is based on 16 spectra from 168–240 nm (Toumadje *et al.*, 1992).
[b]H = helix; A = antiparallel β-sheet; P = parallel β-sheet; T = turn; O = other structure.
[c]The linear model and LLM use the Gauss-Markoff method (Van Stokkum *et al.*, 1990).
[d]LLM-I and LLM-II are the locally linearized model using the sum of fractions closest to 1 and the same combined with smallest standard deviations in the fractions, respectively, as selection critiera.
[e]SVD+VS is singular value decomposition with variable selection (Manavalan and Johnson, 1987).
[f]PROV-I and PROV-II are the Provencher method (Provencher and Glöckner, 1981), with the by-the-program and the by-Manavalan-and-Johnson (1987) suggested solution, respectively.

the criteria suggested by Manavalan and Johnson (1987) provide a significant improvement. In interpreting Table IV it should be considered that, whereas the reference set contains proteins with an α-helical fraction ranging from 0 to 1, the maximum fraction of antiparallel β-sheet, turn, and other structure is 0.5 and that of parallel sheet only is 0.3. Therefore, predictions of the latter structures can have a large relative error. Currently, the reference set of Johnson and co-workers has been expanded to 33 proteins. Whether this improves the results is not yet tested. An advantage of the locally linearized model, along with its ability to produce an estimation of the uncertainties in the secondary structures, is that it requires only 10% of the calculation time of the Provencher and 0.1% of that of the variable selection method (Van Stokkum *et al.*, 1990). Variable selection is the easiest and most transparent of the three methods. It should be stressed that the errors in Table

IV are root mean square residuals, i.e., the average of all errors found. This implies that some predictions will be considerably better and some will be worse. This once again shows the advantage of the Gauss-Markoff model used by Van Stokkum *et al.* (1992), which provides a measure for the uncertainty of a single experiment.

Pancoska and Keiderling (1991) have used the principal component method of factor analysis combined with cluster analysis. This method is comparable, although not identical, to the singular-value decomposition used by Hennessey and Johnson (1981). Nonlinear effects are not taken into account with this method, but it has the advantage that it is not inherently assumed that the complete far-UV CD band is due to secondary structure.

Perczel *et al.* (1991) suggest that convex constraint analysis is a superior method for secondary structure calculation since it avoids the use of X-ray structures as references. However, in order to determine the number of basis spectra and to test the reliability of the method, they use spectra of poly-L-lysine in different conformations. As pointed out above, the reliability of these spectra for real proteins is questionable. Moreover, the method is rather sensitive to the set of proteins used in the analysis. Comparison with X-ray structures gave correlation coefficients that are significantly lower than obtained for the methods in Table IV.

Sreerama and Woody (1993) have developed a self-consistent procedure for singular-value decomposition with variable selection. They include the spectrum of the protein being analyzed in the reference set. Initially, a guess is made for the structure of that protein, and subsequently the predicted structure is used in the computation until self-consistency is reached. Significant improvement over other methods is seen in the prediction of β-turn. A computer program is available from the authors.

A structure present in some proteins, which is not included in any of the programs, is the collagen structure. This shows a minimum in $[\theta]_\lambda$ of $-50{,}000$ at 198 nm and a maximum of $+6{,}000$ at 220 nm (Campbell and Dwek, 1984).

3.5. Advantages, Limitations, and Conclusions

Far-UV CD provides a fast method for monitoring the secondary structure content of proteins. Although it is less accurate than X-ray diffraction or nuclear magnetic resonance (NMR) spectroscopy, its relative ease and speed are major advantages. Moreover, it enables the measurement of proteins in solution and allows a fast scanning of the effect of different conditions on protein structures. However, as Table IV shows, the uncertainties in the determined structure fractions may be rather large. The Gauss-Markoff method of van Stokkum *et al.* (1990) provides a tool to estimate these uncertainties. Results may be misleading when the spectra are not measured down to 184 nm at least, although the α-helical content is reasonably reliable when deduced from spectra truncated at 200 nm and no constraints are used (Manavalan and Johnson, 1985; Johnson, 1986). Spectra of "pure" secondary structures deduced by singular-value decomposition are shown in Fig. 2. It appears that the spectra for α-helix and antiparallel β-sheet resemble each other to some extent, except for the intensity. As a consequence, small antiparallel sheet fractions are easily overestimated when much α-helix and little sheet is present; conversely, when little helix and much sheet is present, the latter may be underestimated (Van Stokkum *et al.*, 1990). An inherent problem of most methods when deducing secondary structure from CD spectra is that the data are compared with crystallographic X-ray structures. As pointed out by Yang *et al.* (1986), classification of structures between several investigators reveals discrepancies as large as 10%. Moreover, the crystal structure may deviate from the solvent structure (Manning, 1989). A consideration, given by Pancoska *et al.* (1991) for infrared CD but equally valid for far-UV CD, is the question of whether the facility of some proteins to crystallize denotes a structural similarity, thus biasing the reference set of proteins. Of course, identical secondary structure content does not prove an identical secondary structure. It only shows that there

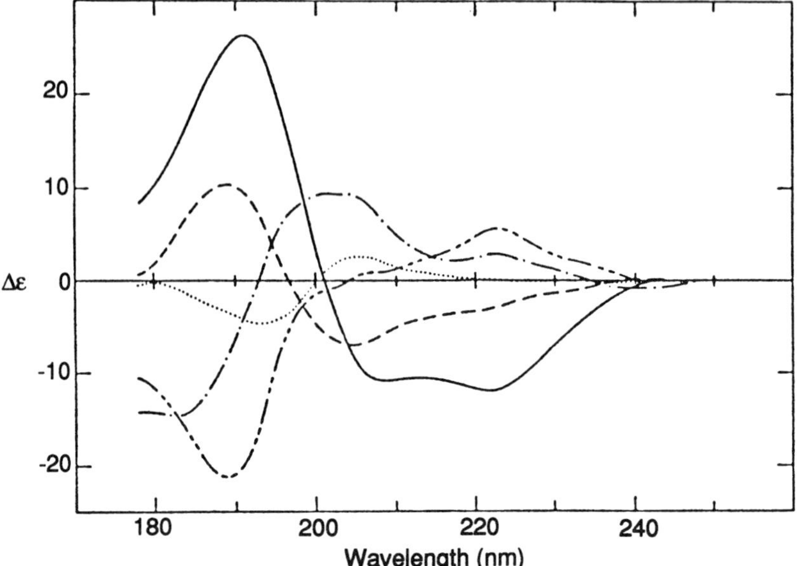

Figure 2. Characteristic circular dichroism spectra for α-helix (solid line), antiparallel β-sheet (– – –), parallel β-sheet (– · – · -), β-turn (– - - - --), and remainder (····). (From Compton and Johnson, 1986, with permission.)

are equivalent amounts of the different types of structure. A minor problem is the fact that the spectral contributions of mirror-imaged structures (e.g. left- and right-handed turns) may cancel each other (Chang *et al.*, 1978; Hennessey and Johnson, 1981). A serious point to be considered before selecting far-UV CD for secondary structure determination is that it can only be measured at relatively low protein concentrations (≈ 1 mg/ml), and that the choice of the buffers is restricted by their absorption in the far-UV. Finally, we stress once more the importance of proper protein concentration determination and maintenance of the equipment for reliable results, the risk of using constraints in deducing secondary structure, and the fact that dihedral angles and not structures per se are measured.

Taking all the points into account, far-UV CD has given valuable contributions to the study of protein structures, especially monitoring structure under varying external conditions.

3.6. Recent Applications: Acid-Induced Structural Changes

The effect of pH on protein structure has been extensively studied by CD spectroscopy (see, e.g., Arakawa *et al.*, 1987; Goto *et al.*, 1990a,b; Jiskoot *et al.*, 1991). In a study on the conformational states of ribulose biphosphate carboxylase (Rubisco) from *Rhodospirillum rubrum*, Van der Vies *et al.* (1992) combined CD with fluorescence measurements. Whereas the native enzyme exists as a dimer (Rubisco-N_2) (Schneider *et al.*, 1990), the site-directed mutant K168E forms a moderately stable folded monomer (Rubisco-N_1). Refolding of urea, Gdn-HCl, or acid-denatured Rubisco is assumed to proceed via an intermediate state (Rubisco-I). The half-time of spontaneous refolding is about 5 hr at 4 °C. Therefore, it is possible to study the properties of the intermediate state before significant reversion to the native state occurs.

The CD spectra of different states of Rubisco are given in Fig. 3. From the similarity of the spectra for Rubisco-N_1 and -N_2 (Fig. 3C), it is concluded that the creation of secondary structure is complete with the formation of the monomer. The effect of acid-induced denaturation in the presence of high and low salt concentration on the CD spectrum is shown in Fig. 3B. The low salt spectrum (curve 4) shows a broad band of negative intensity at 200–203 nm, typical of an unfolded protein. The authors conclude from the CD at 223 nm compared to the Gdn–HCl-denatured species that some residual structure is still present. They quantify this with the method of Chen *et al.* (1972) and conclude that the acid-denatured species contains about 10% α-helix. When at low pH the ionic strength is increased, the far-UV CD spectrum shows a more nativelike form (Fig. 3B, curve 5), which would correspond to the formation of a partially folded protein. The α-helix content of this state (Rubisco-A), calculated according to Chen *et al.* (1972), is 27%, whereas 45% is found for the native state.

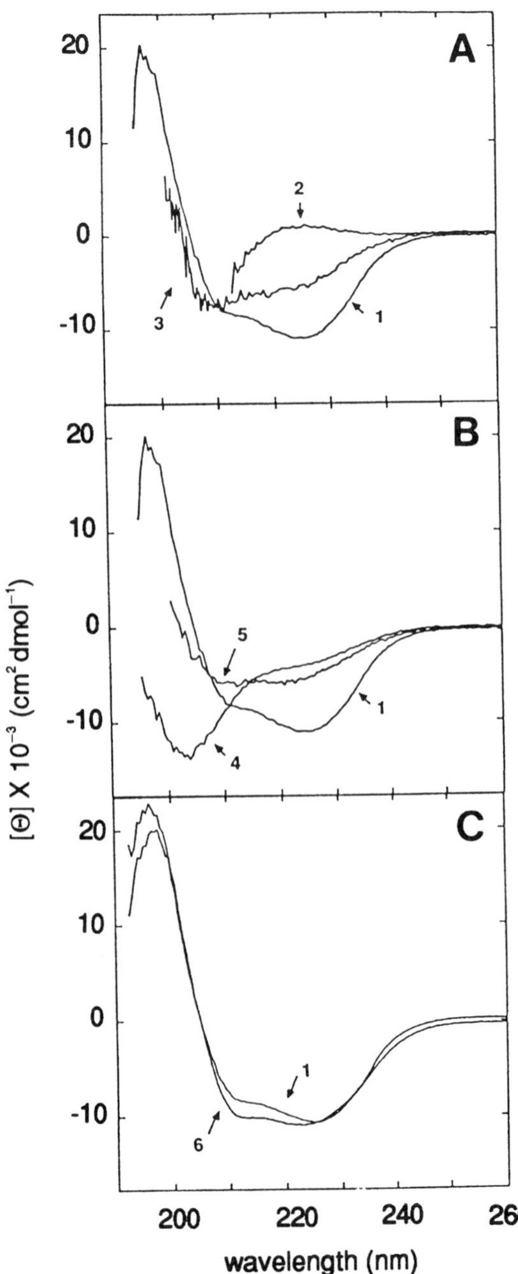

Figure 3. Near-UV CD spectra of Rubisco. (A) Curve 1, Rubisco-N_2; curve 2, Gdn-HCl denatured Rubisco; curve 3, partly refolded Gdn-HCl-denatured Rubisco (Rubisco-I). (B) Curve 1, Rubisco-N_2; curve 4, acid-denatured Rubisco, low salt; curve 5, acid-denatured Rubisco, high salt (Rubisco-A). (C) Curve 1, Rubisco-N_2; curve 6, Rubisco-N_1 (mutant K168E). (From Van der Vies *et al.*, 1992, reproduced with permission.)

The intermediate state was generated by diluting unfolded Rubisco into buffer solutions at 4 °C and by measuring the spectra within 30 min, before significant reversion to the native state had occurred. The far-UV CD spectrum (Fig. 3A, curve 3) is very near the same as that of Rubisco-A (Fig. 3B, curve 5), indicating a similar degree of secondary structure formation, which is clearly less than that of the native state. This, combined with the identical fluorescence spectra and aggregation propensities, leads the authors to suggest that the Rubisco-A state conformation might be closely related to the Rubisco-I state. The former is believed to be structurally similar to the so-called molten globule state (Kuwajima, 1989).

Hence, by using CD the authors could characterize six forms of Rubisco: that is, native dimer (N_2) and native monomer (N_1) with an identical secondary structure; completely or almost completely unfolded by Gdn-HCl or acid at low ionic strength; partially folded by acid at high ionic strength (A); and an intermediate state (I) with secondary structure resembling the A-state. The latter two may be related to the molten globule state.

4. NEAR-UV/VISIBLE CIRCULAR DICHROISM

4.1. Introduction

Whereas the far-UV CD reflects the secondary structure, near-UV and visible (VIS) CD monitors the tertiary structure. The extraction of quantitative information from near-UV/VIS CD is often cumbersome, as will be explained. Therefore, near-UV/VIS CD is mostly used to obtain global information on changes in tertiary structure upon changing conditions (pH, temperature, concentration, and the like) or on differences in conformation between related proteins or mutants. A detailed example of the latter is given by Maune *et al.* (1992). Reviews on near-UV/VIS CD of proteins are given by Strickland (1974) and Kahn (1979). A recent example of quantitative structural analysis from near-UV CD can be found in Scharnagl and Schneider (1991).

4.2. What Does It Measure?

Quantum mechanically, CD originates from the coulomb interaction between the asymmetrically positioned transition dipoles of independent chromophores within a molecule. This interaction will depend on the orientation of the dipole moments and hence on the configuration of the chromophores. Since the interaction decreases rapidly with distance, only interactions between nearby chromophores will yield CD.

Properties of groups showing near-UV/VIS CD are given by Strickland (1974). For Phe and Tyr only a single band, the so-called ^1L_b, shows intense CD. The Trp CD is more complicated, because two dipole moments, ^1L_b and ^1L_a, are significant in this spectral range. The shape and intensity of the near-UV CD band of the cystinyl group (S-S) is highly conformational dependent. In addition, some cations, like Co^{2+}, Cd^{2+}, Zn^{2+}, Hg^{2+}, and Cu^{2+}, show CD when bound to proteins. This has been used by Beltramini *et al.* (1992) to study conformational changes in the active site of hemocyanins (copper-containing proteins) upon ligand binding. Na^+, K^+, and Ca^{2+} have no intrinsic CD, although they may affect spectra by their influence on the conformation of the proteins. For ligands and other groups absorbing in the visible part of the spectrum, VIS CD can provide structural data. The heme group, which has no specific features in its near-UV CD, is an example of this (Singh and Wilson, 1990; Eposti *et al.*, 1991). Several ions show CD in the near-UV above 300 nm or visible due to charge transfer with aromatic residues (Beltramini, 1992).

The position and sharpness of CD bands is influenced by the environment. Hydrogen bonding usually yields a red shift, which for tryptophan can be up to 7 nm (Strickland, 1974). The shifts due to changes in polarity of the environment can be in either direction. Hydrophobic environments tend to sharpen peaks. These phenomena can be used to monitor changes in the equilibrium conformation of near-UV/VIS absorbing chromophores and their surroundings. When an absorbing residue is highly mobile, CD from the different orientations will cancel each other (averaging of asymmetric environment). This can be used to signal flexibility changes in proteins. For epidermolytic toxins, the extraordinarily intense near-UV CD was interpreted as a highly rigid conformation related to the extreme stability of these proteins (Bailey *et al.*, 1982).

It should be noted that a single electronic transition may yield multiple peaks in the CD due to vibrational modes. This complicates the interpretation of near-UV/VIS CD spectra. Moreover, when a chromophore occurs several times in a protein, each individual chromophore may have a different environment, giving rise to different minima and maxima. When the signs of the CD signals are opposite, the net signal may decrease. It has been suggested that generally the average CD per aromatic residue tends to decline as the number of the residues increases. The effect of mobility and multiple occurrence are often difficult to separate (Shire *et al.*, 1991).

4.3. Experimental Details

Since CD signals in the near-UV are weak relative to the normal absorption, signal-to-noise levels may be low and changes sometimes difficult to detect.

However, with modern technology this no longer provides a major obstacle as long as sufficient data are collected to average and proper controls are measured to exclude possible artifacts. The problem of buffer interference is virtually absent in the near-UV. Path lengths are typically between 0.1 and 1.0 cm, depending on the concentration and number of chromophores present, although for very low concentrations even 10-cm cells have been used (Shire *et al.*, 1991). Visible CD bands usually give stronger signals and measurements are more or less straightforward.

4.4. Advantages, Limitations, and Conclusions

Near-UV/VIS CD monitors the tertiary structure of proteins. Although theoretically a quantitative description of it may be obtained (see Kahn, 1979; Scharnagl and Schneider, 1991), this is rarely done, since extensive knowledge on the orientation of the transition dipole moments, their interactions, and the environmental influence on it are not available. However, near-UV/VIS CD is very convenient for a fast qualitative comparison of related structures, such as those found in homologous proteins, related mutants, or a single protein under different conditions. The aromatic residues and occurring ions are proper targets for near-UV/VIS CD studies. This type of CD also provides information on changes in mobility of the chromophores. A technical problem is the weak intensity of near-UV CD signals. Therefore, proper controls and abundant signal averaging are important. Visible spectra show larger signals and are usually quite easy to measure. Due to the low buffer interference in this part of the spectrum, near-UV/VIS CD may be measured over a large range of protein concentrations by the use of cells of different pathlengths (0.1–10 cm) and in almost any buffer.

Especially in combination with other techniques, valuable structural information on protein structure may be obtained from near-UV/VIS CD.

4.5. Recent Applications

4.5.1. AROMATIC RESIDUES DURING ASSOCIATION–DISSOCIATION

Relaxin is a protein hormone that modulates the restructuring of connective tissues in order to generate the required changes in organ structure during pregnancy and parturition. By regulating the softening of cervical tissue and suppressing uterine myometrial contractions, it inhibits premature labor and induces cervical ripening prior to parturition (Shire *et al.*, 1991). Most studies of this protein have been performed on porcine relaxin, since the isolation is well documented and the sow produces large quantities. Shire *et al.* (1991) have compared some properties of the porcine relaxin with that of human source. The former contains no Tyr and two Trps, the latter a single Tyr and also two Trps, one Trp being at an equivalent position in the primary structure of both proteins. Whereas the porcine component is essentially monomeric in solution, the human protein reversibly forms dimers. Far-UV CD (results not shown) showed no difference in secondary structure between human monomeric, human dimeric, and porcine relaxin. The near-UV CD spectra are shown in Fig. 4. Although the protein from both species contains two Trps, the Trp CD (295 nm) is quite small. This might be due to canceling of positive and negative contributions for the Trps or the Trps might be quite mobile. The negative band at 277 nm in the spectrum of human relaxin can be attributed to the lone Tyr, that at 284 nm contains contributions from both Trp and Tyr. However, the spacing of 7 nm is typical for the ^1L_b Tyr band (Strickland, 1974), indicating that it is dominated by Tyr. The porcine compound shows little fine structure in this region, the broad band probably being due to the disulfide CD. Assuming equivalent environment for the Trps and

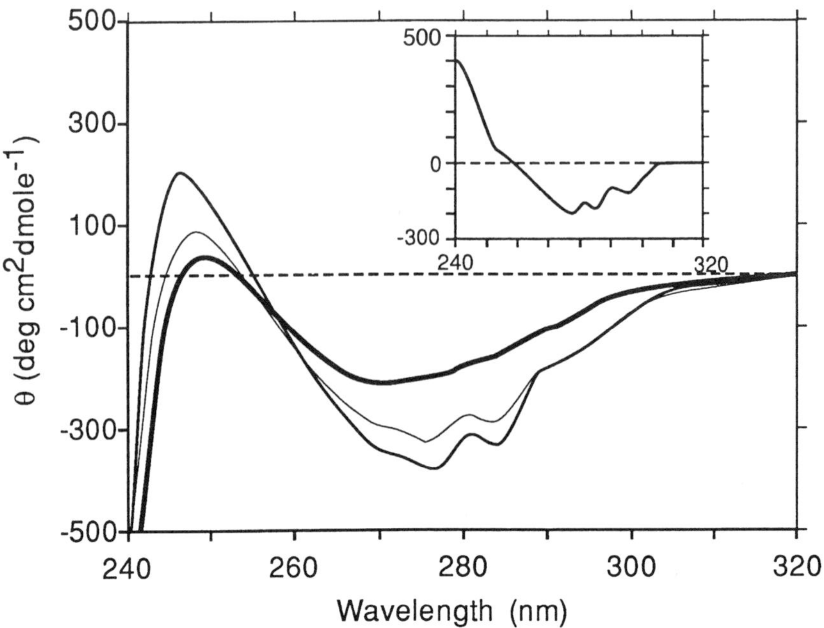

Figure 4. Near-UV CD spectra of porcine relaxin at 0.5 mg/ml (bold line), human relaxin at 0.5 mg/ml (solid line), and human relaxin at 20 μg/ml (thin line). Inset: subtracting the porcine spectrum from that of human at 20 μg/ml. (From Shire *et al.*, 1991, reproduced with permission.)

disulfides in both species, the difference spectrum (inset) should be that of the lone Tyr in the human component. However, the broad band at 293 nm is due to Trp. This points to a different chemical surrounding and/or orientation of at least one Trp in the two species.

When the human compound is diluted, and hence the relative concentration of the monomeric form is increased, the CD at 277 and 284 nm significantly decreases, whereas the CD at 295 nm hardly changes. Apparently the Tyr environment is significantly affected upon dissociation, leading to greater mobility. The secondary structure (far-UV CD, not shown) and Trp CD are more or less unaffected. The Tyr is apparently located closely to the association site.

4.5.2. IRON–SULFUR CLUSTERS

The irons in iron–sulfur clusters that occur in proteins involved in oxidation–reduction cycles are very appropriate for VIS CD analysis (see, e.g., Hirasawa *et al.*, 1989; Stephens *et al.*, 1991; Eposti *et al.*, 1991; Vanoni *et al.*, 1992). In the ammonia assimilation of diazotrophs, L-glutamate reacts with 2-oxyglutamate (2-OG) in the presence of NADPH to produce L-glutamate:

$$\mathrm{NADPH} + \textsc{l}\text{-Gln} + \text{2-OG} \rightleftharpoons 2\,(\textsc{l}\text{-Glu}) + \mathrm{NADP}^+$$

The reaction is catalyzed by glutamate synthase (GltS). This enzyme contains a flavin-adenine, a flavin mononucleotide, and about 8 to 12 nonheme irons (Vanoni *et al.*, 1991,1992). It has been proposed that the reaction is split into two parts occurring at different sites of the enzyme. NADPH would be reduced at one flavin-containing site. Electron transfer, probably via FeS centers, would then occur to the second flavin-containing site where the rest of the reaction takes place (Rendina and Orme-Johnson, 1978; Vanoni *et al.*, 1991). In order to support this model, Vanoni *et al.* (1992) performed sulfite titrations, electron paramagnetic resonance (EPR), and UV/VIS CD spectroscopy on GltS from *Azoaspirillum*. Special anaerobic quartz cuvettes were used for this study. The sulfite reaction proved the nonequivalence of the two flavins and supported the idea of two separate sites for the NADPH and glutamine reaction. The identity of the FeS centers could not be obtained from the absorption spectrum, since the visible spectrum of this enzyme is dominated by the flavin absorption. EPR measurements indicated the presence of three different iron–sulfur clusters ([xFe-yS], x and y denoting the number of Fe and S in the cluster, respectively), with redox potentials following the order $[x\mathrm{Fe}\text{-}y\mathrm{S}]_{\mathrm{I}} > [x\mathrm{Fe}\text{-}y\mathrm{S}]_{\mathrm{II}} > [x\mathrm{Fe}\text{-}y\mathrm{S}]_{\mathrm{III}}$. Whereas $[x\mathrm{Fe}\text{-}y\mathrm{S}]_{\mathrm{I}}$ could be identified unambiguously by EPR as [3Fe-4S], this technique could not prove whether the other clusters were [4Fe-4S] or [2Fe-2S]. The VIS CD spectra of oxidized, partially, and totally reduced GltS are shown in Fig. 5.

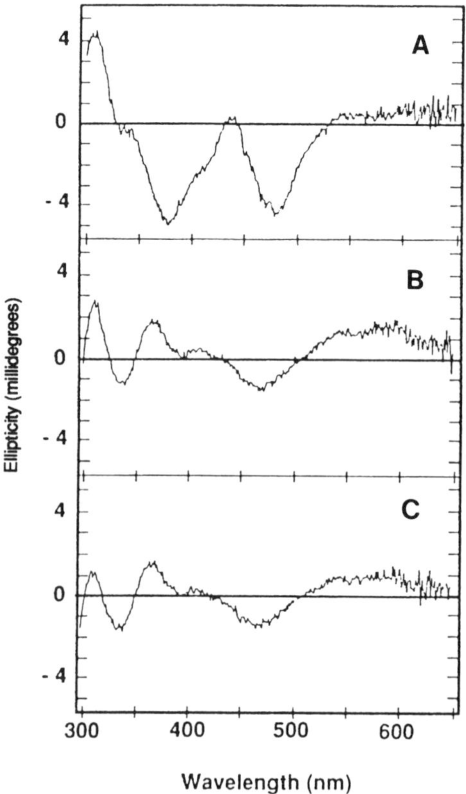

Figure 5. Visible CD spectra of glutamate synthase from *Azospirillum brasilense*: (A) oxidized form, (B) partially reduced (two out of three FeS clusters), and (C) totally reduced form (all FeS clusters). (From Vanoni *et al.*, 1992, reproduced with permission.)

The oxidized form shows two strong negative bands between 350 and 500 nm. For [2Fe-2S] containing proteins, intense positive peaks are observed in this region (Palmer and Massey, 1969; Komai *et al.*, 1969; Hirasawa *et al.*, 1989). Apparently, *Azospirillum* GltS does not contain [2Fe-2S] centers. The fact that both the partially reduced sample [only $[x\text{Fe-}y\text{S}]_{\text{I}}$ and $[x\text{Fe-}y\text{S}]_{\text{II}}$ reduced] and the totally reduced sample show identical CD indicates that [4Fe-4S] clusters do not contribute significantly to the visible CD. Probably the higher symmetry of this group in comparison with that of [2Fe-2S] plays a role.

5. VIBRATIONAL CIRCULAR DICHROISM

5.1. Introduction

Ultraviolet CD monitors electronic transitions, which are restricted in number, often broad and unstructured, overlapping, and mutual interfering. Generally, vibrational transitions (absorbing in the infrared) are better resolved and more straightforward to assign. Therefore, it has been suggested that infrared CD [often called vibrational CD (VCD)] might be a more sensitive and reliable tool for structural studies in biopolymers (Keiderling, 1986; Pancoska *et al.*, 1989; Pancoska and Keiderling, 1991; Gupta and Keiderling, 1992; Baumruk and Keiderling, 1993).

5.2. What Does It Measure?

In principle any infrared-absorbing chromophore in an asymmetric surrounding can be studied with infrared CD. However, due to infrared absorption by water molecules and to a lesser extent by CO_2, in practice the amide I band (1600–1700 cm^{-1}), which reflects secondary structure, and some ligand bands have been used for most work. The latter will be discussed below. The amide II CD band in principle also contains information on secondary structure, but it is rather weak and unstructured (Pancoska *et al.*, 1989). Therefore, it gives poor quantitative results (Gupta and Keiderling, 1992; Baumruk and Keiderling, 1993). Keiderling (1986) and Paterlini *et al.* (1986) have measured infrared CD of the amide I band in D_2O for poly-L-lysine in the α-helix, β-sheet, random coil, and completely denatured form. In the latter case no infrared CD whatsoever was observed (Paterlini *et al.*, 1986). Qualitative theoretical relations between secondary structure and infrared CD for the amide I and amide II bands are given by Snir *et al.* (1975) and Paterlini *et al.* (1986).

In order to calculate secondary structure content from experimentally determined infrared CD spectra of reference proteins, Pancoska and co-workers have used the principal component method of factor analysis combined with cluster analysis (Pancoska *et al.*, 1991; Pancoska and Keiderling, 1991). So far, the nonlinear effects discussed in Section 3.3 have not been considered explicitly for infrared CD on proteins. Pancoska and co-workers (1989, 1991) found that infrared CD reflects shorter-range interactions than far-UV CD, so nonlinear effects may be less important. Indeed, model polypeptides show relatively small chain length dependence for infrared CD spectra of α-helixes (Yasui *et al.*, 1987).

Although both far-UV and infrared CD of the amide I band reflect the secondary structure, the former in fact monitors only dihedral angles, whereas the latter reflects dihedral angles and has a through-bond contribution. Therefore, secondary structures obtained with the two techniques may differ.

Apart from secondary structure, ligand binding to heme proteins also has been studied with infrared CD. In this case the ligand–iron stretch vibration between 2100 and 1950 cm^{-1} is measured (Bormett *et al.*, 1992; Teraoka *et al.*, 1992). Strong negative bands are found for azide as the ligand, while cyanide yields a positive band with double the intensity, and SCN^-, OCN^-, and CO give no observable infrared CD (Teraoka *et al.*, 1992). It appears that sign and magnitude are affected by the symmetry and constraint in the porphyrin ring, by the electronic state and orientation of the electromagnetic dipole moments of the ligand–iron complex, and by the interaction between the ligand and the heme-pocket (Bormett *et al.*, 1992; Teraoka *et al.*, 1992). Theory has not yet been developed to describe infrared CD spectra in terms of the electronic state of the ligand–heme–iron–protein complex (Bormett *et al.*, 1992). Therefore, the present focus of this application of infrared CD is empirical.

5.3. Experimental Details

Infrared CD spectra can be measured on an infrared spectrometer that has been modified to produce modulated circularly polarized light (Diem *et al.*, 1978; Keiderling, 1981, 1986; Lipp and Nafie, 1984; Pancoska *et al.*, 1989; Bormett *et al.*, 1992; Baumruk and Keiderling, 1993). A fast, high sensitivity infrared detector is necessary to measure the modulated intensities. Keiderling (1986) reports a sensitivity of $\Delta A/A$ of 10^{-5}, which is comparable to that of UV CD spectrophotometers. However, since infrared CD signals are smaller, the signal-to-noise ratio is worse. The spectral resolution is 10 to 15 cm^{-1} (Pancoska *et al.*, 1989).

In the rest of this section, we focus on infrared CD for determining secondary structure. For ligand infrared CD, we refer the reader to Bormett *et al.* (1992) and Teraoka *et al.* (1992). Of course, all the problems of ordinary infrared spectroscopy on biomolecules (see, e.g., Jackson *et al.*, 1989) are also encountered in infrared CD. The most severe of these is the strong infrared absorption of water, the natural solvent for most biomolecules. In particular, the HOH bend absorption overlaps strongly with the amide I band. Also, carbonyl groups other than the one in the peptide group may interfere. Thoroughly flushing the spectrometer with dry nitrogen to eliminate CO_2 and using buffers without carbonyls minimize the effects of external carbonyl groups. The water problem can be partially solved by using D_2O instead of H_2O. So far, most quantitative measurements on protein secondary structure in solution have been performed in D_2O. However, in heavy water deuterium exchanges with the protein hydrogens. The rate and extent of this exchange is protein specific and depends on environmental parameters, such as pH (Leichtling and Klotz, 1966; Molday *et al.*, 1972; Delepierre *et al.*, 1987). Therefore, great care must be taken when comparing different systems. Although the problem has been recognized (Pancoska *et al.*, 1989; Baumruk and Keiderling, 1993), it is difficult to take into account. Deuterium-exchanged proteins have been obtained by twice dissolving and lyophilizing the sample in D_2O before final sample preparation. It was hoped that these "carefully controlled sampling conditions" (Pancoska *et al.*, 1989) are sufficient to counteract differences in rate and extent of exchange. Recently, Baumruk and Keiderling (1993) reported first amide I CD spectra in H_2O. These measurements became possible due to improved instrumentation. Although the noise of the measurements is more than in D_2O, the results are promising, since interpretative ambiguities due to deuteration are avoided.

In order to obtain a proper subtraction of the (solvent) water contribution to the spectra, high concentrations of protein (typically 20–50 mg/ml) and very short path lengths (6 μm for H_2O, 20–50 μm for D_2O) are important. Usually, CaF_2 windows separated by a Teflon, tin, or mylar spacer are used as the cell.

An additional problem in infrared CD is the occurrence of an artifact signal correlated to the magnitude of the absorption (Malon and Keiderling, 1988). Subtraction of the infrared CD spectrum of poly-(DL)-lysine measured under identical conditions as the sample spectrum seems to solve this problem, since essentially flat baselines are obtained by this procedure (Pancoska *et al.*, 1989). Alternatively, adaptation of the experimental setup may reduce the artifact problem to a large extent (Lipp and Nafie, 1984; Malon and Keiderling, 1988).

5.4. Advantages, Limitations, and Conclusions

Infrared and far-UV CD of four proteins are compared in Fig. 6. The infrared CD spectra show more differentiation than the UV CD spectra. Whereas all examples shown of far-UV CD have a negative band between 200 and 240 nm and a positive one between 180 and 200 nm, the infrared CD exhibits different signs at comparable wave numbers. Hence, in addition to intensity differences, sign discrimination can be used for the analysis of infrared CD. On the other hand, the signal-to-noise ratio is much better for far-UV CD. Moreover, it remains uncertain as to what extent spectral differences in infrared CD are caused by different levels of deuterium exchange in the proteins. A comparison between secondary structures predicted from far-UV and infrared CD by the principal component method for a set of 13 proteins is given by Pancoska and Keiderling (1991). For the protein concentration determination they assume equal absorption at 1650 cm^{-1} for different structures, a procedure that Lee *et al.* (1990) have questioned. Pancoska

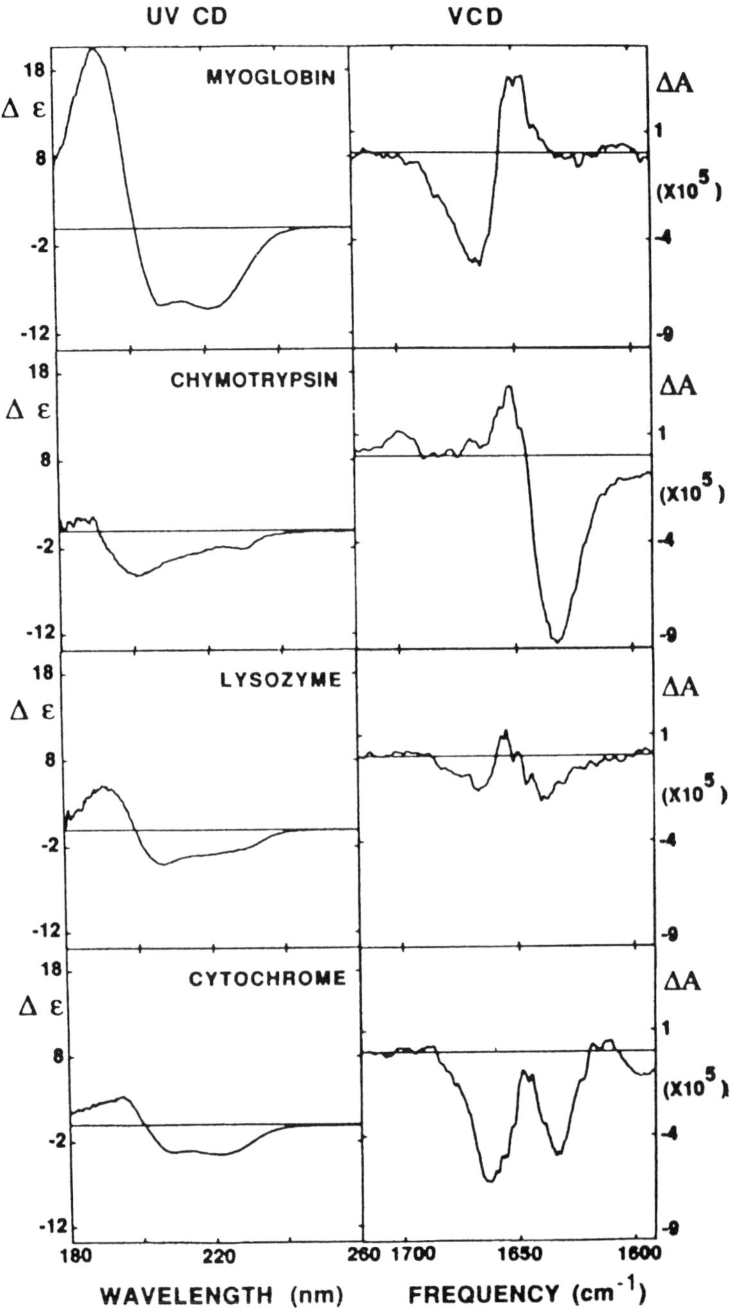

Figure 6. A comparison between UV-CD and IR-CD spectra of some proteins. (From Pancoska *et al.*, 1989, reproduced with permission.)

and Keiderling found a root mean square difference between calculated and X-ray structures of 9% (infrared) versus 8% (UV) for α-helix; 78 (infrared) versus 10% (UV) for β-sheet; 5% (infrared) versus 4% (UV) for a structure called "bend"; and 6% (infrared) versus 3% (UV) for the remainder. Two advantages of infrared CD over far-UV CD are the absence of interference by absorbing aromatic residues (Yasui and Keiderling, 1986) and the absence of CD for totally unordered structure (Paterlini *et al.*, 1986).

Although the series of proteins for which infrared CD measurements have been performed, as well as the available theoretical models, are rather restricted, it appears that in the future infrared CD will have a full place next to far-UV CD in the study of secondary structure. Certainly the recent development to measure the amide I CD in H_2O enhances the possibilities of this technique. Infrared CD measurements require high protein concentrations (20–50 mg/ml), whereas far-UV CD is restricted to low concentrations (~1 mg/ml). In this respect the two techniques are complementary, which may be especially useful for concentration-dependent studies.

The application of infrared CD for the study of ligand binding is not far enough developed to give a proper evaluation of the possibilities and limitations, but there are examples, one of which we will give in the next section.

5.5. Recent Applications

Infrared CD has been used to show that the α-lactalbumin structure in solution differs from that in the crystal (Urbanova *et al.*, 1991). Although α-lact-albumin has a very similar crystal structure to lysozyme and their infrared CD spectra in 33% propanol are similar, the infrared CD spectra of these proteins in aqueous solution are very different. These results suggest that the crystal structure of α-lactalbumin, which is known as a conformationally labile protein, reflects a dehydrated structure.

Bormett *et al.* (1992) measured infrared CD in the azide-heme antisymmetric stretch vibration (2100–1950 cm^{-1}) of a series of azide myo- and hemoglobins in both T- and R-state complexes. The infrared absorption spectra in this region show a double transition, which results from the high and low spin state iron complexes. Only the low spin band is reflected in the infrared CD spectra. The anisotropy ratio, g of Eq. (4), for horse and whale myoglobin is 9.5×10^{-4}. These proteins have identical heme pockets. When the His E7 of the pocket is mutated into a glycine or the Val E11 by asparagine, g is almost zero. Replacement of His E7 by glutamine reduces g by 35%. This shows that infrared CD is highly sensitive for small changes in the weak interactions between the ligand and the heme-pocket residues. All the native hemoglobins studied had a g of 8×10^{-4}. The authors suggest that the decrease (compared to the myoglobin value) is caused by structural changes in the proximal side of the heme or constraint of the porphyrin ring. Also, a mutant monomeric hemoglobin (CTT) had a g of 8×10^{-4}, although the band position was shifted by 20 cm^{-1}. Apparently only the protein–heme interactions (band shift), but not the ligand–protein interaction (constant g) are affected by the quaternary structure.

6. TIME-RESOLVED CIRCULAR DICHROISM

Time-resolved CD (TRCD) is a recent development that allows structural studies of transient intermediates. It has been used to study isomerization (Hasumi, 1980), photoprocesses (Björling *et al.*, 1992; Lewis *et al.*, 1992), and folding intermediates (Labhardt, 1986; Kuwajima *et al.*, 1987, 1991; Elöve *et al.*, 1992).

Lewis *et al.* (1992) have recently reviewed different instrumentation for TRCD. When the required time resolution is milliseconds or more, a conventional CD spectrophotometer can be used, the protein solution being rapidly mixed with a reactant solution in a mixing cell. Mixing chambers can be homemade (Kuwajima *et al.*, 1987) or commercially purchased (Hasumi, 1980; Elöve *et al.*, 1992). Both stopped-flow (Kuwajima *et al.*, 1987, 1991) and continuous flow (Elöve *et al.*, 1992) experiments have been described. The optical path length is between 1.0 and 10 mm, and the dead time 2 msec or higher. In most cases CD signals have been monitored at single wavelengths, which is time-consuming when a complete spectrum is scanned point by point. Alternatively, a multichannel detector can be used. The CD results are subjected to traditional kinetic analyses. Both far- and near-UV/VIS TRCD measurements can be made on these instruments. This allows a detailed study of structural changes in terms of secondary and tertiary structure.

Highly sophisticated TRCD instrumentation allows measurement in the nano- and picosecond timescale (Björling *et al.*; 1992, Lewis *et al.*, 1992). However, only light-induced processes can be followed. Moreover, the measurements are very sensitive to optical artifacts, which have been thoroughly discussed in the literature (Björling *et al.*, 1992; Lewis *et al.*, 1992). Nano- and picosecond technology does not at present allow measurements in the far-UV (Lewis *et al.*, 1992).

Hasumi (1980) resolved a four-step mechanism for the pH-induced isomerization of horse heart ferricytochrome *c* with TRCD. Elöve *et al.* (1992) divided the refolding of cytochrome *c* in four distinct phases, namely, $<$4 msec, 4–100 msec, 100–1000 msec, and $>$ 1 sec. In the first two phases the secondary structure is established, and in the latter two the tertiary structure is formed. It seems that there are no intermediates with partially formed secondary structure. Kuwajima *et al.* (1991) were able to show that, notwithstanding the apparent two-state equilibrium unfolding of dihydrofolate reductase, various folding intermediates are formed during the transition. Using the methods of Greenfield and Fasman

(1969), Chen *et al.* (1974) and Chang *et al.* (1978) Kuwajima and colleagues calculated the secondary structure contents and found that, although a considerable amount of secondary structure is initially formed during folding, this rearranges in later stages. Unfolding and refolding follow different pathways. The signal-to-noise ratio is a problem in refolding studies, because it is necessary to do the measurements at low protein concentrations in order to prevent aggregation during refolding.

The advantage of TRCD over time-resolved absorbance or fluorescence is that both secondary and tertiary structure can be followed with the same instrument. Moreover, the time window in fluorescence is restricted to the nanosecond range. As Kuwajima *et al.* (1991) show, optimal information on secondary structure during folding of proteins is obtained from the combination of TRCD with hydrogen-exchange NMR, since these techniques provide complementary information.

7. CONCLUDING REMARKS

This chapter has illustrated that very different aspects of protein structure can be investigated by CD spectroscopy, ranging from secondary structure (far-UV and infrared), tertiary structure, and flexibility (near-UV), to protein–ligand interaction (magnetic, near-UV/VIS, infrared). Although the information content of CD is significantly lower than that of NMR spectroscopy and crystallographic X-ray diffraction, the speed of the measurements, the simplicity of the spectra obtained, and the relative ease of their interpretation make CD a very attractive approach for the comparison of homologous proteins, the study of a protein at different conditions, or the monitoring of structural changes during a process. For the latter, recent development in TRCD are invaluable. Although much effort has been invested in the theoretical description of CD spectra, the structure of biological molecules is usually too complex for exact solutions and the method is used (semi-)empirically. Although both technology and methodology are well developed to obtain reliable results, improper use can easily lead to erroneous conclusions. We hope the information presented in this chapter will prevent such errors.

ACKNOWLEDGMENT. This work was supported by the Netherlands organization for scientific research (NWO), and NIH grant GM-21479.

REFERENCES

Antanaitis, B. C., Strekas, T., and Aisen, P., 1982, Characterization of pink and purple uteroferrin by resonance Raman and CD spectroscopy, *J. Biol. Chem.* **7**:3766–3770.

Arakawa, T., Hsy, Y-R., and Yphantis, D. A., 1987, Acid induced unfolding and self-association of recombinant *Escherichia coli* derived human interferon γ, *Biochemistry* **26:**5428–5432.

Bailey, J. C., Martin, S. R., and Bayley, P. M., 1982, A circular dichroism study of epidermolytic toxins A and B from *Staphylococcus aureus*, *Biochem. J.* **203:**775–778.

Baumruk, V., and Keiderling, T. A., 1993, Vibrational circular dichroism of proteins in H_2O solution, *J. Am. Chem. Soc.* **115:**6939–6942.

Bayley, P., 1980, Circular dichroism and optical rotation, in: *An Introduction to Spectroscopy for Biochemists* (S. B. Brown, ed.), Academic Press, London, pp. 148–234.

Beltramini, M., Bubacco, L., Salvato, B., Casella, L., Gulotti, M., and Garofani, S., 1992, The aromatic circular dichroism spectrum as a probe for conformational changes in the active site environment of hemocyanins, *Biochim. Biophys. Acta* **1120:**24–32.

Björling, S. C., Zhang, C-F., Farrens, D. L., Song, P-S., and Kliger, D. S., 1992, Time-resolved circular dichroism of native oat phytochrome photointermediates, *J. Am. Chem. Soc.* **114:**4581–4588.

Bokma, J. T., Johnson, W. C., Jr., and Blok, J., 1987, CD of the Li salt of DNA in ethanol/water mixtures: Evidence for the B- to C-form transition in solution, *Biopolymers* **26:**893–909.

Bolotina, T. A., and Lugauskas, V. Y., 1986, Determination of the secondary structure of proteins from the circular dichroism spectra. IV. Consideration of the contribution of aromatic residues to the circular dichroism spectra of proteins in the peptide region, *Mol. Biol. (Moscow)* **19:**1154–1166 (translated from *Molek. Biol.* 1985, **19:**1404–1421).

Bormett, R. W., Asher, S. H., Larkin, P. J., Gustafson, W. G,. Ragunathan, N., Freedman, T. B., Nafie, L. A., Balasubramanian, S., Boxer, S. G., Yu, N-T., Gersonde, K., Noble, R. W., Springer, B. A., and Sligar, S. G., 1992, Selective examination of haem protein azide ligand-distal globin interactions by vibrational circular dichroism, *J. Am. Chem. Soc.* **114:**6864–6867.

Bradford, M. M., 1976, A rapid and sensitive method for quantification of microgram quantities of protein utilizing the principle of protein–dye binding, *Anal. Biochem.* **72:**240–254.

Brahms, S., and Brahms J., 1980, Determination of protein secondary structure in solution by vacuum ultraviolet circular dichroism, *J. Mol. Biol.* **138:**149–178.

Bree, A., and Lyons, L. E., 1956, The intensity of ultraviolet light absorption by monocrystals. Part I. Measurement of thickness of thin crystals by interferometry, *J. Chem. Soc.* **1956:**2658–2662.

Campbell, I. D., and Dwek, R. A. (eds.), 1984, *Biological Spectroscopy*, Benjamin/Cummings, Menlo Park, pp. 255–277.

Cantor, C. R., and Schimmel, P. R. (eds.), 1980, *Biophysical Chemistry*, Vol. 2, Freeman, New York, pp. 409–418.

Chang, C. T., Wu, C.-S., and Yang, J. T., 1978, Circular dichroism analysis of protein conformation: Inclusion of the β-turns, *Anal. Biochem.* **91:**13–31.

Charney, E. (ed.), 1979, *The Molecular Basis of Optical Activity, Optical Rotatory Dispersion and Circular Dichroism*, John Wiley, New York, pp. 1–17, pp. 323–327.

Cheesman, M. R., Greenwood, C., and Thomson, A. J., 1991, Magnetic circular dichroism of hemoproteins, *Adv. Inorg. Chem.* **36:**201–255.

Chen, Y-H., and Yang, J. T., 1977, Two point calibration of circular dichrometers with *d*-10-camphorsulphonic acid, *Anal. Lett.* **10:**1195–1207.

Chen, Y-H., Yang, J. T., and Martinez, H., 1972, Determination of the secondary structures of proteins by circular dichroism and optical rotatory dispersion, *Biochemistry* **11:**4120–4132.

Chen, Y.-H., Yang, J. T., and Chan, K. H., 1974, Determination of the helix and β-form of proteins in aqueous solution by circular dichroism, *Biochemistry* **13:**3350–3359.

Compton, L. A., and Johnson, W. C., Jr., 1986, Analysis of protein circular dichroism spectra for secondary structure using a simple matrix multiplication, *Anal. Chem.* **155:**155–167.

Cotton, M. A., 1896, Recherches sur l'absorption et la dispersion de la lumière par les milieux doués du pouvoir rotatoire, *Ann. Chim. Phys. VII* **8:**347–437.

Delepierre, M., Dobson, C. M., Karplus, M., Poulsen, F. M., States, D. J., and Wedin, R. E., 1987, Electrostatic effects and hydrogen exchange behaviour in proteins: The pH dependence of exchange rates in lysozyme, *J. Mol. Biol.* **197:**111–130.

Diem, M., Gotkin, P. J., Kupfer, J. M., and Nafie, L. A., 1978, Vibrational circular dichroism in amino acids and peptides. 2. Simple alanyl peptides, *J. Am. Chem. Soc.* **100:**5644–5650.

Elöve, G. A., Chaffotte, A. F., Roder, H., and Goldberg, M. E., 1992, Early steps in cytochrome *c* folding probed by time-resolved circular dichroism and fluorescence spectroscopy, *Biochemistry* **31:**6876–6883.

Elwell, M. L., and Schellman, J. A., 1977, Native properties and thermal stability of wild type and two mutant lysozymes, *Biochim. Biophys. Acta* **494:**367–383.

Eglinton, D. G., Johnson, M. K., Thomson, A. J., Gooding, P. E., and Greenwood, C., 1980, Near infra-red magnetic and natural circular dichroism of cytochrome *c* oxidase, *Biochem. J.* **191:**319–331.

Eposti, M. D., Crimi, M., Körtner, C., Kröger, A., and Link, T., 1991, The structure of dihaem cytochrome *b* of fumarate reductase in Wolinella succinogenes: Circular dichroism and sequence analysis studies, *Biochim. Biophys. Acta* **1056:**243–249.

Gans, P., 1980, Vibrational spectroscopy, in: *An Introduction to Spectroscopy for Biochemists*, (S. B. Brown, ed.), Academic Press, London, pp. 148–234.

Goodman, M., Verdini, A. S., Toniolo, C., Phillips, W. D., and Bovey, F. A., 1969, Selective criteria for the critical size for helix formation in oligopeptides, *Proc. Natl. Acad. Sci. USA* **64:**444–450.

Goto, Y., Takahashi, N., and Fink, A. L., 1990a, Mechanism of acid-induced folding of proteins, *Biochemistry* **29:**3480–3488.

Goto, Y., Calciano, L. J., and Fink, A. L., 1990b, Acid-induced folding of proteins, *Proc. Natl. Acad. Sci. USA* **87:**573–577.

Greenfield, N., and Fasman, G. D., 1969, Computed circular dichroism spectra for the evaluation of protein conformation, *Biochemistry* **8:**4108–4116.

Gross, K. P, and Schnepp, O., 1977, Improved circular dichroism instrument in the vacuum ultraviolet, *Rev. Sci. Instrum.* **48:**362–363.

Grosse, R., Malur, J., Meiske, W., and Repke, K. R. H., 1974, Statistical behaviour and suitability of protein derived circular dichroic basis spectra for the determination of globular protein information, *Biochim. Biophys. Acta* **359:**33–46.

Gupta, V. P., and Keiderling, T. A., 1992, Vibrational CD of the amide II band in some model polypeptides and proteins, *Biopolymers* **32:**239–248.

Hadden, J. M., Bloemendal, M., Haris, P. I., Srai, S. K. S., and Chapman, D., 1994, Fourier transform infrared spectroscopy and differential scanning calorimetry of transferrins: Human serum transferrin, rabbit serum transferrin and human lactoferrin, *Biochim. Biophys. Acta* **205:**59–67.

Hasumi, H., 1980, Kinetic studies on isomerization of ferricytochrome *c* in alkaline and acid pH ranges by the circular dichroism stopped-flow method, *Biochim. Biophys. Acta* **626:**265–276.

Hennessey, J. P., Jr., and Johnson, W. C., Jr., 1981, Information content in the circular dichroism of proteins, *Biochemistry* **20:**1085–1094.

Hennessey, J. P., Jr., and Johnson, W. C., Jr., 1982, Experimental errors and their effect on analyzing circular dichroism spectra of proteins, *Anal. Biochem.* **125:**177–188.

Hirasawa, M., Chang, K. T., Morrow, K. J., and Knaff, D. B., 1989, *Biochim. Biophys. Acta* **977:**150–156.

Hvidt, S., and Lehrer, S. S., 1992, Thermally induced chain exchange of frog αβ-tropomyosins, *Biophys. Chem.* **45:**51–59.

Jackson, M., Haris, P. I., and Chapman, D., 1989, Fourier transform infrared spectroscopic studies of lipids, polypeptides and proteins, *J. Mol. Struct.* **214:**329–355.

Jirgensons, B., 1973, Optical activity of proteins and other macromolecules, *Mol. Biol. Biochem.* **5:**1–191.

Jiskoot, W., Bloemendal, M., van Haeringen, B., van Grondelle, R., Beuvery, E. C., Herron, J. N., and Crommelin, D. J. A., 1991, Non-random conformation of a mouse IgG_{2a} monoclonal antibody at low pH, *Eur. J. Biochem.* **201:**223–232.

Johnson, W. C., Jr., 1971, A circular dichroism spectrometer for the vacuum ultraviolet, *Rev. Sci. Instrum.* **42:**1283–1286.

Johnson, W. C., Jr., 1986, Extending circular dichroism spectra into the vacuum UV and its application to proteins, *Photochem. Photobiol.* **44:**307–313.

Johnson, W. C., Jr., 1988, Secondary structure of proteins through circular dichroism spectroscopy, *Annu. Rev. Biophys. Biophys. Chem.* **17:**145–166.

Johnson, W. C., Jr., 1990, Protein secondary structure and circular dichroism: A practical guide, *Proteins* **7:**205–214.

Kahn, P. C., 1979, The interpretation of near-ultraviolet circular dichroism, in: *Methods in Enzymology*, Vol. 61 (C. H. W. Hirs and S. N. Timasheff, eds.), Academic Press, San Diego, CA, pp. 339–377.

Keiderling, T. A., 1981, Vibrational circular dichroism, *Appl. Spectrosc. Rev.* **17:**189–226.

Keiderling, T. A., 1986, Vibrational CD of biopolymers, *Nature* **322:**851–852.

Komai, H., Massey, V., and Palmer, G., 1969, The preparation and properties of deflavoxanthineoxidase, *J. Biol. Chem.* **244:**1692–1700.

Kuwajima, K., 1989, The molten globule state as a clue for understanding the folding and cooperativity of globular protein structure, *Proteins* **6:**87–103.

Kuwajima, K., Yamaya, H., Miwa, S., Sugai, S., and Nagamura, T., 1987, Rapid formation of secondary structure framework in protein folding studied by stopped-flow circular dichroism, *FEBS Lett.* **221:**115–118.

Kuwajima, K., Garvey, E. P., Finn, B. E., Matthews, C. R., and Sugai, S., 1991, Transient intermediates in the folding of dihydrofolate reductase as detected by far-ultraviolet circular dichroism spectroscopy, *Biochemistry* **30:**7693–7703.

Labhardt, A. M., 1986, Folding intermediates studied by circular dichroism, in: *Methods in Enzymology*, Vol. 131 (C. H. W. Hirs and S. N. Timasheff, eds.), Academic Press, San Diego, CA, pp. 126–135.

Lee, D. C., 1985, Infrared spectroscopic studies of biological and model membranes, Ph.D. thesis, Royal Free Hospital School of Medicine, London, pp. 62, 71.

Lee, D. C., Haris, P. I., Chapman, D., and Mitchell, R. C., 1990, Determination of protein secondary structure using factor analysis of infrared spectra, *Biochemistry* **29:**9185–9193.

Leichtling, B. H., and Klotz, I. M., 1966, Catalysis of hydrogen-deuterium exchange in polypeptides, *Biochemistry* **5:**4026–4037.

Lewis, J. W., Goldbeck, R. A., Kliger, D. S., Xie, X., Dunn, R. C., and Simon, J. D., 1992, Time-resolved circular dichroism spectroscopy: Experiment, theory, and applications to biological systems, *J. Phys. Chem.* **96:**5243–5254.

Lewis, P. N., Momany, F. A., and Scheraga, H. A., 1973, Chain reversals in proteins, *Biochim. Biophys. Acta* **303:**211–229.

Lipp, E. D., and Nafie, L. A., 1984, Fourier transform infrared vibrational circular dichroism: Improvement in methodology and mid-infrared spectral results, *Appl. Spectr.* **38:**20–26.

Lowry, O. H., Rosebrough, N. J., Farr, A. L., and Randall, R. J., 1951, Protein measurement with the folin phenol reagent, *J. Biol. Chem.* **193:**265–275.

Lowry, T. M., and French, H. S., 1932, The rotatory dispersive power of organic compounds. Part XX. Rotatory dispersion and circular dichroism of camphor-β-sulphonic acid in the region of absorption, *J. Chem. Soc.* **1932:**2655–2658.

Madison, V., and Schellman, J., 1972, Optical activity of polypeptides and proteins, *Biopolymers* **11:**1041–1076.

Malon, P., and Keiderling, T. A., 1988, A solution to the artifact problem in Fourier transform vibrational circular dichroism, *Appl. Spectrosc.* **42:**32–38.

Manavalan, P., and Johnson, W. C., Jr., 1985, Protein secondary structure from circular dichroism spectra, *J. Biosci.* **8**(Suppl)**:**141–149.

Manavalan, P., and Johnson, W. C., Jr., 1987, Variable selection method improves the prediction of protein secondary structure from circular dichroism spectra, *Anal. Biochem.* **67:**76–85.

Manning, M. C., 1989, Underlying assumptions in the estimation of secondary structure content in proteins by circular dichroism spectroscopy—a critical review, *J. Pharmac. Biomed. Anal.* **7:**1103–1119.

Maune, J. F., Beckingham, K., Martin, S. R., and Bayley, P. M., 1992, Circular dichroism studies on calcium binding in two series of Ca^{2+} binding site mutants of *Drosophila melanogaster* calmodulin, *Biochemistry* **31:**7779–7786.

Molday, R. S., Englander, S. W., and Kallen, R. G., 1972, Primary structure effects on peptide group hydrogen exchange, *Biochemistry* **11:**150–158.

Nölting, B., Jung, C., and Snatzke, G., 1992, Multichannel circular dichroism investigations of the structural stability of bacterial cytochrome *p*-450, *Biochim. Biophys. Acta* **1100:**171–176.

Osborne, G. A., Cheng, J. C., and Stephens, P. J., 1973, A near-infrared circular dichroism and magnetic circular dichroism instrument, *Rev. Sci. Instrum.* **44:**10–15.

Palmer, G., and Massey, V., 1969, Electron paramagnetic resonance and circular dichroism studies on milk xanthine oxidase, *J. Biol. Chem.* **244:**2614–2620.

Pancoska, P., and Keiderling, T. A., 1991, Systematic comparison of statistical analyses of electronic and vibrational circular dichroism for secondary structure prediction of selected proteins, *Biochemistry* **30:**6885–6895.

Pancoska, P., Yasui, S. C., and Keiderling, T. A., 1989, Enhanced sensitivity to conformation in various proteins. Vibrational circular dichroism results, *Biochemistry* **28:**5917–5923.

Pancoska, P., Yasui, S. C., and Keiderling, T. A., 1991, Structural analyses of the vibrational circular dichroism of selected proteins and relationship to secondary structure, *Biochemistry* **30:**5089–5103.

Pasteur, M. L., 1848, Sur les relations qui peuvent exister entre la forme cristalline, la composition chimique et le sens de la polarisation rotatoire, *Ann. Chim. Phys. III* **24:**442–459.

Paterlini, M. G., Freedman, T. B., and Nafie, L. A., 1986, Vibrational circular dichroism spectra of three conformationally distinct states and an unordered state of poly(-L-lysine) in deuterated aqueous solution, *Biopolymers* **25:**1751–1765.

Perczel, A., Hollósi, M., Tusnády, G., and Fasman, G. D., 1991, Convex constraint analysis: A natural deconvolution of circular dichroism curves of proteins, *Protein Eng.* **4:** 669–679.

Pribic, R., Van Stokkum, I. H. M., Chapman, D., Haris, P. I., and Bloemendal, M., 1993, Protein secondary structure from Fourier transform infrared and/or circular dichroism spectra, *Anal. Biochem.* **214:**336–378.

Provencher, S. W., 1982a, A constrained regularization method for inverting data represented by linear algebraic or integral equations, *Comput. Phys. Commun.* **27:** 213–227.

Provencher, S. W., 1982b, CONTIN, a general purpose constrained regularization program for inverting noisy linear algebraic and integral equation, *Comput. Phys. Commun.* **27:**229–242.

Provencher, S. W., and Glöckner, J., 1981, Estimation of globular protein secondary structure from circular dichroism, *Biochemistry* **20:**33–37.

Rendina, A. R., and Orme-Johnson, W. A., 1978, Glutamate synthase: On the kinetic mechanism of the enzyme from *Escherichia coli* W, *Biochemistry* **17:**5388–5393.

Saxena, V. P., and Wetlaufer, D. B., 1971, A new basis for interpreting the circular dichroic spectra of proteins, *Proc. Natl. Acad. Sci. USA* **68:**969–972.

Scharnagl, C., and Schneider, S., 1991, UV-visible absorption and circular dichroism spectra of the subunits of c-phycocyanin. II. A quantitative discussion of the chromophore-protein and chromophore–chromophore interactions in the β-subunit, *J. Photochem. Photobiol. B* **8:**129–158.

Schneider, G., Lindqvist, Y., and Lindqvist, T., 1990, Crystallographic refinement and structure of ribulose-1,5-biphosphate carboxylase from *Rhodospirillum rubrum* at 1.7 Å resolution, *J. Mol. Biol.* **211:**989–1008.

Sharonov, Yu. A., 1991, Room temperature and low-temperature magnetic circular dichroism of hemoproteins and related compounds, *Sov. Sci. Rev. D Physicochem. Biol.* **10:**1–118.

Shindo, Y., and Nakagawa, M., 1985, Circular dichroism measurements. I. Calibration of a circular dichroism spectrometer, *Rev. Sci. Instrum.* **56:**32–39.

Shire, S. J., Holladay, L. A., and Rinderknecht, E., 1991, Self-association of human and porcine relaxin as assessed by analytical ultracentrifugation and circular dichroism, *Biochemistry* **30:**7703–7711.

Siegel, J. B., Steinmetz, W. E., and Long, G. L., 1980, A computer-assisted model for estimating protein secondary structure from circular dichroic spectra: Comparison of animal lactate dehydrogenases, *Anal. Biochem.* **104:**160–167.

Singh, H. K., and Wilson, M. T., 1990, Characterization of haem disorder in cytochrome b_5 by circular dichroism, *Biochem. Soc. Trans.* **18:**1272–1273.

Smith, P. K., Krohn, R. I., Hermanson, G. T., Mallia, A. K., Gartner, F. H., Provenzano, M. D., Fujimoto, E. K., Goeke, N. M., Olson, B. J., and Klenk, D. C., 1985, Measurements of protein using bicinchonic acid, *Anal. Biochem.* **150:**76–85.

Snir, J., Frankel, R. A., and Schellman, J. A., 1975, Optical activity of polypeptides in the infrared. Predicted CD of the amide I and amide II bands, *Biopolymers* **14:**173–196.

Sreerama, N., and Woody, R. W., 1993, A self-consistent method for the analysis of protein secondary structure from circular dichroism, *Anal. Biochem.*, **209:**32–44.

Stephens, P. J., Jenson, G. M., Devlin, F. J., Morgan, T. V., Stout, C. D., Martin, A. E., and Burgens, B. K., 1991, Circular dichroism and magnetic circular dichroism of *Azobacter vinelandii* ferredoxin I, *Biochemistry* **30:**3200–3209.

Strickland, E. H., 1974, Aromatic contribution to circular dichroism spectra of proteins, *CRC Crit. Rev. Biochem.* **2:**113–175.

Sutherland, J. C., and Holmqvist, B., 1980, Magnetic circular dichroism of biological molecules, *Annu. Rev. Biophys. Bioeng.* **9:**293–326.

Teraoka, J., Nakamura, K., Nakahara, Y., Kyogoku, Y., and Sugeta, Y., 1992, Extraordinarily intense vibrational circular dichroism of a metmyoglobin cyanide complex, *J. Am. Chem. Soc.* **114:**9211–9213.

Toumadje, A., Alcorn, S. W., and Johnson, W. C., Jr., 1992, Extending CD spectra of proteins to 168 nm improves the analysis for secondary structure, *Anal. Biochem.* **200:** 321–331.

Urbanova, M., Dukor, R. K., Pancoska, P., Gupta, V. P., and Keiderling, T. A., 1991, Comparison of α-lactalbumin and lysozyme using vibrational circular dichroism. Evidence for a difference in crystal and solution structures, *Biochemistry* **30:**10479–10485.

Van der Vies, S. M., Viitanen, P. V., Gatenby, A. A., Lorimer, G. H., and Jaenicke, R., 1992, Conformational states of ribulosebiphosphate carboxylase and their interaction with chaperonin 60, *Biochemistry* **31:**3635–3644.

Vanoni, M. A., Nuzzi, L., Rescigno, M., Zanetti, G., and Curti, B., 1991, The kinetic mechanism of the reactions catalyzed by the glutamate synthase from *Azospirillum brasilense*, *Eur. J. Biochem.* **202:**181–189.

Vanoni, M. A., Edmondson, D. E., Zanetti, G., and Curti, B., 1992, Characterization of the flavins and the iron-sulphur centres of glutamate synthase from *Azospirillum brasilense* by absorption, circular dichroism, and electron paramagnetic resonance spectroscopy, *Biochemistry.* **314**:4613–4623.

Van Stokkum, I. H. M., Spoelder, H. J. W., Bloemendal, M., Van Grondelle, R., and Groen, F. C. A., 1990, Estimation of protein secondary structure and error analysis from circular dichroism spectra, *Anal. Biochem.* **191**:110–118.

Wu, Y., Huang, H. W., and Olah, G. A., 1990, Method of oriented circular dichroism, *Biophys. J.* **57**:796–806.

Yang, J. T., Wu C-S. C., and Martinez, H. B., 1986, Calculation of protein conformation from circular dichroism, in: *Methods in Enzymology*, Vol. 130 (C. H. W. Hirs and S. N. Timasheff, eds.), Academic Press, San Diego, CA, pp. 208–269.

Yasui, S. C., and Keiderling, T. A., 1986, Vibrational circular dichroism of polypeptides. VI. Polytyrosine α-helical and random coil results, *Biopolymers* **25**:5–15.

Yasui, S. C., Keiderling, T. A., and Katakai, R., 1987, Vibrational CD on polypeptides. X. A study of α-helical oligopeptides in solution, *Biopolymers* **26**:1407–1420.

3

Fourier Transform Infrared Spectroscopy Investigations of Protein Structure

E. A. Cooper and K. Knutson

1. INTRODUCTION

It has been recognized since 1950 (Elliott and Ambrose, 1950) that infrared spectroscopy has the potential to provide information regarding protein secondary structure. Krimm lists nine protein-sensitive infrared active vibrations arising from the amide backbone linkage (Krimm and Bandekar, 1986). The majority of protein structural information, however, has been obtained from one absorbance originating primarily from the amide C=O stretching vibrations: the amide I band. The sensitivity to variations in both geometric arrangement of atoms and hydrogen bonding enables infrared spectroscopy to discriminate between the various secondary structures, i.e., helical, extended sheet, unordered, and turns, incorporated within the three-dimensional organization of peptides and proteins. Many studies have demonstrated, both theoretically and experimentally, that infrared spectroscopy can be used to identify specific secondary structures.

One of the strengths of infrared spectroscopy is that it is amenable to a variety of sample forms including solid films or powders, solutions, liquids, and so forth. Crystal structures are not necessary nor are external molecular probes required,

E. A. Cooper • Department of Bioengineering, University of Utah, Salt Lake City, Utah 84112. *K. Knutson* • Department of Pharmaceutics and Pharmaceutical Chemistry, University of Utah, Salt Lake City, Utah 84112.

Physical Methods to Characterize Pharmaceutical Proteins, edited by James N. Herron *et al.*, Plenum Press, New York, 1995.

which would supply information only about the surrounding microenvironment. Infrared spectroscopy not only provides information about protein structure in native environments, it can also contribute insight into conformational alterations associated with changing environmental conditions such as pH, temperature, pressure, and solvent.

Until two decades ago, it appeared that infrared spectroscopy was becoming obsolete due to the low sensitivity of dispersive instruments (Griffiths and de Haseth, 1986). However, the advent of commercially available Fourier transform interferometers, as well as personal computer systems, has enabled infrared spectroscopy to emerge as one of the most rapidly growing spectroscopic techniques. Because of these advances, it is becoming more straightforward to extract secondary structural information from a protein spectrum. One hindrance in obtaining aqueous protein spectra has been the interference of overlapping H_2O bands. Absorbance subtraction procedures now enable H_2O to be used regularly as an infrared solvent. Resolution-enhancement techniques have provided the means to obtain quantifiable data from a protein infrared spectrum, even with H_2O as the solvent (Surewicz and Mantsch, 1988).

2. INFRARED SPECTROSCOPY

Vibrational spectroscopy is the measurement of the vibrational energy changes of a molecule incited by electromagnetic radiation of specific wavelength. It is vibrational energy that keeps the covalently bonded atoms of a molecule constantly oscillating about an equilibrium position. The movements of these vibrations oscillate with frequencies that comprise the infrared spectral region. Thus, when exposed to infrared radiation, energy is absorbed by the molecule, which results in an increase of vibrational motion. The physical effects of increasing the vibrational energy include changes in interatomic distances as well as bond angles. The radiation wavelengths that are capable of exciting the vibrational levels include the infrared region, 7.8×10^{-5} to 1×10^{-1} cm. The region between 2.5×10^{-4} to 1.5×10^{-3} cm, the mid-infrared region, is most commonly studied. The single vibrational energy changes are accompanied by a number of rotational energy changes, which result in an absorption band rather than a discrete line. Since the wavelength of infrared radiation is significantly larger than most molecules, the electric field of the photon can be assumed to be uniform for the entire molecule. The electric field generates forces on the charges present within the molecule or molecular group. Because the electric field oscillates, the forces exerted on the molecular charges alternate direction and cause a dipole moment to oscillate at the same frequency as the electromagnetic radiation. Only when the frequency of the oscillating dipole moment matches the frequency of the molecular vibration is energy allowed to be absorbed. Thus, only those vibrations that result in a change of dipole moment are able to absorb infrared radiation. This is the requirement or selection rule for a vibrational mode to be infrared active.

Molecular systems consisting of N atoms with a fixed orientation are defined as having 3N-6 (3N-5 for linear molecules) degrees of freedom that are associated with normal modes of vibration. Each of the normal vibrational modes has characteristic frequencies dependent on the masses of the atoms, interatomic forces holding the atoms about equilibrium positions, and geometry of the atoms comprising the molecule. The description of this system is based on the assumption that the motion of the atoms involved in a molecular vibration behaves as a harmonic oscillator.

The complexity of molecular vibrations arises from numerous atoms of a molecule contributing to a particular vibration. This can complicate the analysis of polymeric molecules, including proteins, which are themselves complicated and often ill-defined. However, the ability of a molecular group to influence a molecular vibration within a polymeric molecule is dependent on the distance between the groups. Thus, it is possible to consider the motions of just a few atoms within the polymer chain, i.e., the amide linkage of a protein, without taking into consideration the remainder of the chain for a good approximation of the ensuing molecular vibration. Also, polymer chains containing at least one dimension of crystallinity can be evaluated in terms of a unit cell. A specific chemical repeat unit does not necessarily comprise the unit cell, but rather the geometrical arrangement of a molecule contributes to the definition of a unit cell. For example, the number of chemical repeat units contained in a one-dimensional polypeptide unit cell is 2, 3, and 18 for a planar zigzag, a 3_1 helix, and an α-helix, respectively (Koenig, 1979).

The infrared spectrum of a nonlinear molecule does not exhibit exactly 3N-6 fundamental vibrational absorption bands for a number of reasons, the most obvious at this point being the failure to uphold the selection rule. Also, not all vibrations oscillate at a frequency within the mid-infrared region. Furthermore, the intensity of an infrared band is dependent on how effectively infrared energy can be transferred to the molecule. Effective energy transfer is directly dependent on the square of the changing dipole moment magnitude. Thus, because the changing dipole moment may be small, some bands are too weak to be observed. Other bands are not separated sufficiently in wavelengths to be resolved.

Additional bands may appear that do not directly arise from fundamental vibrations. Combination bands result from the excitation of two different fundamental vibrations. The sum of the frequencies results as the combination band. Difference bands appear at the frequency of the difference between two fundamental vibrational frequencies when the molecule, already in an excited state, absorbs enough energy to raise it to an additional energy level. Overtones result when there is anharmonicity within a fundamental vibration. These bands appear as multiples of the fundamental vibration band. Most of these additional bands are quite weak. However, the frequency of an overtone or combination band may nearly coincide with the frequency of a fundamental vibration and result in an interaction, known as Fermi resonance, which produces two relatively strong bands instead of the expected single strong fundamental band. Both bands involve the fundamental as well as the overtone contributions, and the positions are observed at slightly higher and slightly lower frequencies than the expected positions of the fundamental and overtone bands.

Since inter- or intramolecular interactions, conformation, crystallinity, and orientation can influence the force constants of chemical bonds, vibrational modes can be affected resulting in differences in an infrared spectrum. For example, hydrogen bonding is an interaction that can alter the force constants of both the proton donor and proton acceptor groups, thus affecting the frequencies of the associated stretching and bending vibrations. Hydrogen bonding is important in distinguishing between the different secondary structures present within the three-dimensional structure of a protein molecule. The amide C=O and N-H groups form the hydrogen bonds that stabilize regular secondary structures such as the α-helix and the β-sheet conformations. The strength of the hydrogen bond is dependent on the distance between donor and acceptor groups as well as the angle between the proton donor group and the axis of the proton acceptor's single pair of electrons. Therefore, since the force constant depends on the strength of the hydrogen bond, the absorbance frequency can detect differences in the hydrogen-bonding character of proteins and peptides.

3. BAND ASSIGNMENTS

An early empirical investigation demonstrated the ability of three infrared active bands to distinguish between either the α- or β-conformers of several synthetic polypeptides (Elliott and Ambrose, 1950). Since then, numerous studies have been undertaken to establish the validity of infrared spectroscopy for determining protein secondary structure (Beer *et al.*, 1959; Miyazawa, 1960; Miyazawa and Blout, 1961; Susi *et al.*, 1967; Susi, 1969; Chirgadze *et al.*, 1973, 1976; Chirgadze and Nevskaya, 1976a,b; Koenig and Tabb, 1980; Painter *et al.*, 1982; Parker, 1983; Byler and Susi, 1985; Krimm and Bandekar, 1986; Olinger *et al.*, 1986; Dong *et al.*, 1990; Lee *et al.*, 1990; Surewicz *et al.*, 1990). Many of the early investigations focused on synthetic polypeptides and fibrous proteins because of the difficulties associated with aqueous solution studies (Susi *et al.*, 1967). Today, with the increased sophistication and availability of instrumentation and data analysis methods, aqueous solution studies have become routine.

As previously mentioned, for every molecule there exist 3N-6 normal vibrational modes that may be infrared active. Therefore, a simple peptide molecule, such as polyglycine in the extended form, has about 50 infrared active bands (Krimm and Bandekar, 1986). These bands could provide a wealth of information about the polypeptide's three-dimensional structure, including details about intermolecular as well as intramolecular interactions. However, primarily nine backbone amide bands have been the most informative concerning protein secondary structure. The amide bands (Table I) were initially identified from studies using the amide bond model *N*-methylacetamide (NMA) (Susi, 1969). Although the vibrations are highly localized in the backbone peptide linkage, they cannot be simply described in terms of individual group stretching or bending modes. Other interactions such as intrachain coupling through the α-carbon atoms and intra- and interchain coupling through hydrogen bonds contribute to the composition of the vibrational modes. Krimm and Bandekar (1986) have suggested that transition dipole coupling, a resonance interaction occurring between two nearby oscillators when one of them is in an excited state, also contributes to the vibrational modes. Table I summarizes the absorbance bands arising from the NMA amide vibrations and their corresponding potential energy distributions. These distributions are indicative of the composition of a polypeptide or protein's amide vibrations. However, the potential energy distributions are not quantitatively equivalent for either polypeptides or proteins and may even differ between the various structural classes (Krimm and Bandekar, 1986).

Although the frequency and intensity of the absorption bands arising from the backbone CONH grouping inherent in all proteins are similar, variations do

Table I. Characteristic Absorption Bands of Amide Linkage[a]

Absorbance band	Wavenumber position (cm^{-1})	Vibrational modes[b–e]
Amide A	~3300	NH stretch in plane, 2 × amide II in Fermi resonance
Amide B	~3100	NH stretch in plane, 2 × amide II in Fermi resonance
Amide I	1597–1672	CO s (83)[e], CN s (15), CCN d (11)
Amide II	1480–1575	NH ib (49), CN s (33), CO ib (12), CC s (10), NC s (9)
Amide III	1229–1301	NH ib (52), CC s (18), CN s (14), CO ib (11)
Amide IV	625–767	CO ob (44), CC s (34), CNC d (11)
Amide V	640–800	CN t (75), NH ob (38)
Amide VI	537–606	CO ob (85), CN t (13)
Amide VII	~200	NH ob (64), CN t (15), CO ob (12)

[a]Based on *N*-methylacetamide spectrum.
[b]Susi (1969) and references therein.
[c]Krimm and Bandekar (1986) and references therein.
[d]s, stretch; d, deformation; t, torsion; ib, in-plane bend; ob, out-of-plane bend.
[e]Number in parenthesis is potential energy distribution.

occur. These differences can be used to draw conclusions about the structure and conformation of a protein or peptide. Whereas conformational changes are accompanied by changes in hydrogen bonding, absorbance bands that are sensitive to hydrogen bonding should also reflect changes in secondary structure of the protein molecule. Since the amide I, II, and III (Fig. 1) modes are composed primarily of vibrations originating in the C=O and/or N-H bonds, then these vibrations should be sensitive to hydrogen bonding and protein secondary structure. However, it is the amide I band that has been used most frequently to provide secondary structural information.

Early studies set the stage for the qualitative basis of secondary structural identification. However, it was recognized that while the amide I and II bands of polypeptides containing a single structural component are sharp and well-defined (Fig. 2), the corresponding bands in proteins are "broad, ill-defined and often contain several subsidiary maxima" (Beer *et al.*, 1959, p. 155) (Fig. 3). Also, the "investigation of the amide I and II bands of fibrous proteins suggests that a uniform conformation is the exception rather than the rule.... This presents a complication as well as a challenge" (Susi, 1969, p. 606). Because of the heterogeneous nature of protein conformations, the protein amide bands are actually composites of individual absorbances arising from the different structural classes. Current advances in resolution-enhancement techniques (Section 5.2) have aided in separating the various components of the amide bands, especially the amide I

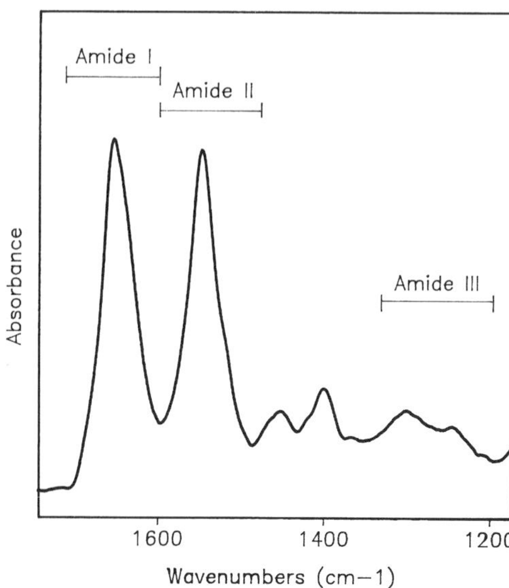

Figure 1. Amide I, II, and III bands of bovine serum albumin, pH = 7.4, after PBS buffer subtraction.

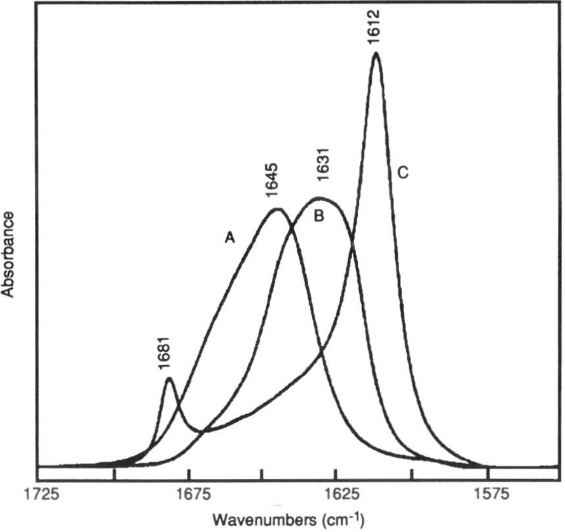

Figure 2. Amide I and II bands of poly(L-lysine) in (A) random, (B) helical, and (C) β-sheet conformations. (From Carrier *et al.*, 1990, with permission.)

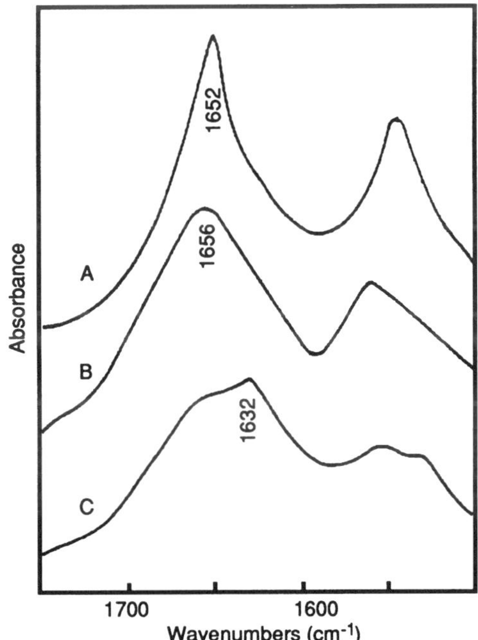

Figure 3. Amide I and II bands of aqueous protein solutions: (A) β-lactoglobulin, a β sheet protein, (B) α_s casein, a random protein, and (C) myoglobin, an α-helical protein. (Adapted from Susi *et al.*, 1967.)

band, making qualitative identification of secondary structure more straightforward. Quantitative determination of secondary structure is becoming a reality.

To better understand and interpret polypeptide–protein spectral data, two theoretical approaches have been employed. The first method utilizes a first-order perturbation treatment that considers the peptide backbone as a weakly coupled oscillator model (Miyazawa, 1960; Miyazawa and Blout, 1961; Krimm, 1962; Chirgadze and Nevskaya, 1976a). The fundamental repeat unit is designated for each structural class. These studies provide insight into the interpretation of the polypeptide and protein spectra, particularly the amide I region. One assumption is that the system approximates an infinite array of identical subunits. Homopolypeptides may closely approximate this model, but proteins, especially globular proteins, do not closely approximate the model. Globular protein structure is heterogeneous in nature; the substructure present is interspersed with and separated by other classes of secondary structure. Also, the length of the repeat structures cannot be considered infinite, and in many cases the structure is less than ideal.

The second method is the normal coordinate analysis of polypeptide models representing specific classes of secondary conformations (Krimm and Bandekar, 1986), i.e., α-helix, β-sheet, and turns. This method appears to be a good predictor of how small conformational changes will affect the vibrational spectrum (Dwivedi and Krimm, 1984). Using normal mode analysis in combination with experimental results from both infrared and Raman spectroscopies, especially from the amide I, II, III, and V modes, can allow fine distinctions between different conformations (Naik and Krimm, 1986a,b). Although proteins contain ordered secondary structure, the order may be distorted and not ideal. Therefore, the theoretical and experimental predictions made from synthetic polypeptide models may not directly compare to proteins.

Investigations of aqueous solutions are undertaken with either H_2O or D_2O. D_2O is more widely utilized since the D-O-D stretching vibrations do not overlap the amide I band, unlike the H-O-H stretching vibrational modes of H_2O (Section 4.2). In some literature the amide I and II bands obtained in D_2O are referred to as amide I′ and II′. The amide hydrogens exchange with deuterium atoms, resulting in the bands associated with N-H stretching and hydrogen bonding shifting to lower wavenumbers, the extent of the shift being dependent on the pattern of hydrogen bonding. The frequency shift and rate of hydrogen–deuterium exchange can provide information about the structural composition of a protein (Haris *et al.*, 1986; Krimm and Bandekar, 1986; Olinger *et al.*, 1986).

Current compilations of secondary structure-frequency correlations compare both experimental results and theoretical calculations (Byler and Susi, 1986; Surewicz and Mantsch, 1988; Dong *et al.*, 1990). The remainder of this section will attempt to collate what is currently understood about the relationships between the amide bands and protein structure.

3.1. Amide A and B

The amide A band at ~3300 cm^{-1} is associated with the NH stretching vibration of hydrogen-bonded amide groups (as opposed to the nonhydrogen-bonded NH stretching modes that absorb at 3400–3460 cm^{-1}). In amide-containing molecules, the amide A band is usually accompanied by the amide B band, which absorbs near 3100 cm^{-1}. In NMA, this doublet arises from a Fermi resonance with the overtone of the amide II (Susi, 1969; Krimm and Bandekar, 1986). However, for certain conformations of a polypeptide or protein, the doublet is in resonance with a combination of the amide II modes (Krimm and Bandekar, 1986).

Because the N-H stretching vibrational mode is highly localized in the NH bond, the associated bands are not likely to be directly sensitive to the chain conformation. However, the frequency is dependent on the strength of the N-H ··· O=C hydrogen bond. Thus, these vibrational modes provide some indication about structural variations. For example, Bendit (1966b) observed two bands in the amide A region of α-keratin, one at 3286 cm^{-1} and the other at 3310–3320 cm^{-1}. The bands were assigned to helical and nonordered structures, respectively. The frequency difference is attributed to stronger hydrogen bonding within the helical phase than the nonhelical phase. With hydration, the nonhelical band shifts to lower frequencies (Bendit, 1966b), indicating that the H_2O molecules form hydrogen bonds with the amide groups of nonordered structure. Krimm and Bandekar (1986) report that the β forms of polyglycine I and polyalanine have different amide A frequencies (3272 cm^{-1} and 3242–3250 cm^{-1}, respectively) as a result of differing hydrogen bond lengths. The α-helical form of polyalanine also has a higher amide A frequency than its β counterpart, reflecting the weaker hydrogen bonding in the helical structure (Krimm and Bandekar, 1986). The amide A band can also be modified by Fermi resonance. Therefore, it is difficult to draw explicit conclusions about the hydrogen-bond geometry (Krimm and Bandekar, 1986), although attempts have been made (Fraser and MacRae, 1973; Krimm and Bandekar, 1986).

Early protein studies utilized the amide A and B bands as indicators of the types of structure present based on the dichroism of the absorbances (Elliott and Ambrose, 1950; Beer *et al.*, 1959; Bendit, 1966b). Both the amide A and B bands for oriented helix containing proteins exhibit parallel dichroism, whereas the two bands exhibit perpendicular dichroism for β-structures (Beer *et al.*, 1959). The stability of the structure is reflected by the bandwidth. A broader band indicates a greater distribution of peptide hydrogen bonds, suggesting a decrease in the stability of the structure as observed in α-helical structures by Chirgadze and co-workers (1976).

The practicality of using the amide A and B bands for structural determination in solution studies is compromised by the strong absorbance of the aqueous OH stretching vibrational mode at approximately 3400 cm^{-1}. This band greatly obstructs the amide A absorbance, rendering the band useless unless the protein is in the dry state, under humid conditions (Bendit, 1966b), or in D_2O base solutions.

3.2. Amide I

The amide I band, arising from about 1620 to 1700 cm^{-1}, has been the most important source of polypeptide–protein secondary structural information provided by infrared spectroscopy. Through theoretical and systematic experimental studies, strong correlations between the frequency of absorbance and the secondary structure have been determined. Because the primary components of the amide I band are C=O stretching vibrations (Table I), the band is influenced by hydrogen bonding, and thus reflects variations in secondary structure.

Homopolypeptide secondary structure is fairly straightforward to identify from the amide I frequency since only one structural class tends to be present. However, most proteins contain several types of secondary structure of which one type may predominate. Therefore, the amide I band is actually a composite of absorbances arising from the various structural types, resulting in a broad and "distorted" band shape. However, the composite frequency is not informative about secondary structure composition. With the use of Fourier transform technology, as well as the development of sophisticated methods for spectral analysis and data reduction, previously unresolvable bands have been separated into their underlying structural components (Fig. 4).

Dong *et al.* (1990) collated a set of frequency structure correlations from 12 globular proteins in H_2O (Table II) and concluded that their 11 assignments are "essentially the same for all proteins." Byler and Susi (1986) also identified 11 frequencies that represent the various secondary structures for proteins in D_2O (Table II). They determined that the amide I band for most proteins contains six to nine of these components. Exceptions include Met aporepressor in the native form, which exhibits five amide I bands (Yang *et al.*, 1987), and casein, a structurally nonordered protein, which exhibits three bands (Byler and Susi, 1986). Sometimes, not all amide I bands are unequivocally assignable to a specific structure, yet the spectral differences are suggestive of conformational variations. For example, under four different environmental conditions, β-lactoglobulin exhibits four different amide I bands (Purcell and Susi, 1984), indicating that one native and three denatured forms exist. Differences in the hydrogen–deuterium exchange rates (Section 4.2.2) for various amide I bands indicate differences in the openness and flexibility of a protein, as noted between ribonuclease A and S (Haris *et al.*, 1986).

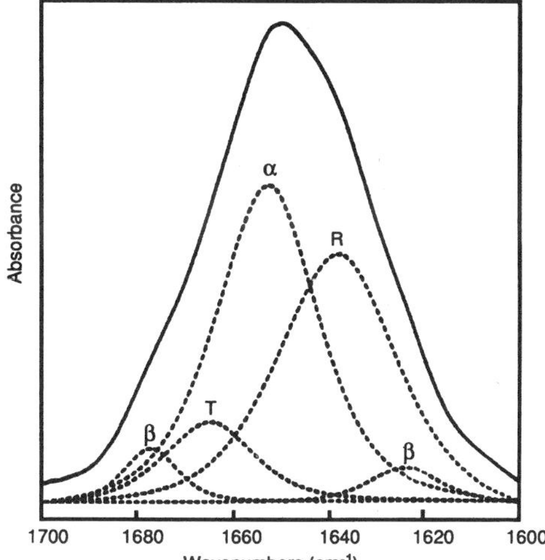

Figure 4. Original and curve-fitted amide I band of native Met aporepressor in D_2O. α, β, T, and R correspond to α-helices, β-structures, turns, and random conformations, respectively. (From Yang *et al.*, 1987, with permission.)

Table II. Amide I Component Frequencies (cm^{-1}) and Assignments

Secondary structure	Experimental in D_2O[a]	Experimental in H_2O[b]	Fibrous proteins (solid-state)[c]	Theoretical
Extended chains	1624 (2.4)[d]	1624.0 (0.5)[e]		
		1627.0 (1.0)		
	1631 (2.5)	1632.0 (1.0)	1630–1634	1630 (AP)[f]
	1637 (1.4)	1638.0 (1.0)		1637 (P)[g]
		1642.0 (1.0)		
	1675 (2.6)			1670 (AP)[f]
			1695–1697	1695 (AP)[f]
Helix	1654 (1.5)	1656.0 (2.0)	1650–1653	1653[h]
Random	1645 (1.5)	1650.0 (1.0)	1660–1664	
Turns and bends	1663 (2.2)	1666.0 (1.0)		
	1670 (1.4)	1672.0 (1.0)		
	1683 (1.5)	1680.0 (1.0)		
	1688 (1.1)	1688.0 (1.0)		
	1694 (1.7)			

[a]Byler and Susi (1986).
[b]Dong *et al.* (1990).
[c]Krimm (1962).
[d]Rounded average (RMS deviation).
[e]Mean frequency (SD).
[f]Antiparallel (Chirgadze and Nevskaya, 1976a).
[g]Parallel (Chirgadze and Nevskaya, 1976b).
[h]Nevskaya and Chirgadze (1976).

3.2.1. HELICAL CONFORMATIONS

The amide I frequency for α-helical conformations is generally accepted as absorbing in the 1650 to 1658 cm^{-1} region (Byler and Susi, 1986; Surewicz *et al.*, 1987c; Dong *et al.*, 1990). Table II lists some representative frequency values for helical structures. The theoretically determined absorbance for an infinite right-handed α-helix (1653 cm^{-1}) agrees with experimental values for proteins in the solid state (1650–1653 cm^{-1}), D_2O (1654 cm^{-1}), and H_2O (1656 cm^{-1}). In collating experimental values for α-helical polypeptides, Nevskaya and Chirgadze (1976) determined that the absorbances arising from this particular conformation are almost independent of such factors as physical state and type of side-chain moieties. This is corroborated by Susi *et al.* (1967), who noted that the amide I frequency for myoglobin in H_2O solution is identical to the crystalline form and mineral oil dispersion, as well as similar to values for solid fibrous proteins.

Departures from the "accepted" frequency range have been documented for polylysine and polyglutamic acid whose absorbances in D_2O are near 1635 cm^{-1} and 1640 cm^{-1}, respectively (Susi *et al.*, 1967). This departure has been explained by the strong interaction between the charged side groups within the α-helix (Nevskaya and Chirgadze, 1976). The heterogeneity of side-chain composition in proteins appears to have less affect on the position of the amide I frequencies than does the side chain composition in homopolypeptides. However, the distorted form of the α-helix has been reported to shift up to 1662 cm^{-1} (Lee *et al.*, 1985), supporting the thought that small conformational differences lead to large wavenumber shifts. In D_2O, the helical component of the amide I band has been observed as low as 1647 cm^{-1} (Purcell and Susi, 1984). The standard $\alpha(\alpha_I)$-helix structure is not the only helical structure possible in peptides and proteins. The theoretically predicted absorbance for a 3_{10} helix is 1665 cm^{-1}; however, experimental studies have provided a band assignment at 1638 cm^{-1} (Holloway and Mantsch, 1989; Prestrelski *et al.*, 1991) for the same structure. It has also been proposed that there exists a helical structure (α_{II}) that differs from the α_I-helix (Dwivedi and Krimm, 1984). The plane of the peptide group has more tilt with respect to the helix axis, resulting in weaker hydrogen bonding. The amide I band of the α_{II}-helix is predicted to shift from 1658 cm^{-1} to 1667 cm^{-1} (Dwivedi and Krimm, 1984). This band could, however, be mistaken for random (Section 3.2.3) or turn (Section 3.2.4) conformations.

Not only is frequency an indicator of secondary structure, but bandwidths are strongly related to the degree of order present (Chirgadze *et al.*, 1976): the more ordered, the narrower the band; and the lower the thermodynamic stability of the helix, the greater the bandwidth, about twice as large, indicating a greater distribution of peptide hydrogen bonding. Chirgadze *et al.* (1976) explain that the broadening of the amide band of α-helical proteins is not affected by either the solvent or the nature of the side chains. Thus, the broadening of the amide band appears to be due to properties of the peptide or protein backbone, not the amino acid sequence. The bandwidth of the unordered forms of the polypeptide and protein is even greater than that of the unstable helix. Yang *et al.* (1987) monitored the bandwidth of Met aporepressor as a function of temperature and determined that the increase of bandwidth correlated with a change from helical to random conformations.

3.2.2. EXTENDED CHAIN CONFORMATIONS

Band assignments for the extended chain conformations are not unique such as for the helical structures. There are several variations in the extended structure, including regular long β-strands as opposed to short irregular β-like strands. Bandekar and Krimm (1988) have theoretically determined that the sheet arrangement, as well as intersheet distance in multisheet structures, affects the frequency of both the amide I and II bands. The distortions and structural geometries available to the β-strands make it possible for more than one absorbance band to arise, conceivably providing more details about the structure than would a single band. Some proteins such as concanavalin A, elastase, and papain may exhibit as many as three low-frequency components arising from β-structures (Byler and Susi, 1986). The absorbances arising from β-structures span the range of 1624 to 1642 cm^{-1} (Table II) and may include a high wavenumber component (due to antiparallel structure) absorbing at 1670–1680 cm^{-1} in D_2O (Byler and Susi, 1986) or 1689–1687 cm^{-1} in H_2O (Haris *et al.*, 1986; Olinger *et al.*, 1986). Calculations have been made to predict the absorbances of both the parallel and antiparallel sheet structures (Table II). However, it has been argued that it is impossible to distinguish between these two structures on the basis of the presence of a band at high wavenumbers (Olinger *et al.*, 1986; Surewicz *et al.*, 1987c). The high wavenumber frequency is less intense and can also overlap with the turn absorbances (Section 3.2.4), making it difficult to conclusively identify the antiparallel structure. Another indication that a high-frequency band may arise from β-structure is a slow hydrogen–deuterium exchange rate (Haris *et al.*, 1986).

It is also expected that the β-structures in globular proteins are of finite dimensions as well as distorted. This leads to broadening of the bands as well as a decrease in the frequency splitting (Chirgadze *et al.*, 1973; Chirgadze and Nevskaya, 1976a). It has been suggested that proteins containing distorted β-structures can exhibit amide I absorbances as low as 1615 to 1618 cm^{-1} (Purcell and Susi, 1984; Muga *et al.*, 1990). Others have attributed bands absorbing between 1625 to 1615 cm^{-1} to aggregates of intermolecular β-structure as opposed to intramolecular β-sheets (Surewicz *et al.*, 1990; Jackson *et al.*, 1991; Jackson and Mantsch, 1992).

3.2.3. RANDOM CONFORMATIONS

For random conformations (conformations with no well-defined repetitive structure), the characteristic amide I band component features a large half-bandwith about two or three times greater than those arising from the stable forms of helices or β-structures (Chirgadze *et al.*, 1973, 1976; Byler and Susi, 1986). In unordered structures, the nature of the coupling between neighboring amide groups, as well as the geometry of hydrogen bonding, is irregular and largely distributed, which can result in a broad band. Not only are these bands broad, but they also exhibit amide I frequencies that are dependent on state, i.e., solid or solvated in D_2O or H_2O (Table II). It is thought that in the solid state the peptide linkages forming the unordered regions do not hydrogen bond, yet polar solvents can hydrogen bond with the unordered protein as a result of solvation. The degree of hydrogen bonding affects the amide I band position. For example, the amide I band frequency of poly-*S*-carbobenzoxymethyl-L-cysteine solvated in mixtures of 1,2-dichloroethane and trifluoroacetic acid shifted from 1659 cm^{-1} to 1646 cm^{-1} with increasing acid content (and solvent polarity) from 4% to 9% (Chirgadze *et al.*, 1973). Because the absorbance frequency is dependent on the environment, the amide I absorbance is not well established. Olinger *et al.* (1986) suggest that the band absorbing as high as 1667 cm^{-1} for ribonuclease could arise from either the random structure or turn structures.

In H_2O solutions, current opinion has indicated that the unordered component cannot be distinguished from the α-helical component (Villalain *et al.*, 1989) because of the small frequency difference between the two (Table II). However, Dong *et al.* (1990) determined from systematic studies that the absorbance arising from unordered structures appears as an asymmetric shoulder on the low-frequency side of the α-helix band. In proteins with little helical structure, the random band is well-defined (Dong *et al.*, 1990). Others claim the band at higher wavenumbers (1657 cm^{-1} vs. 1650 cm^{-1}) is due to unordered structures and the lower wavenumber absorbance arises from helical structures (Surewicz *et al.*, 1987d; Wasacz *et al.*, 1987). In D_2O solutions, because the separation between absorbances is greater, the absorbance contributions arising from the two structures can in most cases be sufficiently resolved using deconvolution or derivative methods (Surewicz *et al.*, 1988).

3.2.4. TURN CONFORMATIONS

Turn conformations have been determined to be quite common in globular proteins, about 32% of the residues reside in these structures for 29 proteins (Chou and Fasman, 1977). In a separate survey of 38 nonhomologous proteins of known crystal structure, it was determined that 29% of the residues reside in β-turns (Bandekar and Krimm, 1979). The reverse turns are composed of either three or four amino acid residues. At least 11 turn types have been identified, and these turns are also expected to vary in proteins with respect to the "standard angles" (Krimm and Bandekar, 1986). Therefore, "while α-helix and β-sheet components of a protein may be expected to have relatively constant amide I frequencies, the same is not likely to be true of a β-turn component" (Krimm and Bandekar, 1986, p. 301). Until recently, these types of structure had not been considered in the assignment of amide band absorbances. Dong and co-workers (1990) have assigned the highly reproducible bands at 1666, 1672, 1680, and 1688 cm^{-1} to the β-turn structures, "since α-helix, unordered, and β-sheet can be assigned to other bands" (p. 3306). These assignments correlate well with the assignments for proteins in D_2O made by Byler and Susi (1986) (Table II). Experimentally, more specific assignments have not been made that distinguish between the subclasses of β-turns. Krimm and Bandekar (1986) have theoretically predicted amide I band positions for the standard forms of β_I-, β_{II}-, and β_{III}-turns to absorb at 1690, 1666 and ~1656, and 1686 cm^{-1} respectively. As mentioned previously, however, it is not likely that the standard dihedral angles will exist in the folded protein or polypeptide molecules. On varying the turn dihedral angles, large frequency shifts are predicted (Krimm and Bandekar, 1986), thus making specific band assignments difficult. It is also possible that the turn absorbances could overlap with helical or antiparallel β-sheet absorbances.

3.3. Amide II

The amide II band is more complex than the amide I band. It is composed largely of NH bending vibrations coupled with CN stretching and minor contributions from CO bending and CC and NC stretching vibrations (Table I). However, the potential energy provided by the NH bend can be variable even within the same structural class (Krimm and Bandekar, 1986), leading to a shift in absorbance frequencies.

The early work with polypeptides and fibrous proteins established the dichroism of some of the amide II components and attempted to establish frequency–structure correlations (Beer *et al.*, 1959; Miyazawa and Blout, 1961; Krimm, 1962; Bendit, 1966a). The strongly absorbing amide II components of the α-helix and β-sheet components exhibit perpendicular and parallel dichroism, respectively. The frequency–structure correlations for fibrous proteins are: α-helix, 1515–1550 cm^{-1}; extended chain, 1523–1555 cm^{-1}; and random, 1520 cm^{-1} (Krimm, 1962). Others indicate that the random conformation absorbs around 1535 cm^{-1} (Miyazawa and Blout, 1961; Koenig and Tabb, 1980). However, it is argued that these assignments are not as distinctive as the amide I band assignments, and the bands are difficult to differentiate for globular proteins since the frequency differences are small or nonexistent between the various structural components (Koenig and Tabb, 1980; Yang *et al.*, 1985; Haris *et al.*, 1986; Olinger *et al.*, 1986). The amide II bands arising from β-turns overlap with absorbances arising from both helix and sheet structures and therefore do not appear to be distinctive (Krimm and Bandekar, 1986). The amide II band is generally not considered amenable to structural studies. Thus, it is not surprising that the amide II band has not been extensively used for analyzing the structure of globular proteins. However, the amide II band appears to be more sensitive to changes in hydrogen bonding than the amide I band. Shifts to higher frequencies can indicate stronger hydrogen bonding (Koenig and Tabb, 1980; Krimm and Bandekar, 1986) and suggest the occurrence of conformational changes (Alvarez *et al.*, 1987).

Other factors inhibit the use of the amide II band for structural studies. It is strongly influenced by side-chain contributions. For example, tyrosine, glutamate, and aspartate absorb at ~1515, 1567–1569, and ~1584 cm^{-1}, respectively (Chirgadze *et al.*, 1975). Also, because of such large contributions from the NH stretching vibrations, deuteration causes the amide II bands arising from the structural features to rapidly disappear (Section 4.2.2).

The amide II band has been used to quantify the amount of protein adsorbed to polymer surfaces by measuring the absorbance at 1550 cm^{-1} (Gendreau *et al.*, 1982; Fink *et al.*, 1987; Young *et al.*, 1988; Giroux and Cooper, 1991) or the amide II band area (Castillo *et al.*, 1984; Pitt *et al.*, 1987) (Section 5.3).

3.4. Amide III

The amide III band arising in the region of 1330–1200 cm^{-1} is a complex mode (Table I) consisting of NH bending and CN stretching vibrations with additional contributions from CC stretching and CO bending (Krimm and Bandekar, 1986). The fact that the potential energy distribution can be different at different frequencies provides additional complexity (Krimm and Bandekar, 1986). Yet, the amide III band has two advantages over the amide I and II bands, the first being that the absorbance range is broader, and therefore the bands associated with the various secondary structures are better separated. This is especially useful for distinguishing between random and helical conformations. The second advantage is that the H-O-H bending vibration of liquid or vapor H_2O does not interfere with the amide III absorbances. It has been suggested that this band can be a sensitive indicator of denaturation (Anderle and Mendelsohn, 1987; Kaiden *et al.*, 1987) due to the fact that the absorbances are very sensitive to changes in angular geometry (Hsu *et al.*, 1976; Krimm and Bandekar, 1986). Thus, the amide III band can qualitatively verify or even clarify amide I data (Kaiden *et al.*, 1987; Wasacz *et al.*, 1987). In spite of the potential advantages of the amide III band, it must be recognized that the absorbances in this region are much less intense than either the amide I or II bands. In order to obtain a signal that is not significantly affected by noise, the protein concentration must be high.

Many frequency–structure correlations for the amide III region have been based on Raman studies. Because of differences in infrared and Raman selection rules, intensity and frequency differences are present (Krimm and Bandekar, 1986; Anderle and Mendelsohn, 1987), causing some ambiguity in the range of frequency assignments for structural correlations. However, there is general agreement that absorbances arising between 1295 and 1260 cm^{-1} are due to helical structures, bands between 1245 and 1230 cm^{-1} are attributed to β-sheet structure, and fairly broad medium-intensity bands from 1260 to 1240 cm^{-1} arise from the random conformations (Hsu *et al.*, 1976; Parker, 1983; Krimm and Bandekar, 1986; Anderle and Mendelsohn, 1987; Kaiden *et al.*, 1987). β-turn absorbances are "consistently" predicted above 1290 cm^{-1} (Krimm and Bandekar, 1986); however, lower absorbances have been observed (Krimm and Bandekar, 1986). Also, it has been suggested that the vibrational modes active in this region are sensitive to side-chain composition (Hsu *et al.*, 1976).

3.5. Other Amide Bands

The amide V band arising from vibrational contributions of NH bend and CN torsion (Table I) appears to be sensitive to polypeptide conformation (Krimm and Bandekar, 1986). The spectra of the β-forms of polyglycine I, polyalanine, and polyalanylglycine exhibit a band in the region of 705–708 cm^{-1}. However, in the β-form of polyglutamate, the band is absent in the 700 cm^{-1} region, but a band is observed at 653 cm^{-1}. Thus there is some variability in the amide V region that could be dependent on hydrogen bond strength as well as side-chain composition. Two bands in the amide V region (658 and 618 cm^{-1}) are documented for the helical form of polyalanine (Krimm and Bandekar, 1986). The helical form of polyglutamic acid gives rise to medium-weak bands at 670 and 618 cm^{-1} (Krimm and Bandekar, 1986). Turn absorbances are predicted in the frequency regions 575–570 cm^{-1} for the β_I turn, 607–571 cm^{-1} for the β_{II} turn, and 589–573 cm^{-1} for the β_{III} turn (Krimm and Bandekar, 1986).

Krimm and Bandekar (1986) have theoretically calculated the absorbances of polypeptide structures arising in the other amide band regions, but find that these absorbances are not defined as distinctively as those assigned to NMA (Table I). Combined with additional normal-mode analysis, this region could provide additional information about structural differences.

3.6. Other Protein Bands

For structural purposes, the vibrations arising from the backbone conformations are important. However, contributions from side chains must be considered so that these vibrations are not confused with the backbone conformational vibrations. Chirgadze *et al.* (1975) indicate that the absorbance contributions from side chains can account for 10–30% of the overall absorption. However, the contributions to the protein spectra will depend strongly on the protein amino acid composition. The side-chain groups of tyrosine, arginine, glutamine, asparagine, glutamic and aspartic acids, lysine, histidine, and phenylalanine residues are the substantial contributors to the amino acid absorption in the 1400–1800 cm^{-1} region (Chirgadze *et al.*, 1975; Venyaminov and Kalnin, 1990). The absorption positions for the amino acid side chains are listed in Table III.

4. SAMPLING METHODS

The ability of infrared spectroscopy to accommodate a variety of sample forms makes this technique more than suitable for probing a protein's conformation, especially since proteins are found in a variety of environments. In general, membrane proteins tend to exist in a hydrophobic lipid environment, fibrous proteins in a dry state, and globular proteins in an aqueous medium. It is preferred to study biological systems in their native environments, and with infrared spectroscopy it is possible in many cases to study proteins *in situ* or at least in environments that model their native surroundings. Also, understanding how proteins react to changes in environment (i.e., a change in pH, a change in solvent, a change in temperature) may provide some insight into protein stability. Basic principles of sample preparation are discussed in introductory texts such as *Infrared Spectroscopy* by Conley (1972) or *Introduction to Infrared and Raman Spectroscopy* by Colthup *et al.* (1975). Parker (1983), in *Applications of Infrared, Raman, and Resonance Raman Spectroscopy in Biochemistry*, focuses on the use of infrared spectroscopy for studying biological systems.

4.1. Solid State

There are several sampling techniques for studying solid-state protein samples, including powder, mull, alkali halide pellet, and film. Powder samples can be difficult due to the high incidence of scattered light, which results in a loss of energy transmitted to the detector. To reduce scattering, samples should be ground

Table III. Characteristic Amino Acid Side-Chain Absorbances

Amino acid residue	Band assignment (cm^{-1})[a]	Band assignment (cm^{-1})[b]	Band assignment (cm^{-1})[c]
Alanine			~1465
Valine			~1450
Leucine			~1375
Serine			1350–1250
Aspartic acid	1713 (COOD)	1716 (COOH)	1720
	1584 (COO^-)	1574 (COO^-)	
Glutamic acid	1706 (COOD)	1712 (COOH)	
	1567 (COO^-)	1560 (COO^-)	1560
			~1415
Asparagine	1648	1678	~1650
		1622	
Glutamine	1635	1670	~1615
		1610	
Arginine	1608	1673	
	1586	1633	
Lysine		1629	1640–1610
		1526	1550–1485
			~1160
			~1100
Histidine		1494	
Phenylalanine			1602
		1494	~1450
			760
			700
Tyrosine	1615		~1600
	1515	1518	~1450
	1603	1602	
	1500	1498	
Amine end		1631	
		1560	
		1515	
Carboxyl end	1720 (COOH)	1740 (COOH)	
	1592 (COO^-)	1598 (COO^-)	

[a] D_2O solution (Chirgadze *et al.*, 1975).
[b] H_2O solution (Venyaminov and Kalnin, 1990).
[c] Adapted from Krimm and Bandekar (1986), and references therein.

to a powder of 5-μm particle size or less (Parker, 1983), smaller than the radiation wavelengths (Hacskaylo, 1954).

To further minimize scattering, samples can be dispersed in a medium of similar refractive index such as in a mineral oil mull or alkali halide (KBr) pellet. Preparation of particles fine enough for inclusion in mulls and pellets requires vigorous grinding in order to achieve small and homogenous particles. Because preparation of both pellets and mulls requires some degree of mechanical processing, the conformations of the protein may be affected. It is known that proteins are marginally stable at room temperature (Dill, 1990); thus, any degree of processing should be considered when analyzing protein secondary structure. In preparing the pellet, pressures on the order of about 5.5 kbar (Parker, 1983) are required to compress the alkali halide and sample to form a transparent disk. Wong and Heremans (1988) concluded that chymotrypsinogen in aqueous solutions undergoes irreversible conformational changes at pressures as low as 3.7 kbar. Buchet *et al.* (1989) observed an irreversible loss of Ca^{2+}-ATPase activity in its native sarcoplasmic reticulum environment after exposure to 1.5 kbar of pressure. The loss of activity was attributed to irreversible conformational changes within the enzyme. Differences between two forms of solid-state samples, the pellet and film, were noted in the infrared spectra of both myoglobin and bovine serum albumin (Kaiden *et al.*, 1987).

Probably the best form in which to study "dried" protein samples is as a film deposited from solution. There are no interfering bands from either solvents or dispersion media and the sample preparation does not require extremes of mechanical processing. However, care must be taken to avoid precipitation of salts or buffers as large crystals or aggregates if at high concentrations. Also, films of proteins can often be examined under polarized radiation in order to learn more about the orientation of infrared active functional groups that in turn can provide insight into molecular orientation (Section 4.4).

4.2. Solution

In many cases, a biological system implies an aqueous environment. Both H_2O and D_2O have been employed, thus providing more reliable structural information about the protein (Haris *et al.*, 1986; Krimm and Bandekar, 1986; Olinger *et al.*, 1986). Protein solutions are also advantageous for studying the effects of systematic changes in environment, i.e., pH, temperature, solvent, and pressure on protein conformations.

4.2.1. H_2O

It has been only in the last decade that investigations regarding protein structure in the presence of H_2O rather than D_2O environments have become practical. Koenig and Tabb (1980) demonstrated how the H_2O influence could be digitally subtracted from the spectra of aqueous protein solutions. Prior to their study, H_2O influences had been removed using time-consuming differential techniques that are not easily adopted for routine investigations (Susi *et al.*, 1967). The overwhelming problem of examining aqueous solutions of a protein in the past has been due to the fact that the H-O-H bending vibrations of H_2O absorb strongly near 1640 cm^{-1}, the same wavenumber at which the amide I band (Section 3.2) absorbs. Thus informative data, especially data that could be useful for quantitative purposes, in the past have been difficult to obtain. With the introduction and commercialization of the Fourier transform infrared (FTIR) spectrometer to improve the signal-to-noise ratio and computerized techniques to subtract out the influence of the water, it has become possible to collect informative data.

In order for protein information to be transmitted through to the detector, it is necessary that the water contribution not obscure the protein signal by overabsorbing. It is best that the combined protein–water absorbance at ~1640 cm^{-1} not exceed 1 absorbance unit in order to conserve the linear relationship between absorbance and concentration (Beer's Law), especially if quantitative information is desired. This necessitates the use of short pathlengths on the order of 6–10 μm and fairly high protein concentrations, generally on the order of 10 to 50 mg/ml.

Liquid cells are widely available. The most common windows used with aqueous protein solutions are either CaF_2 or BaF_2. These materials have been chosen because they have refractive indices close to H_2O, thus minimizing and possibly eliminating interference fringes (White and Ward, 1965; Trewhella *et al.*, 1989). However, these windows are no longer infrared transparent below 1100 cm^{-1} and 770 cm^{-1}, respectively. Below these wavenumbers, the infrared radiation is totally absorbed by the window, and amide bands (Table I) may not be detected.

4.2.2. D_2O

In early protein solution investigations, D_2O was employed as the aqueous solvent because of the strong overlapping H-O-H bending vibration of H_2O. The D-O-D bending vibration is shifted to 1220 cm^{-1}, eliminating its absorbance contribution in the amide I region. However, the amide II band, composed largely of NH bending, is shifted markedly to lower wavenumbers. Unfortunately, it is very difficult to avoid small amounts of HOD, and since this bending frequency interferes with the amide II frequency of the COND groups, the amide II information is essentially lost (Susi, 1969).

To prepare a sample in D_2O, it is necessary to allow the amide hydrogen and deuterium atoms to exchange. The time frame for exchange will depend on the degree of folding and structure within the protein as well as the temperature of exchange. It is generally accepted that the more buried the amide hydrogen atom, the less accessible it will be to the solvent, thus slowing down exchange. The accessibility is controlled by the degree of hydrogen bonding as well as how buried the N-H groups are in the molecule's three-dimensional structure. The exchange can then be followed by the disappearance of the amide II band (Gregory and Rosenberg, 1986). It appears that most exchanges reach equilibrium within 48 hr. However, it has been noted that exchange in the stable α-helix has a time of half-exchange of over 30 days (Chirgadze *et al.*, 1976). Questions arise, however, about the effects of exchange on the native protein structure, as well as the completeness of exchange.

Because there is no interference from the D_2O band in the amide I region, the sample pathlength can be longer, generally up to 50 μm, thereby making it possible to lower protein concentrations. Perkins *et al.* (1989) have utilized a pathlength of 100 μm to study the complement protein properdin at a concentration of 1.5 mg/ml.

Not only is D_2O transparent in the amide I region of the protein spectrum, but it is also useful in studying protein conformations, conformational changes, as well as conformational dynamics by following the rate of the isotopic exchange between hydrogen and deuterium (Hvidt and Wallevik, 1972; Venyaminov *et al.*, 1976; Olinger *et al.*, 1986; Wong and Heremans, 1988; Wantyghem *et al.*, 1990). D_2O exchange is sensitive to the "local chemical environment of the exchange sites and to the overall conformational dynamics of the proteins" (Wantyghem *et al.*, 1990, p. 6606).

4.3. Attenuated Total Reflection

In conventional transmission infrared spectroscopy, the infrared radiation passes directly through the sample (Fig. 5). However, some samples (in the form of solids, films, or solutions) do not lend themselves to transmission infrared spectroscopy or are required to remain intact. These samples may benefit from being examined by attenuated total reflection infrared (ATR-IR) spectroscopy (Fig. 5). Due to the nature of the optical path, interference fringes that can be a problem in normal transmission methods are not present when using the ATR method (Fujiyama *et al.*, 1970). The two techniques of internal reflection and infrared spectroscopy were coupled in the early 1960s by Fahrenfort (1961) and Harrick (1960) independently of one another. The basis for ATR-IR spectroscopy is that a beam of infrared radiation enters an optically transparent medium [the internal reflection element (IRE)] of higher refractive index than the contacting sample, at an angle greater than the critical angle. The radiation is essentially totally reflected at the sample–IRE interface; however, an electromagnetic field having the same frequency as the totally reflected radiation is established within the sample near the interface and decays logarithmically as an evanescent wave

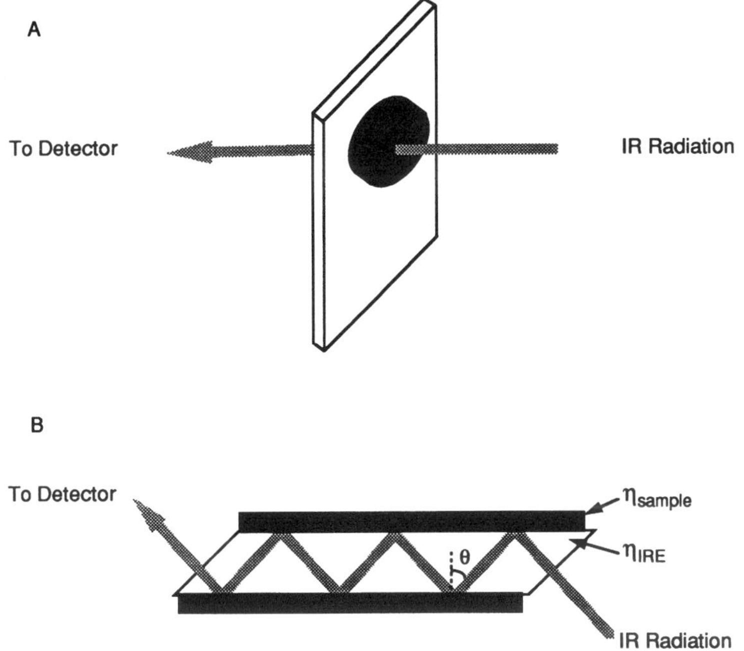

Figure 5. Schematic of (A) transmission and (B) ATR sampling methods.

into the sample medium. An absorbing sample will selectively absorb the energy, thus resulting in a reflectivity less than unity. The resultant infrared spectrum of the sample is similar to its transmission spectrum and characteristic of the sample surface within a finite distance of the interface. For a more detailed description of this technique, the reader is referred to Harrick (1967).

The ATR spectrum differs from the transmission spectrum in that it is wavelength dependent. A symmetric band in a transmission spectrum will be asymmetric in the ATR spectrum with higher absorbance at lower wavenumbers. This reflects the decay function of the evanescent wave into the sample. The decay function is termed as depth of penetration (d_p) and has been defined differently by various investigators, as discussed in the review by Knutson and Lyman (1985). For example, the distance into the sample required for the evanescent wave of the electric field vector to decay to $1/e$ its initial value at the interface (Harrick, 1967) or the intensity of the electric field to decay to $1/e$ its original value (Hirschfeld, 1977) is defined as depth of penetration and is described by Eq. (4.1) or (4.2), respectively:

$$d_p = \frac{\lambda}{2\pi\eta_{IRE}\left[\sin^2\theta - \left(\frac{\eta_{sample}}{\eta_{IRE}}\right)^2\right]^{1/2}} \tag{4.1}$$

$$d_p = \frac{\lambda}{4\pi\eta_{IRE}\left[\sin^2\theta - \left(\frac{\eta_{sample}}{\eta_{IRE}}\right)^2\right]^{1/2}} \tag{4.2}$$

In either case, the decay is a function of the indexes of refraction of the sample and IRE, angle of the incident radiation at the internal reflection interface, and wavelength. The surface sensitivity increases with higher incident angles and higher IRE indexes of refraction. Thus, by systematically varying experimental conditions, it is possible to profile the sample as a function of depth up to ~5 μm (Knutson and Lyman, 1985). It is also possible to combine this surface technique with polarization studies in order to learn more about molecular orientation (Section 4.4). Under certain experimental conditions, the influence of the aqueous phase may be less intrusive than transmission experiments (Robinson and Vinogradov, 1964; Fringeli and Günthard, 1981). However, when studying protein solutions, the influence of any adsorbed protein and subsequent conformational changes must be considered. Because of the capability to increase sampling area due to the multiple reflections (30–100 for standard size ATR elements), sample concentrations can possibly be decreased relative to transmission infrared. Swedberg *et al.* (1990) used enzyme concentrations as low as 2.5 mg/ml to form thin films from which they obtained structural information concerning the *trans*-cinnamoyl-α-chymotrypsin acyl-enzyme intermediate. They suggest that this technique can be used to obtain structural information of enzyme–substrate or enzyme–inhibitor complexes using dilute protein solutions. This particular technique has been well suited for studying the protein conformational changes that accompany adsorption onto either the internal reflection element itself or onto polymer surfaces (Castillo *et al.*, 1984; Lenk *et al.*, 1989; Giroux and Cooper, 1991).

4.4. Dichroic Measurements

The transition moment (changing dipole moment) of a molecular vibration is directional and nearly always fixed with respect to the nuclear configuration of the molecular group. Theoretically, if a particular vibration is infrared active, its band intensity will be proportional to the square of the transition moment. Because the transition moment is directional and interacts with the corresponding directional component of the electromagnetic radiation, it is possible to gain information regarding band assignments or orientation of a sample when using polarized radiation. For example, small frequency shifts occur between absorbances arising from the various secondary structures in the amide I band. However, in oriented solid samples of polypeptides and proteins the absorbance arising from the α-helix shows strong parallel dichroism, whereas the low wavenumber band for the β-extended form exhibits strong perpendicular dichroism (Elliott and Ambrose, 1950; Miyazawa, 1960; Krimm, 1962; Fraser and MacRae, 1973), allowing the structural components to be more easily distinguished. The amide II (Section 3.3) and amide A and B (Section 3.1) bands also show strong dichroism corresponding to the α and β forms.

Polarization studies can also reveal information about the directionality of the vibrational transition moment, thus providing some indication about a functional group or molecular orientation. The dichroic ratio, $R = A_{\parallel}AL_{\perp}$, determined by ratioing the absorbances arising from the electric vector parallel and perpendicular to the sample, is necessary to calculate the orientation angle of the transition moment (Fringeli and Günthard, 1981). However, because the transition moment does not always have the same directionality as the bond, molecule, or unit cell itself, prior knowledge of the value is required. For example, the transition moment of an α-helix C=O stretch (amide I) vibration lies close to parallel to the helix axis (~27°), whereas the antiparallel β-sheet C=O stretch vibration lies close to perpendicular to the axis (Cabiaux *et al.*, 1989).

Rothschild and Clark (1979) and Rath *et al.* (1991) have used polarized FTIR in the transmission mode to determine the orientations of bacteriorhodopsin of the purple membrane of *Halobacterium halobium* and the membrane-bound channel-forming COOH-terminal peptide of colicin E1, respectively. Polarized light can also be combined with ATR studies when well-defined conditions are maintained to provide information about three-dimensional orientation. Fringeli and Günthard (1981) have reviewed the application of ATR spectroscopy to membrane research, and Fraser and McRae (1973) have described infrared dichroism and orientation density functions in greater detail. Among those who have incorporated ATR with orientational studies using Fringeli and Günthard's method are Gremlich *et al.* (1983), who determined the orientation of corticotropin-(1-24)-tetracosapeptide in dioleoylphosphatidylcholine bilayer systems, and Cabiaux *et al.* (1989), who studied the orientation of diphtheria toxin fragments CB1 and CB4 with respect to phospholipid acyl chains.

5. DATA ANALYSIS

As mentioned previously, early protein studies were only able to provide gross qualitative information due to practical limitations of instrumentation such as low sensitivity of infrared instruments, spectral interference of surrounding environment such as H_2O, and probably most importantly, inability to separate the overlapping information contained within the amide bands especially characteristic of condensed-phase samples. The development of computerized FTIR spectrometers aided in improving the signal-to-noise ratio, allowing increased sensitivity and data manipulations. This made it possible to remove or subtract spectral interferences such as aqueous media from the sample spectrum. Additionally, the improved spectra allowed the development and application of computational procedures for the "resolution enhancement" of broad infrared bands that are unable to be instrumentally resolved. The following section presents some of the computer-assisted developments and mathematical techniques that have enabled infrared spectroscopy to develop as a valuable tool for protein structural evaluation.

5.1. Subtraction

As mentioned in Section 4.2, H_2O subtraction has only become experimentally feasible in the past 15 years with the growth of sophisticated computerized techniques. H_2O is the desirous solvent to use because of its naturally occurring role in biological systems. Even though D_2O is environmentally similar and has found widespread acceptance as a H_2O substitute, the completeness of the hydrogen–deuterium exchange must be considered, as well as the fact that complete exchange may occur at the expense of native structure. Also, only in an aqueous environment is it possible to obtain spectral information from the amide I, II, and III bands. However, it is important to understand the particular absorbance subtraction procedure since it is not a straightforward, unbiased procedure. Figure 6 presents the spectra of a protein in an aqueous buffer, the aqueous buffer itself, and the resultant protein spectrum with the aqueous phase subtracted. There are no unique individual bands in either the protein or H_2O spectra to use as a reference for subtraction. Currently, the method most commonly used to evaluate the subtraction of liquid H_2O is the presence of a straight baseline from 2000 to 1750 cm^{-1} and the absence of negative side lobes (Haris *et al.*, 1986; Olinger *et al.*, 1986; Surewicz *et al.*, 1987d; Mitchell *et al.*, 1988; Dong *et al.*, 1990), although some have tried to remove the bias by incorporating a computer algorithm (Powell *et al.*, 1986) to automatically subtract out the H_2O spectrum. H_2O vapor can also present a problem if the infrared path is not completely dry during data collection. The vapor characteristically appears in the amide I region as a series of sharp spikes occurring at 1684, 1670, 1662, 1653, 1646, and 1617 cm^{-1} superimposed on the protein band. The importance of subtracting any residual H_2O vapor has been recognized since the extra "peaks" dramatically influence the resolution enhancement of the spectra (Dong *et al.*, 1990; Mantsch *et al.*, 1986). Another considera-

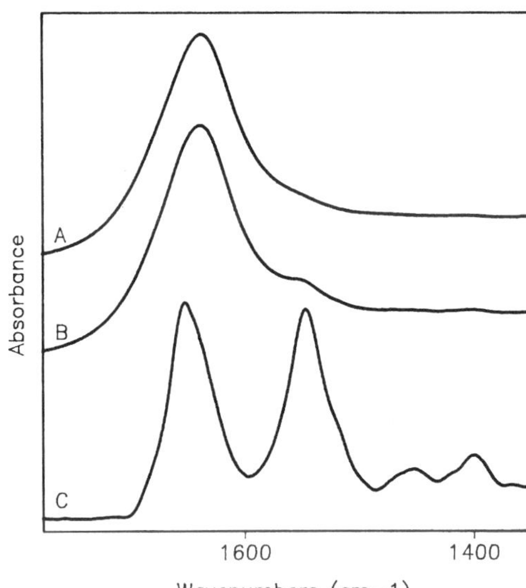

Figure 6. Spectra of (A) aqueous PBS alone, (B) aqueous PBS and bovine serum albumin, pH = 7.4, and (C) bovine serum albumin after buffer subtraction.

tion is that the interaction between protein and water could change the absorbance characteristics of the water band due to deformation of the water hydrogen bonding network, therefore affecting the quality or adequacy of the water subtraction (Susi and Byler, 1986). Also, the subtraction of a very strong water band adversely affects the signal-to-noise ratio. And because the choice of water subtraction factor is still subjective, there can be a large degree of uncertainty when the subtraction factor is manually chosen and appears to be dependent on operator experience with the subtraction procedure (Powell *et al.*, 1986).

5.2. Resolution Enhancement

Increasing instrumental resolution will not resolve the naturally broad infrared bands of condensed-phase spectra but will instead adversely affect the signal-to-noise ratio. Several techniques have been developed to aid in distinguishing overlapping components. The instrumental resolution is not actually increased, but rather the techniques minimize the effects of the instrument transform function on the experimentally measured band to reveal a band more closely resembling an intrinsic infrared bandshape.

Deconvolution was first applied to the analysis of infrared spectra in 1981 (Kauppinen *et al.*, 1981) and has become the most common method used to distinguish the amide I component bands. Fine details of the procedure are beyond the scope of this chapter and are discussed in greater detail elsewhere (Kauppinen *et al.*, 1981; Cameron and Moffatt, 1984; Mantsch *et al.*, 1986; Susi and Byler, 1986; Markovich and Pidgeon, 1991). Basically, an infrared band can be expressed as the convolution of a line shape (Lorentzian, Gaussian, or a mixture of both) with a line position (Dirac delta function). These procedures are usually carried out in the Fourier domain because better control of variables is obtained. In the Fourier domain, the convolution process becomes the multiplication of a decaying exponential function (the transform of the line shape) with a cosine function (the transform of the line position), which results in an exponentially decaying cosine function. This latter function is transformed back to the wavenumber domain and results in the infrared band. The rate of cosine decay is determined by the bandwidth. The broader the band, the faster the decay. If the decay is decreased by multiplying by an increasing exponential function, the band can be "mathematically" narrowed. Thus, the contributions of the narrow bands overlapping under a broadband envelope can be distinguished. This is the process of deconvolution (Cameron and Moffatt, 1984; Markovich and Pidgeon, 1991) as illustrated in Fig. 7.

However, it should be emphasized that although the procedure cannot create finer detail than is inherent in the spectrum, artifacts can arise from the amplification of random noise and water vapor as well as the side lobes that are generated due to the finite nature of the spectrum. Surewicz and Mantsch (1988) suggest that the best results are obtained if the signal-to-noise ratio is better than 500:1. Apodization functions are often used to smooth the data by alleviating some of the noise amplification (Cameron and Moffatt, 1984). Practically, the region being deconvolved should contain components with similar bandwidths, although the bandwidths and the amount of noise is unknown (Cameron and Moffatt, 1984). Thus an empirical approach to deconvolution must be taken. Proceed by maintaining a constant degree of smoothing while increasing the bandwidth to the point just before side lobes begin to appear (once the chosen bandwidth is greater than the width of the narrowest band in the spectrum, the negative lobes will appear). Using this bandwidth, decrease the amount of smoothing to the point at which the noise begins to rapidly increase. Although the parameters are controlled by the inherent bandwidth and amount of noise, the procedure is still subjective. By practicing on a synthetic spectrum, the influence these variables have on the resultant deconvolved spectrum can be observed.

Derivative procedures are similar to deconvolution in that both utilize weighting functions in the Fourier domain to achieve their results. The second and fourth derivatives are produced by weighting the data in the Fourier domain, using quadratic and quartic functions, respectively (Cameron and Moffatt, 1984), as opposed to the exponential function used in deconvolution. The fourth derivative results in narrower bands than the second derivative but also requires a higher resolution and better signal-to-noise ratio in order to gain the additional information.

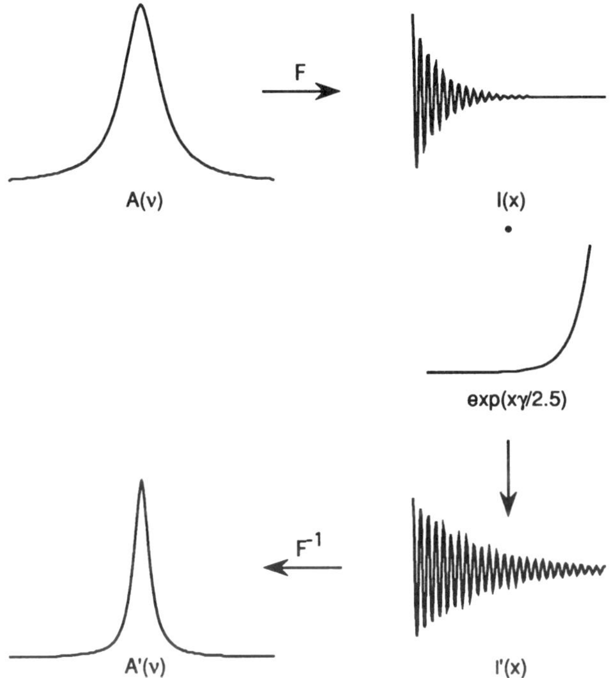

Figure 7. Illustration of deconvolution: The original band A(υ) is cosine Fourier transformed to yield the function $I(x)$ in which the decay is determined by the original bandwidth γ. It is then multiplied in the Fourier domain by an increasing exponential function to yield the new function $I'(x)$, and after reverse Fourier transformation, results in the narrowed band, $A'(\upsilon)$.

Because derivative and deconvolution methods preserve the band position, both have been used extensively to resolve and identify the position of overlapped protein amide I bands and can be used in concert to corroborate the results of the other method. Although the band height is lost with both techniques, deconvolution theoretically retains the integrated intensity of the component bands. Both techniques are equally sensitive to noise, i.e., a degradation of the signal-to-noise ratio occurs. Figure 8 illustrates the spectrum of β-lactoglobulin deconvolved, before resolution enhancement, and after second-derivative calculations.

5.3. Quantitation

The Beer-Lambert Law states that there exists a relationship between the transmitted radiation intensity and the sample concentration:

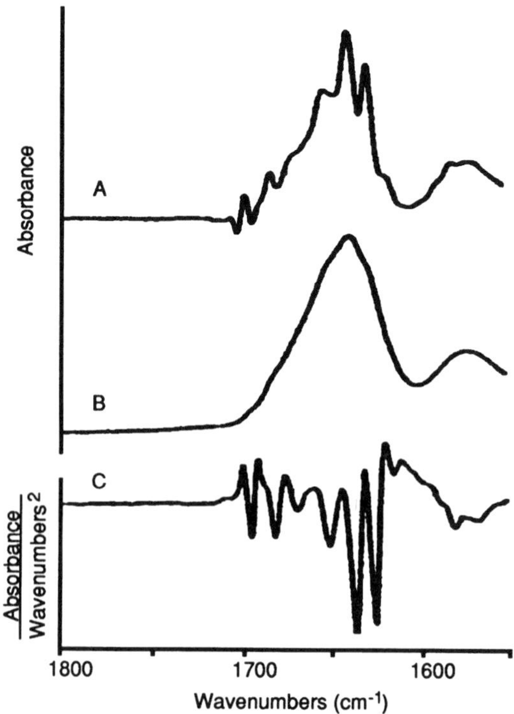

Figure 8. Spectrum of β-lactoglobulin: (A) deconvolved, (B) before resolution enhancement, and (C) after second-derivative procedures. (Adapted from Purcell and Susi, 1984.)

$$\log \frac{I_o}{I} = \log \frac{1}{T} = A = abc$$

where I_o and I denote the incident and transmitted radiation intensity, respectively, T is the transmittance, A the absorbance, c the concentration, b the pathlength, and a the absorptivity that is specific for every substance at each frequency (analogous to an extinction coefficient). Resolution-enhancement techniques have improved the identity and separation of the individual absorbances arising from specific secondary structures, making it possible for quantitative analysis. Currently, the most common method of quantitation is to first identify the frequency and number of individual amide I components using resolution enhancement techniques and then employ these values as input parameters for synthesizing a curve to fit the bands. Band heights are not conserved in the derivative and deconvolved spectra and therefore should not be used as a quantitative indicator of structural composition.

There are several assumptions implied with quantitative evaluations. The absorbances due to the secondary structural classes are distinct and well defined, there is an absence of contributions from side chains in the amide I region, and the absorptivities of the integrated bands are equivalent for all structures as well as all proteins. Krimm and Bandekar (1986) point out that it is possible for absorbances arising from turn structures to overlap with helical modes, and several studies indicate that the amide I region is not completely void of side-chain absorbances (Chirgadze *et al.*, 1975; Venyaminov and Kalnin, 1990). Among the evidence that the latter assumption is not valid is that there exist intensity differences between the random forms of silk fibroin and poly-*S*-carboxymethyl-L-cysteine. Chirgadze *et al.* (1973) attributed this observation to differences in electronic properties between amide groups of different molecules. In the same study, molar absorptivities were found to be quite different between the random and β-structures, ranging from 275 to 380 liter/mole per cm^{-1} and 960 to 1060 liter/mole per cm^{-1}, respectively. Even so, there is a good correlation for the curve fitting of the deconvolved spectra in D_2O (Byler and Susi, 1986) and the derivative spectra in H_2O (Dong *et al.*, 1990) with X-ray data.

It is important to keep in mind that quantitative studies do not provide absolute values but rather estimates of protein secondary structure. Some investigators caution against placing high expectations on infrared spectroscopy for the determination of secondary structural content (Surewicz *et al.*, 1993). Other investigators have attempted to eliminate the subjectivity of assigning various bands to specific secondary structure by incorporating a least-squares curve-fitting analysis of proteins in D_2O (Eckert *et al.*, 1977), a partial least-squares analysis in H_2O (Dousseau and Pézolet, 1990), and factor analysis (Kalnin *et al.*, 1990; Lee *et al.*, 1990; Sarver and Krueger, 1991), with varying degrees of success. These techniques also have their own shortcomings (Surewicz *et al.*, 1993).

Another facet of quantitation is the estimation of the amount of protein adsorbed to a surface. Various studies have shown the applicability of the amide II band for the quantitation of protein either by measuring the integrated band area (Castillo *et al.*, 1984; Pitt *et al.*, 1987) or the band absorbance at a particular wavenumber or maximum band absorbance (Gendreau *et al.*, 1982; Fink *et al.*, 1987; Giroux and Cooper, 1991). The amide II band's relevance for quantitative means stems from its relative stability toward structural changes.

6. LITERATURE EXAMPLES

There are numerous investigations in the literature in which infrared spectroscopy has been utilized to aid in elucidating protein structure. The ability of this technique to encompass a variety of sample forms and mixtures has made it invaluable for studying various aspects of protein molecular structure. The following sections will highlight some of the many studies incorporating infrared spectroscopy for protein structural determination.

6.1. Solution Studies

Water is synonymous with biological systems. It is therefore of interest to study the structure of biological molecules in their native environments. Solution studies can provide insight into the similarities between crystal structure and solution structure (Wei *et al.*, 1991), the structural composition of a protein if the crystal structure is unknown (Perkins *et al.*, 1989), structural changes important in the mechanistic actions of proteins (Surewicz *et al.*, 1990), effects of intermolecular interactions such as dimerization and self-assembly on structure (Jakobsen *et al.*, 1983; Wei *et al.*, 1992), the structure–function relationships between homologous proteins (Trewhella *et al.*, 1989; Jackson *et al.*, 1991), the role of water in protein folding, unfolding, and stabilization, the mechanisms of protein folding, and so forth. Various studies have probed the stability of proteins to changing environmental conditions such as pH (Casal *et al.*, 1988), temperature (Surewicz *et al.*, 1987d; Casal *et al.*, 1988), pressure (Wong and Heremans, 1988), and solvent (Purcell and Susi, 1984; Wasacz *et al.*, 1987; Jackson and Mantsch, 1992) in order to assess the mechanisms of protein folding and stability.

Jackson and Mantsch (1992) studied the helicogenic effects of halogenated alcohols on proteins. The amide I band of native concanavalin A, a β-sheet protein, has major component bands at 1636 and 1625 cm^{-1} (as determined from deconvolution techniques) in D_2O. In pure bromoethanol the frequency of the amide I band shifts to 1654 cm^{-1}, indicating the formation of α-helical structure. The pure solvents, dichloroethanol > bromoethanol > trifluoroethanol > chloroethanol > fluoroethanol, can induce helical conformations. However, low alcohol concentrations appear to have no effect on the structure of concanavalin A, yet amide I temperature profiles indicate that the halogenated solvent (0.026 mole chloroethanol) induces thermal destabilization at 60 to 65 °C, whereas the native structure remains stable up to 80 °C. Based in part on amide I data, it is proposed that with increasing alcohol concentrations, the alcohols disrupt the hydrogen bonding, resulting in unfolding of the native structure and leading to aggregation. Higher concentrations lead to complete unfolding followed by refolding into helical conformations. These pure solvents induce helical conformations in other nonhelical proteins as well. Jackson and Mantsch (1992) suggest that helix formation is dependent on the properties of the solvent and not necessarily the polypeptide chain.

Because of the close sequence homologies between the L/M subunits of purple bacterial reaction center and the D1/D2 subunits of higher plant photosystem II (PSII) reaction centers, the structure of the latter is proposed to be similar to the crystallographically determined structure of the former (He *et al.*, 1991). However, the PSII reaction center (including α- and β-subunits of cytochrome b_{559} and the product of the *bspI* gene) is sensitive to photodamage, whereas the bacterial center (including the H subunit) is not. By quantitatively analyzing the amide I band using a factor analysis method that is independent of band assignments, He *et al.* (1991) determined that the PSII reaction center contains 67% α-helix. The L/M subunits of bacterial reaction centers themselves contain similar percentages of helices. The PSII reaction center when left in Triton X-100, a solubilization enhancer, even under dark conditions undergoes conformational changes within 30 min as revealed by the relative increase in the absorbance band at 1627 cm^{-1} with respect to the helical absorbance at 1656 cm^{-1}. By exchanging Triton X-100 with dodecyl maltoside, the reaction center becomes stable as indicated by the stability of the amide I band contours. However, with exposure to light (2000 $\mu E\ m^{-2}s^{-1}$) in the presence of O_2 at room temperature or 60 °C temperatures, spectral changes occur that are similar to those observed when the reaction center was destabilized by Triton X-100. These changes also correspond to a loss of photochemical activity. The conformational changes occurring are all indicative of the formation of intermolecular β-sheet structures as suggested by the appearance of bands at 1624 and 1686 cm^{-1}.

Prestrelski *et al.* (1991) studied the effects of Ca^{2+}, Mn^{2+}, Zn^{2+}, and Na^+ binding on the secondary structure of bovine α-lactalbumin to obtain insight into conformation and stability of the molecule. The amide I band visualized by second-derivative and deconvolution procedures demonstrates that both the apo and Ca^{2+} holo forms of α-lactalbumin in D_2O undergo conformational changes with dissolution as indicated by relative intensity changes, shifts in band positions, and the appearance of a new band. These changes are complete within 24 hr, suggesting that full dissolution is a time-dependent process. The apo and Ca^{2+} holo forms exist in different conformations as indicated by the differences in the amide I bands. The most apparent difference is the presence of a band at 1645 cm^{-1} in the apo form that is absent in the holo form. The presence of this band results in a 0.16 fractional loss of intensity from those components arising from the α-helix. Shifts in the positions of the turn absorbances also indicate conformational alterations. These interpretations agree with previous circular dichroism (CD) and fluorescence data. Prestrelski *et al.* (1991) suggest that the rigid, coordination complex resulting from the bound Ca^{2+} undergoes a conformational change to an essentially orderless loop with removal of the Ca^{2+} ion. The binding of Zn^{2+}, Mn^{2+}, and Na^+ cause slight alterations in the amide I band that are not as substantial as those caused by Ca^{2+} binding, indicating that the former metal ions do not induce a structure similar to the Ca^{2+} holo form. This is in disagreement with aromatic CD and fluorescent studies. However, infrared vibrations are sensitive to the backbone of the protein, whereas the other experimental techniques are more sensitive to local environments. Thus, infrared spectroscopy can provide complementary information that is unavailable from other techniques.

6.2. Membrane Studies

Membrane proteins isolated from their hydrophobic lipid environments do not provide comprehensive insight into their natural states. With infrared spectroscopy, it is possible to study the proteins in an intact biological membrane and even obtain information about the lipid structural state (although description of lipid structure is beyond the scope of this chapter) without relying on environmental probes (Krimm and Dwivedi, 1982; Lee *et al.*, 1985, 1987; Arrondo *et al.*, 1987; Mitchell *et al.*, 1988; Buchet *et al.*, 1989; Haris *et al.*, 1989). Infrared studies of proteins reconstituted with model membranes have also been undertaken (Surewicz *et al.*, 1987b,c; Rothschild *et al.*, 1990); however, care must be exercised in choosing the model lipid so as not to interfere with the protein bands. For example, the amide bands arising from sphingolipids and the carboxyl absorbances of phosphatidylserine interfere with the amide I absorbance of the protein (Surewicz and Mantsch, 1988).

Ca^{2+}-transporting ATPase of sarcoplasmic reticulum (SR) has been widely studied by infrared spectroscopy. The structure in its native environment has been established, indicating that the major components are α-helix and random conformations with minor contributions from β-sheet structures (Arrondo *et al.*, 1987). In the absence of the SR, as well as reconstituted with a model phospholipid, a similar structure is predicted (Lee *et al.*, 1985). In contrast, when making quantitative measurements, Villalain *et al.* (1989) determined that a high β-sheet content (32%) is present with very little random structure. They attribute their contradictory results to ambiguous band assignments as well as the assumptions made when using band-fitting techniques for quantitative analysis. The same study revealed that in the presence of various ligands such as Ca^{2+}, vanadate, ATP, and phosphate, no marked spectral changes were observed, contrary to the conclusions of Arrondo *et al.* (1987), who examined the differences between the enzyme's E_1 (Ca^{2+} stabilized) state and E_2 (Ca^{2+}-free, vanadate environment) state. Arrondo *et al.* (1987) documented small but significant differences in the deconvolved and second-derivative amide I spectra. Both groups, however, indicate that the results could simplistically suggest the rearrangement of the domains as opposed to significant unfolding of the protein. Buchet *et al.* (1989) took their study a step further and investigated the effects of dimethyl sulfoxide (DMSO), a known ATPase inhibitor, on the structure of the enzyme in SR vesicles. There were no significant alterations in the amide I band up to 40% DMSO concentrations. This correlates with reversible ATPase inhibition if the solvent concentration is dropped to 4%. Between 40 and 60% DMSO, however, inhibition becomes irreversible and is accompanied by the appearance of a shoulder at about 1630 cm^{-1} and increased intensity at 1678–1682 cm^{-1}, suggesting the formation of new β-structures. Increasing solvent concentrations above 60% caused the appearance of a sharp band at 1622 cm^{-1} and a distinct band at 1682 cm^{-1}, while the maximum shifted to 1639 cm^{-1}. These investigators propose that the 1622 cm^{-1} band arises from the unionized -COOH vibrations. Rehydration of the enzyme resulted in the disappearance of the 1622 cm^{-1} band and the maximum shifting back to 1645 cm^{-1}, yet the 1630 cm^{-1} shoulder remained and activity was not restored.

An early study of the purple membrane of *H. halobium* suggests that the helical portion of the membrane protein, bacteriorhodopsin, does not assume an α-helical conformation but rather an α_{II}-type helix. The major structural difference in the α_{II}-type helix is that the plane of the peptide group is tilted versus lying parallel to the helix axis. This results in a different transition dipole coupling as well as an increase in the C=O ··· H-N hydrogen bond length. Predictions indicate that the amide I and A frequencies for the α_{II}-helix should be significantly higher, which agrees well with observed values (Krimm and Dwivedi, 1982). Employing derivative and deconvolution for the amide I and II band analyses (Lee *et al.*, 1987) revealed that the predominant structure is helical and agrees with the calculated value for an α_{II} structure. β-structure and some antiparallel and random structures were also identified. After a 2-hr digestion with papain, the amide I band maximum absorbance of the purple membrane shifts from 1661 to 1663–1664 cm^{-1} and the bandwidth becomes narrower. This indicates that the C-tail (which is removed by the papain) probably has a random structure. Further treatment (24 hr) altered relative intensities but did not indicate the depletion of any structure. The interpretation is that some of the β-structure remains imbedded in the membrane and is not all contained within the C-tail, N-terminus, or the transmembrane loop between segments 2 and 3.

6.3. Other Studies

Peptides have biologically active roles, whether as antigens or as hormones, for example, in which structural characteristics are important in their ensuing activities. Infrared spectroscopy has proven useful in determining structural characteristics. For example, Surewicz *et al.* (1987a) studied the 24-amino acid peptide atriopeptin III, an atrial natriuretic factor that is thought to play a regulatory role in balancing electrolyte and water levels. They identified from the amide I band that the monomeric form of the peptide in solution consists primarily of random structure, whereas the self-associated form as well as the lipid-associated form contain a large fraction of β-sheet. After deconvolution it was observed that the lipid-complexed and self-associated forms, though both containing β-structures, did not contain identical structures. The lipid-bound peptide contained a larger fraction of random structure yet fewer turns than the self-associated forms. Also, the β-components absorbed at different wavenumbers, indicating that the component was not identical. Another example of the contributions of infrared spectroscopy in determining peptide secondary structure is demonstrated in the study of peptide M, the synthetic 18-residue peptide derived from the highly immunogenic sequence of retinal S-antigen (Muga *et al.*, 1990). The peptide demonstrated conformational adaptability to environmental changes as expected for small peptides. Deconvolution of the amide I band indicates that three different pH-dependent conformational states are present. The structures present at low pH (<4), intermediate pH (4–9.5), and high pH (>9.5) correspond to residual β-structure and possibly distorted helix or turns, predominantly β-sheet structure, and nonordered structure, respectively. Concentration also plays a role in the secondary structure. At concentrations above 2.5 mM, the predominant structure is β-sheet, the 1612 cm^{-1} absorbance most likely arising from intermolecular association. Below 2.5 mM, the relative band intensities arising from β-structure, especially the 1612 cm^{-1} band, decrease, indicating a concentration-dependent self-association. Elevated hydrostatic pressures are known to promote the dissociation of protein aggregates. At elevated pressures, distinct changes in the amide I band occur, again indicating that the intermolecular sheet structure is destroyed while the monomeric form adopts an intramolecular β-extended structure. The extended structure of peptide M could play an important role in its recognition by the immune system.

Infrared spectroscopy has also contributed to the understanding of how proteins respond to adsorption to various surfaces. For example, Castillo *et al.* (1984) adsorbed human serum albumin onto soft contact lenses and determined that β-sheet and random conformations are gained at the expense of helical structure. In another study (Giroux and Cooper, 1991), human fibronectin and human fibrinogen were adsorbed under static conditions onto thin polyurethane films cast directly onto the germanium internal reflection element. Using the absorbance intensity at 1550 cm^{-1} as an estimation of surface concentration, the adsorption kinetics were followed. The more hydrophobic, treated polymer adsorbed less human fibrinogen but more human fibronectin than the untreated polyurethane. β-Structure content of both proteins increased with adsorption to the treated polyurethane. Lenk *et al.* (1989) studied bovine serum albumin adsorption from a flowing solution onto a series of polyetherurethanes. They concluded that there was a loss of helical conformations and a gain of β-structure.

7. SUMMARY

Infrared spectroscopy can provide insight into protein structure. This technique is sensitive to the backbone amide arrangement of peptide and protein molecules. In many cases, complementary as well as more expansive information is obtained as opposed to information obtained by other methods that examine the molecule's environmental surroundings, require molecular probes, or perhaps cannot investigate the molecule in its native environment. The foundation for spectroscopic differences between the various secondary structures arises not only from geometrical differences and hydrogen bond variations but also transition dipole coupling between neighboring oscillators. Theoretical predictions of protein spectra have been made using normal mode analysis and combined with experimental data. At present the amide I band has provided the most insight into secondary structure. Even more convincing results are obtained when both H_2O and D_2O are used as solvents.

Recent advances in computerized technology and mathematical techniques have expanded the potential contributions of infrared spectroscopy in the area of protein structural determination. However, the limitations of resolution enhancement and curve-fitting techniques must be taken into consideration. The parameters must be carefully and optimally chosen and evaluated on a case-by-case basis. The subjectivity of these techniques makes a thorough understanding of the algorithms necessary, especially those commercially available. Infrared spectroscopy continues to provide insight into protein and peptide structures under biologically relevant conditions that enable the structure–function relationships for such molecules to be better understood.

ACKNOWLEDGMENTS. This work was supported in part by the NIH Biotechnology Training Grant, Number GM08393 and NIH-HD R01-23000.

REFERENCES

Alvarez, J., Haris, P. I., Lee, D. C., and Chapman, D., 1987, Conformational changes in concanavalin A associated with demetallization and α-methylmannose binding studied by Fourier transform infrared spectroscopy, *Biochim. Biophys. Acta* **916:**5–12.

Anderle, G., and Mendelsohn, R., 1987, Thermal denaturation of globular proteins Fourier transform-infrared studies of the amide III spectral region, *Biophys. J.* **52:**69–74.

Arrondo, J. L. R., Mantsch, H. H., Mullner, N., Pikula, S., and Martonosi, A., 1987, Infrared spectroscopic characterization of the structural changes connected with the E_1–E_2 transition in the Ca^{2+}-ATPase of sarcoplasmic reticulum, *J. Biol. Chem.* **262:**9037–9043.

Bandekar, J., and Krimm, S., 1979, Vibrational analysis of peptides, polypeptides, and proteins: Characteristic amide bands of β-turns, *Proc. Natl. Acad. Sci. USA* **76:** 774–777.

Bandekar, J., and Krimm, S., 1988, Normal mode spectrum of the parallel-chain β-sheet, *Biopolymers* **27:**909–921.

Beer, N., Sutherland, G. B. B. M., Tanner, K. N., and Wood, D. L., 1959, Infra-red spectra and structure of proteins, *Roy. Soc. London Proc.* **249:**147–172.

Bendit, E. G., 1966a, Infrared absorption spectrum of keratin. I. Spectra of α-, β-, and supercontracted keratin, *Biopolymers* **4:**539–559.

Bendit, E. G., 1966b, Infrared absorption spectrum of keratin. II. Deuteration studies, *Biopolymers* **4:** 561–577.

Buchet, R., Jona, I., and Martonosi, A., 1989, Correlation of structure and function in the Ca^{2+}-ATPase of sarcoplasmic reticulum: A Fourier transform infrared spectroscopy (FTIR) study on the effects of dimethyl sulfoxide and urea, *Biochim. Biophys. Acta* **983:**167–178.

Byler, D. M., and Susi, H., 1985, Protein structure by FTIR self-deconvolution, *Proc. SPIE Int. Soc. Opt. Eng.* **553:**289–290.

Byler, D. M., and Susi, H., 1986, Examination of the secondary structure of proteins by deconvolved FTIR spectra, *Biopolymers* **25:**469–487.

Cabiaux, V., Brasseur, R., Wattiez, R., Falmagne, P., Ruysschaert, J.-M., and Goormaghtigh, E., 1989, Secondary structure of diphtheria toxin and its fragments interacting with acidic liposomes studied by polarized infrared spectroscopy, *J. Biol. Chem.* **264:**4928–4938.

Cameron, D. G., and Moffatt, D. J., 1984, Deconvolution, derivation, and smoothing of spectra using Fourier transforms, *J. Test. Eval.* **12:**78–85.

Carrier, D., Mantsch, H. H., and Wong, P. T. T., 1990, Protective effect of lipidic surfaces against pressure-induced conformational changes of poly(L-lysine), *Biochemistry* **29:**254–258.

Casal, H. L., Köhler, U., and Mantsch, H. H., 1988, Structural and conformational changes of β-lactoglobulin B: An infrared spectroscopic study of the effect of pH and temperature, *Biochim. Biophys. Acta* **957:**11–20.

Castillo, E. J., Koenig, J. L., Anderson, J. M., and Lo, J., 1984, Characterization of protein adsorption on soft contact lenses, *Biomaterials* **5:**319–325.

Chirgadze, Y. N., and Nevskaya, N. A., 1976a, Infrared spectra and resonance interaction of amide-I vibration of the antiparallel-chain pleated sheet, *Biopolymers* **15:**607–625.

Chirgadze, Y. N., and Nevskaya, N. A., 1976b, Infrared spectra and resonance interaction of amide-I vibration of the parallel-chain pleated sheet, *Biopolymers* **15:**627–636.

Chirgadze, Y. N., Shestopalov, B. V., and Venyaminov, S. Y., 1973, Intensities and other spectral parameters of infrared amide bands of polypeptides in the β- and random forms, *Biopolymers* **12:**1337–1351.

Chirgadze, Y. N., Fedorov, O. V., and Trushina, N. P., 1975, Estimation of amino acid residue side-chain absorption in the infrared spectra of protein solutions in heavy water, *Biopolymers* **14:**679–694.

Chirgadze, Y. N., Brazhnikov, E. V., and Nevskaya, N. A., 1976, Intramolecular distortion of the α-helical structure of polypeptides, *J. Mol. Biol.* **102:**781–792.

Chou, P. Y., and Fasman, G. D., 1977, β-turns in proteins, *J. Mol. Biol.* **115:**135–175.

Colthup, N. B., Daly, L. H., and Wiberley, S. E., 1975, *Introduction to Infrared and Raman Spectroscopy*, 2nd ed., Academic Press, New York.

Conley, R. T., 1972, *Infrared Spectroscopy*, 2nd ed., Allyn and Bacon, Boston.

Dill, K. A., 1990, Dominant forces in protein folding, *Biochemistry* **29:**7133–7154.

Dong, A., Huang, P., and Caughey, W. S., 1990, Protein secondary structures in water from second-derivative amide I infrared spectra, *Biochemistry* **29:**3303–3308.

Dousseau, F., and Pézolet, M., 1990, Determination of the secondary structure content of proteins in aqueous solutions from their amide I and amide II infrared bands. Comparison between classical and partial least-squares methods, *Biochemistry* **29:**8771–8779.

Dwivedi, A. M., and Krimm, S., 1984, Vibrational analysis of peptides, polypeptides, and proteins. XVIII. Conformational sensitivity of the α-helix spectrum: α_I- and α_{II}-poly(L-alanine), *Biopolymers* **23:**923–943.

Eckert, K., Grosse, R., Malur, J., and Repke, K. R. H., 1977, Calculation and use of protein-derived conformation-related spectra for the estimate of the secondary structure of proteins from their infrared spectra, *Biopolymers* **16:**2549–2563.

Elliott, A., and Ambrose, E. J., 1950, Structure of synthetic polypeptides, *Nature* **165:**921–922.

Fahrenfort, J., 1961, Attenuated total reflection: A new principle for the production of useful infra-red reflection spectra of organic compounds, *Spectrochim. Acta* **17:**698–709.

Fink, D. J., Hutson, T. B., Chittur, K. K., and Gendreau, R. M., 1987, Quantitative surface studies of protein adsorption by infrared spectroscopy II. Quantification of adsorbed and bulk proteins, *Anal. Biochem.* **165:**147–154.

Fraser, R. D. B., and MacRae, T. P., 1973, *Conformation in Fibrous Proteins*, Academic Press, New York.

Fringeli, U. P., and Günthard, H. H., 1981, Infrared membrane spectroscopy, in: *Membrane Spectroscopy*, Vol. 31 (E. Grell, ed.), Springer-Verlag, Berlin, pp. 270–332.

Fujiyama, T., Gerrin, J., Bryce, L., and Crawford, J., 1970, Vibrational intensities XXV: Some systematic errors in infrared absorbtion spectrophotometry of liquid samples, *Appl. Spectrosc.* **24:**9–15.

Gendreau, R. M., Leininger, R. I., Winters, S., and Jakobsen, R. J., 1982, Fourier transform infrared spectroscopy for protein-surface studies, in: *Biomaterials: Interfacial Phenomena and Applications* (S. L. Cooper and N. A. Peppas, eds.), American Chemical Society, Washington, D.C., pp. 371–394.

Giroux, T. A., and Cooper, S. L., 1991, FTIR/ATR studies of protein adsorption on plasma derivatized polyurethanes, *J. Colloid Interface Sci.* **146:**179–194.

Gregory, R. B., and Rosenberg, A., 1986, Protein conformational dynamics measured by hydrogen isotope exchange, *Methods Enzymol.* **131:**448–508.

Gremlich, H.-U., Fringeli, U.-P., and Schwyzer, R., 1983, Conformational changes of adrenocorticotropin peptides upon interaction with lipid membranes revealed by infrared attenuated total reflection spectroscopy, *Biochemistry* **22:**4257–4264.

Griffiths, P. R., and de Haseth, J. A., 1986, *Fourier Transform Infrared Spectrometry*, John Wiley and Sons, New York.

Hacskaylo, M., 1954, Preparation of compounds for infrared spectrometry, *Anal. Chem.* **26:**1410–1412.

Haris, P. I., Lee, D. C., and Chapman, D., 1986, A Fourier transform infrared investigation of the structural differences between ribonuclease A and ribonuclease S, *Biochim. Biophys. Acta* **875:**255–265.

Haris, P. I., Coke, M., and Chapman, D., 1989, Fourier transform infrared spectroscopic investigation of rhodopsin structure and its comparison with bacteriorhodopsin, *Biochim. Biophys. Acta* **995:**160–167.

Harrick, N. J., 1960, Study of physics and chemistry of surfaces from frustrated total internal reflections, *Phys. Rev. Lett.* **4:**224–226.

Harrick, N. J., 1967, *Internal Reflection Spectroscopy*, Interscience Publishers, New York.

He, W.-Z., Newell, W. R., Haris, P. I., Chapman, D., and Barber, J., 1991, Protein secondary structure of the isolated photosystem II reaction center and conformational changes studied by Fourier transform infrared spectroscopy, *Biochemistry* **30:**4552–4559.

Hirschfeld, T., 1977, Subsurface layer studies by attenuated total reflection Fourier transform spectroscopy, *Appl. Spectrosc.* **31:**289–292.

Holloway, P. W., and Mantsch, H. H., 1989, Structure of cytochrome b_5 in solution by Fourier-transform infrared spectroscopy, *Biochemistry* **28:**931–935.

Hsu, S. L., Moore, W. H., and Krimm, S., 1976, Vibrational spectrum of the unordered polypeptide chain: A Raman study of feather keratin, *Biopolymers* **15:**1513–1528.

Hvidt, A., and Wallevik, K., 1972, Conformational changes in human serum albumin as revealed by hydrogen-deuterium exchange studies, *J. Biol. Chem.* **247:**1530–1535.

Jackson, M., and Mantsch, H. H., 1992, Halogenated alcohols as solvents for proteins: FTIR spectroscopic studies, *Biochim. Biophys. Acta* **1118:**139–143.

Jackson, M., Haris, P. I., and Chapman, D., 1991, Fourier transform infrared spectroscopic studies of Ca^{2+}-binding proteins, *Biochemistry* **30:**9681–9686.

Jakobsen, R. J., Brown, L. L., Hutson, T. B., Fink, D. J., and Veis, A., 1983, Intermolecular interactions in collagen self-assembly as revealed by Fourier transform infrared spectroscopy, *Science* **220:**1288–1290.

Kaiden, K., Matsui, T., and Tanaka, S., 1987, A study of the amide III band by FT-IR spectrometry of the secondary structure of albumin, myoglobin, and γ-globulin, *Appl. Spectrosc.* **41:**180–184.

Kalnin, N. N., Baikalov, I. A., and Venyaminov, S. Y., 1990, Quantitative IR spectrophotometry of peptide compounds in water (H_2O) solutions. III. Estimation of the protein secondary structure, *Biopolymers* **30:**1273–1280.

Kauppinen, J. K., Moffatt, D. J., Mantsch, H. H., and Cameron, D. G., 1981, Fourier self-deconvolution: A method for resolving intrinsically overlapped bands, *Appl. Spectrosc.* **35:**271–276.

Knutson, K., and Lyman, D. J., 1985, Surface infrared spectroscopy, in: *Surface and Interfacial Aspects of Biomedical Polymers*, Vol. 1 (J. D. Andrade, ed.), Plenum Press, New York, pp. 197–247.

Koenig, J. L., 1979, Vibrational spectroscopy of polypeptides and proteins, in: *Infrared and Raman Spectroscopy of Biological Molecules* (T. M. Theophanides, ed.), D. Reidel Publishing Company, Dordrecht, Holland, pp. 109–124.

Koenig, J. L., and Tabb, D. L., 1980, Infrared spectra of globular proteins in aqueous solution, in: *Analytical Applications of FT-IR to Molecular and Biological Systems* (J. R. Durig, ed.), D. Reidel Publishing Company, Dordrecht, Holland, pp. 241–255.

Krimm, S., 1962, Infrared spectra and chain conformation of proteins, *J. Mol. Biol.* **4:** 528–540.

Krimm, S., and Bandekar, J., 1986, Vibrational spectroscopy and conformation of peptides, polypeptides, and proteins, *Adv. Prot. Chem.* **38:**181–364.

Krimm, S., and Dwivedi, A. M., 1982, Infrared spectrum of the purple membrane: Clue to a proton conduction mechanism? *Science* **216:**407–408.

Lee, D. C., Hayward, J. A., Restall, C. J., and Chapman, D., 1985, Second-derivative infrared spectroscopic studies of the secondary structures of bacteriorhodopsin and Ca^{2+}-ATPase, *Biochemistry* **24:**4364–4373.

Lee, D. C., Herzyk, E., and Chapman, D., 1987, Structure of bacteriorhodopsin investigated using Fourier transform infrared spectroscopy and proteolytic digestion, *Biochemistry* **26:**5775–5783.

Lee, D. C., Haris, P. I., Chapman, D., and Mitchell, R. C., 1990, Determination of protein secondary structure using factor analysis of infrared spectra, *Biochemistry* **29:**9185–9193.

Lenk, T. J., Ratner, B. D., Gendreau, R. M., and Chittur, K. K., 1989, IR spectral changes of bovine serum albumin upon surface adsorption, *J. Biomed. Mater. Res.* **23:**549–569.

Mantsch, H. H., Casal, H. L., and Jones, R. N., 1986, Resolution enhancement of infrared spectra of biological systems, in: *Spectroscopy of Biological Systems*, Vol. 13 (R. J. H. Clark and R. E. Hester, eds.), John Wiley and Sons, Chichester, pp. 1–46.

Markovich, R. J., and Pidgeon, C., 1991, Introduction to Fourier transform infrared spectroscopy and applications in the pharmaceutical sciences, *Pharm. Res.* **8:**663–675.

Mitchell, R. C., Haris, P. I., Fallowfield, C., Keeling, D. J., and Chapman, D., 1988, Fourier transform infrared spectroscopic studies on gastric H^+/K^PL-ATPase, *Biochim. Biophys. Acta* **941:**31–38.

Miyazawa, T., 1960, Perturbation treatment of the characteristic vibrations of polypeptide chains in various configurations, *J. Chem. Phys.* **32:**1647–1652.

Miyazawa, T., and Blout, E. R., 1961, The infrared spectra of polypeptides in various conformations: Amide I and II bands, *J. Am. Chem. Soc.* **83:**712–719.

Muga, A., Surewicz, W. K., Wong, P. T. T., Mantsch, H. H., Singh, V. K., and Shinohara, T., 1990, Structural studies with the uveopathogenic peptide M derived from retinal S-antigen, *Biochemistry* **29:**2925–2930.

Naik, V. M., and Krimm, S., 1986a, Vibrational analysis of the structure of gramicidin A. I. Normal mode analysis, *Biophys. J.* **49:**1131–1145.

Naik, V. M., and Krimm, S., 1986b, Vibrational analysis of the structure of gramicidin A. II. Vibrational spectra, *Biophys. J.* **49:**1147–1154.

Nevskaya, N. A., and Chirgadze, Y. N., 1976, Infrared spectra and resonance interactions of amide-I and II vibrations of α-helix, *Biopolymers* **15:**637–648.

Olinger, J. M., Hill, D. M., Jakobsen, R. J., and Brody, R. S., 1986, Fourier transform infrared studies of ribonuclease in H_2O and ^2H_2O solutions, *Biochim. Biophys. Acta* **869:**89–98.

Painter, P. C., Coleman, M. M., and Koenig, J. L., 1982, *The Theory of Vibrational Spectroscopy and Its Application to Polymeric Materials*, John Wiley & Sons, New York.

Parker, F. S., 1983, *Applications of Infrared, Raman, and Resonance Raman Spectroscopy in Biochemistry*, Plenum Press, New York.

Perkins, S. J., Nealis, A. S., Haris, P. I., Chapman, D., Goundis, D., and Reid, K. B. M., 1989, Secondary structure in properdin of the complement cascade and related proteins: A study by Fourier transform infrared spectroscopy, *Biochemistry* **28:**7176–7182.

Pitt, W. G., Spiegelberg, S. H., and Cooper, S. L., 1987, Adsorption of fibronectin to polyurethane surfaces: Fourier transform infrared spectroscopic studies, in: *Proteins at Interfaces: Physiochemical and Biochemical Studies* (J. A. Brash and T. A. Horbett, eds.), American Chemical Society, Washington, D.C., pp. 324–338.

Powell, J. R., Wasacz, F. M., and Jakobsen, R. J., 1986, An algorithm for the reproducible spectral subtraction of water from the FT-IR spectra of proteins in dilute solutions and adsorbed monolayers, *Appl. Spectrosc.* **40:**339–344.

Prestrelski, S. J., Byler, D. M., and Thompson, M. P., 1991, Effect of metal ion binding on the secondary structure of bovine α-lactalbumin as examined by infrared spectroscopy, *Biochemistry* **30:**8797–8804.

Purcell, J. M., and Susi, H., 1984, Solvent denaturation of proteins as observed by resolution-enhanced Fourier transform infrared spectroscopy, *J. Biochem. Biophys. Methods* **9:**193–199.

Rath, P., Bousché, O., Merrill, A. R., Cramer, W. A., and Rothschild, K. J., 1991, Fourier transform infrared evidence for a predominantly alpha-helical structure of the membrane bound channel forming COOH-terminal peptide of colicin El, *Biophys. J.* **59:**516–522.

Robinson, F. P., and Vinogradov, S. N., 1964, Infrared attenuated total reflection spectra of aqueous solutions of some amino acids, *Appl. Spectrosc.* **18:**62–63.

Rothschild, K. J., and Clark, N. A., 1979, Polarized infrared spectroscopy of oriented purple membrane, *Biophys. J.* **25:**473–488.

Rothschild, K. J., He, Y.-W., Mogi, T., Marti, T., Stern, L. J., and Khorana, H. G, 1990, Vibrational spectroscopy of bacteriorhodopsin mutants: Evidence for the interaction of proline-186 with the retinylidene chromophore, *Biochemistry* **29:**5954–5960.

Sarver, R. W., Jr., and Krueger, W. C., 1991, Protein secondary structure from Fourier transform infrared spectroscopy: A data base analysis, *Anal. Biochem.* **194:**89–100.

Surewicz, W. K., and Mantsch, H. H., 1988, New insight into protein secondary structure from resolution-enhanced infrared spectra, *Biochim. Biophys. Acta* **952:**115–130.

Surewicz, W. K., Mantsch, H. H., Stahl, G. L., and Epand, R. M., 1987a, Infared spectroscopic evidence of conformational transitions of an atrial natriuretic peptide, *Proc. Natl. Acad. Sci. USA* **84:**7028–7030.

Surewicz, W. K., Moscarello, M. A., and Mantsch, H. H., 1987b, Fourier transform infrared spectroscopic investigation of the interaction between myelin basic protein and dimyristoylphosphatidylglycerol bilayers, *Biochemistry* **26:**3881–3886.

Surewicz, W. K., Moscarello, M. A., and Mantsch, H. H., 1987c, Secondary structure of the hydrophobic myelin protein in a lipid environment as determined by Fourier-transform infrared spectrometry, *J. Biol. Chem.* **262:**8598–8602.

Surewicz, W. K., Szabo, A. G., and Mantsch, H. H., 1987d, Conformational properties of azurin in solution as determined from resolution-enhanced fourier-transform spectra, *Eur. J. Biochem.* **167:**519–523.

Surewicz, W. K., Stepanik, T. M., Szabo, A. G., and Mantsch, H. H., 1988, Lipid-induced changes in the secondary structure of snake venom cardiotoxins, *J. Biol. Chem.* **263:** 786–790.

Surewicz, W. K., Leddy, J. J., and Mantsch, H. H., 1990, Structure, stability, and receptor interaction of cholera toxin as studied by Fourier-transform infrared spectroscopy, *Biochemistry* **29:**8106–8111.

Surewicz, W. K., Mantsch, H. H., and Chapman, D., 1993, Determination of protein secondary structure by Fourier transform infrared spectroscopy: A critical assessment, *Biochemistry* **32:**389–394.

Susi, H., 1969, Infrared spectra of biological macromolecules and related systems, in: *Structure and Stability of Biological Macromolecules* (S. N. Timasheff and G. D. Fasman, eds.), Marcel Dekker, New York, pp. 575–663.

Susi, H., and Byler, D. M., 1986, Resolution-enhanced Fourier transform infrared spectroscopy of enzymes, *Methods Enzymol.* **130:**290–311.

Susi, H., Timasheff, S. N., and Stevens, L., 1967, Infrared spectra and protein conformations in aqueous solutions I. The amide I band in H_2O and D_2O solutions, *J. Biol. Chem.* **242:**5460–5466.

Swedberg, S. A., Pesek, J. J., and Fink, A. L., 1990, Attenuated total reflectance Fourier transform infrared analysis of an acyl-enzyme intermediate of α-chymotrypsin, *Anal. Biochem.* **186:**153–158.

Trewhella, J., Liddle, W. K., Heidorn, D. B., and Strynadka, N., 1989, Calmodulin and troponin C structures studied by Fourier transform infrared spectroscopy: Effects of Ca^{2+} and Mg^{2+} binding, *Biochemistry* **28:**1294–1301.

Venyaminov, S. Y., and Kalnin, N. N., 1990, Quantitative IR spectrophotometry of peptide compounds in water (H_2O) solutions. I. Spectral parameters of amino acid residue absorption bands, *Biopolymers* **30:**1243–1257.

Venyaminov, S. Y., Rajnavölgyi, É., Medgyesi, G. A., Gergely, J., and Závodszky, P., 1976, The role of interchain disulphide bridges in the conformational stability of human immunoglobulin G1 subclass. Hydrogen–deuterium exchange studies, *Eur. J. Biochem.* **67:**81–86.

Villalain, J, Gomez-Fernandez, J. C., Jackson, M., and Chapman, D., 1989, Fourier transform infrared spectroscopic studies on the secondary structure of the Ca^{2+}-ATPase of sarcoplasmic reticulum, *Biochim. Biophys. Acta* **978:**305–312.

Wantyghem, J., Baron, M.-H., Picquart, M., and Lavialle, F., 1990, Conformational changes of *Robinia pseudoacacia* lectin related to modifications of the environment: FTIR investigation, *Biochemistry* **29:**6600–6609.

Wasacz, F. M., Olinger, J. M., and Jakobsen, R. J., 1987, Fourier transform infrared studies of proteins using nonaqueous solvents. Effects of methanol and ethylene glycol on albumin and immunoglobulin G, *Biochemistry* **26:**1464–1470.

Wei, J., Lin, Y.-Z., Zhou, J.-M., and Tsou, C.-L., 1991, FTIR studies of secondary structures of bovine insulin and its derivatives, *Biochim. Biophys. Acta* **1080:**29–33.

Wei, J., Xie, L., Lin, Y.-Z., and Tsou, C.-L., 1992, The pairing of the separated A and B chains of insulin and its derivatives, FTIR studies, *Biochim. Biopys. Acta* **1120:**69–74.

White, J. U., and Ward, W. M., 1965, Effects of interference fringes in infrared absorption cells, *Anal. Chem.* **37:**268–270.

Wong, P. T. T., and Heremans, K., 1988, Pressure effects on protein secondary structure and hydrogen–deuterium exchange in chymotrypsinogen: A Fourier transform infrared spectroscopic study, *Biochim. Biophys. Acta* **956:**1–9.

Yang, P. W., Mantsch, H. H., Arrondo, J. L. R., Saint-Girons, I., Guillou, Y., Cohen, G. N., and Barzu, O., 1987, Fourier transform infrared investigation of the *Escherichia coli* methionine aporepressor, *Biochemistry* **26:**2706–2711.

Yang, W.-J., Griffiths, P. R., Byler, D. M., and Susi, H., 1985, Protein conformation by infrared spectroscopy: Resolution enhancement by Fourier self-deconvolution, *Appl. Spectrosc.* **39:**282–287.

Young, B. R., Pitt, W. G., and Cooper, S. L., 1988, Protein adsorption on polymeric biomaterials II. Adsorption kinetics, *J. Colloid Interface Sci.* **125:**246–260.

4

Mass Spectrometry in Protein Structural Analysis

Peter Roepstorff

1. INTRODUCTION TO MASS SPECTROMETRY OF PROTEINS

Mass spectrometric protein analysis was pioneered in the 1960s and early 1970s by K. Biemann and co-workers at MIT, E. Lederer's group at the Institute for Natural Product Chemistry in Gif sur Yvette, France, in collaboration with M. Barber at AEI, and at the Shemyakin Institute for Natural Product Chemistry in Moscow. The mass spectrometric techniques at that time required volatile samples that, because of the zwitter ionic nature of peptides, required derivatization. Ionization was performed by electron impact, resulting in high excitation of the formed ions and subsequent extensive fragmentation. As a result, only rather small peptides were amenable to mass spectrometric analysis. The development of new ionization methods in the 1970s such as chemical ionization and field desorption, although promising, did not improve the perspectives for mass spectrometric protein analysis. In spite of these limitations, mass spectrometric research, carried out in a rather limited number of groups, made significant contributions to protein research, especially in characterization of posttranslationally modified amino acid residues and sequencing of N-terminally blocked peptides (reviewed, for example, in Arpino and McLafferty, 1976).

In 1974, a new mass spectrometric method, plasma desorption mass spectrometry (PDMS), was introduced by Macfarlane and co-workers (Torgerson

Peter Roepstorff • Department of Molecular Biology, Odense University, DK-5230 Odense M, Denmark.

Physical Methods to Characterize Pharmaceutical Proteins, edited by James N. Herron *et al.*, Plenum Press, New York, 1995.

et al., 1974). PDMS was based on simultaneous desorption and ionization of the molecules from the solid state by bombardment with high energy (MeV) primary ions. It was quickly demonstrated to allow analysis of underivatized peptides, and a real breakthrough for the application in protein chemistry came after observation for the first time of molecular ions of insulin (Håkansson *et al.*, 1982) and commercialization of the instrument in 1984.

Independently, Barber and co-workers in 1981 discovered a new ionization technique based on bombardment with neutral atoms (e.g., Ar or Xe) of the analyte molecules dissolved in a suitable matrix (Barber *et al.*, 1981). The method was termed fast atom bombardment mass spectrometry (FABMS). It has later been demonstrated that the neutral atoms could be replaced by Cs^+ ions. Many authors in this case called the method liquid secondary ion mass spectrometry (LSIMS). For convenience, the term FABMS will be used in this chapter independently of the use of neutral atoms or ions as primary particles. FABMS was quickly demonstrated to be a very soft ionization method that, like PDMS, allowed observation of molecular ions of insulin (Dell and Morris, 1982; Barber *et al.*, 1982).

FABMS was quickly introduced in many mass spectrometry facilities because this methodology could be installed by adapting existing sector and quadrupole mass spectrometers. PDMS commercial instruments were not immediately available. But, with a little delay, PDMS gained footing in a number of biochemical laboratories due to its simplicity and relatively low price. In the last half of the 1980s, both techniques paved the way for the general use of mass spectrometry in protein chemistry (for recent reviews, see, for example, Desiderio, 1991). However, the use of mass spectrometry also led to demands for better performance, i.e., increased mass range, sensitivity, and mass accuracy. The answer to these demands came in the summer of 1988, when two new mass spectrometric methods were demonstrated to fulfill these demands. Thus, molecular ions of proteins in excess of 200 kDa were obtained with matrix-assisted laser desorption ionization mass spectrometry (MALDIMS) on a time-of-flight mass spectrometer (Karas and Hillenkamp, 1988) and series of highly charged molecular ions was demonstrated to be formed upon electrospray ionization mass spectrometry (ESIMS) of proteins (Fenn *et al.*, 1989). The high charge state obtained with the latter technique eliminated the need for instruments with high mass range and the first ESI spectra of proteins were indeed recorded on a 1600 mass range quadrupole instrument.

At present, mass spectrometry must be considered an established technique in protein chemistry. There is a selection of techniques available including FABMS, PDMS, MALDIMS, and ESIMS. Each of these techniques have their advantages and shortcomings. This chapter describes the principles of the different techniques with special emphasis on their relative strengths and weaknesses for practical protein studies. It further illustrates the type of information that can be obtained by mass spectrometry and a number of examples of the use of this information in practical protein studies.

2. THE CONTEMPORARY MASS SPECTROMETRIC TECHNIQUES

Mass spectrometry in principle include three steps: (1) the ionization step, i.e., formation of gas phase ions of the analyte; (2) separation of the formed ions according to their mass to charge ratio (m/z) in the analyzer; and (3) detection of the formed ions after separation. In contemporary mass spectrometry a wealth of different techniques is available for all three steps. In this chapter only the first two steps will be dealt with. A further restriction is that only techniques relevant for the application to protein studies will be considered. The requirements of the technique to sample conditions such as purity and concentration, its tolerance to the presence of contaminants and salts, as well as its compatibility with procedures traditionally used in biochemistry are also important for its practical use and will be dealt with in this section.

2.1. Plasma Desorption Mass Spectrometry

PDMS is a desorption method, i.e., the secondary gaseous ions are generated in a sputtering process by bombardment with primary particles. In PDMS the primary particles are highly charged heavy ions with energies in the megavolt range. These ions are formed in commercially available mass spectrometers by spontaneous fission of ^{252}Cf placed behind or in front of the sample. The principle of the instrument is illustrated in Fig. 1. The sample is deposited on a thin 0.5- to 1-μm thick aluminized polyester foil. In the fission event two colinear fission fragments are formed. One hits the start detector and triggers the time measurement, whereas the other penetrates the sample and causes desorption of a number of secondary ions. These are accelerated between the sample foil at 10- to 20-keV potential and the acceleration grid at ground potential and allowed to drift through the field free-flight tube to the stop detector. The flight times are measured by the time-to-digital converter. Each fission event only results in formation of a few ions. Therefore, in order to obtain sufficient ion statistics, spectra are accumulated for a large number of fission events, typically 5×10^5–10^7. The lower number, being the most frequently used, corresponds to 3- to 8-min accumulation time depending on the age of the californium source. The flight time T is correlated with the m/z by the equation $T = a(m/z)^{1/2} + b$, where a and b are instrument-dependent constants. Thus, an internal mass calibration can be performed based on two known peaks in the spectrum (typically H^+ and NO^+ or Na^+).

A specific advantage of PDMS for protein analysis is the possibility to use a nitrocellulose support. The presence of salts is frequently ruinous for PDMS. To prepare salt-free samples, Jonsson *et al.* (1986) introduced a procedure involving adsorption of the protein sample to a thin layer of nitrocellulose deposited on the same foil followed by removal of the salt by washing with ultrapure water. This

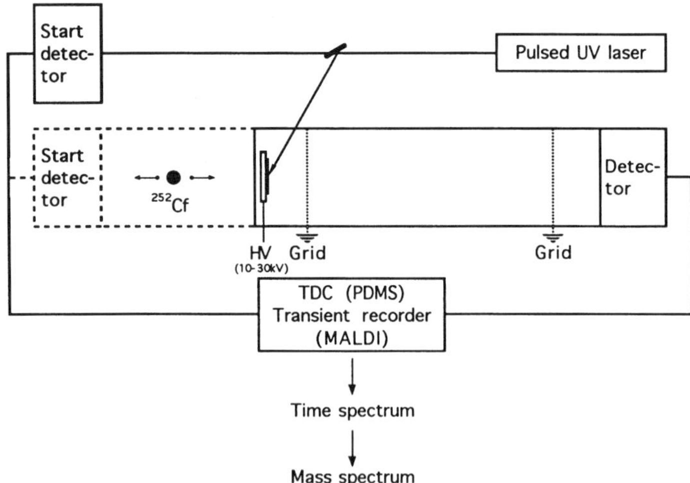

Figure 1. Principle of the time-of-flight mass spectrometer used for PDMS and MALDIMS. The sample is placed at high voltage (HV). As a PD-instrument, the components drawn with a dotted line are needed, while the laser is omitted.

procedure allows application of samples from salt-containing buffers, urea solutions, or directly from reaction mixtures. The use of nitrocellulose has several other advantages such as increased sensitivity, and it offers the possibility for subsequent execution of *in situ* reactions (e.g., reduction of disulfide bonds and enzymatic cleavages) and extraction or transfer of the sample to, for example, a polyvinyl difluoride (PVDF) membrane used in automatic sequencing (reviewed by Roepstorff *et al.*, 1990) because only a very small part of the sample is consumed during recording of a spectrum. The use of nitrocellulose supports is now generally accepted for PDMS of proteins because of its compatibility with a large number of relevant biochemical procedures and suitable protein solvents. An experimental protocol for preparation of the nitrocellulose support and for execution of different *in situ* reactions has been published by Roepstorff (1990).

2.2. Fast Atom Bombardment Mass Spectrometry

In FABMS, the sample is dissolved in an appropriate liquid matrix. Desorption and ionization are induced by bombardment with 8–10 keV Ar or Xe atoms or 20–40 keV Cs^+ ions. The atom or ion current is continuous. For a solid sample this would result in rapid destruction of the surface with concomitant reduction of the secondary ion yield. By using a liquid matrix in which the analyte has a certain surface activity, the surface is continuously replenished, resulting in a continuous production of secondary ions. Therefore, FAB is well suited for scanning mass analyzers such as magnetic sectors and quadrupoles and is now available for most instruments of this type. For analysis of large peptides and proteins it has been necessary to extend the mass range of the analyzers (>10,000 for magnetic sectors and 4,000 for quadrupoles are presently available), which unfortunately also results in high instrument costs. The wide mass range has also created a need for improved calibration procedures. Although calibration is usually performed in an external calibration protocol with a known compound or compound mixture, the best mass accuracy is obtained with an internal calibration compound. FAB is a soft ionization method and only few fragment ions are formed. It is, however, possible to enhance fragmentation by collision-induced dissociation, i.e., to add internal energy to a selected ion by collision with neutral gas molecules, followed by analysis of the fragment ions. This has led to the development of tandem mass spectrometry (MS/MS) using either large four sector instruments or triple quadrupole mass spectrometers (for recent reviews, see Biemann, 1990, and Hunt *et al.*, 1990, respectively).

Several matrices have been described. The most frequently used for peptide analysis are glycerol, thioglycerol, and a mixture of dithiothreitol and dithioerythritol often containing traces of trifluoro- or trichloroacetic acid to enhance positive ion formation. The secondary ion yield depends on the surface activity of the sample in the matrix. This, unfortunately, may lead to pronounced suppression effects when analyzing mixtures. An alternative to direct mixture analysis is to use a hyphenated system, in the case of peptide analysis, combined liquid chromatography and mass spectrometry (LCMS). Due to the liquid matrix, FAB is well suited for LCMS using continuous flow FAB in which the effluent from the chromatograph is led through a capillary directly to the tip of the FAB probe (reviewed by Caprioli and Suter, 1992). The flow rate is limited to 2–10 μl/min because it must correspond with the removal of solvent by the combined evaporation and sputtering processes. In general, FABMS requires salt-free samples, and highly purified samples that are needed as an *in situ* purification like in PDMS is not possible.

2.3. Matrix-Assisted Laser Desorption Ionization Mass Spectrometry

MALDIMS was first described at the International Mass Spectrometry Conference in 1988 (Hillenkamp, 1989) where spectra showing molecular ions of proteins up to 200 kDa were presented. The principle is an extension of laser desorption mass spectrometry in which infrared lasers are used to produce rapid heating of solid samples. In the original MALDI experiments, protein solutions were mixed with a 100- to 1000-fold molar excess of nicotinic acid and dried. The solid mixture was then irradiated with a short laser pulse at 260 nm from a frequency-doubled Nd YAK-laser and the desorbed ions analyzed in a time-of-flight mass spectrometer. Surprisingly, high-intensity molecular ions of the proteins were observed. It was suggested that a prerequisite for successful molecular ion formation was that the matrix compound exhibited strong absorption at the laser wavelength. Since then, several alternative matrices have been developed (Table I), many of which have their absorption maximum at 330–350 nm, the wavelength of the frequency-tripled Nd YAK-laser or the much cheaper nitrogen laser (for a review, see Beavis, 1992a). The commonly used laser instruments are time-of-flight instruments either equipped with a straight flight tube like the PDMS instrument (Fig. 1) or with an electrostatic mirror. The ion production upon each laser shot is remarkably high, and frequently excellent spectra can be obtained upon a single laser shot, although 10 to 50 spectra are often added to improve ion statistics. This high ion production has necessitated replacement of the single ion counting time to digital converter used in the plasma desorption instrument by a transient recorder (Fig. 1).

MALDIMS has many attractive features. The size of molecules that can be desorbed seems unlimited, although detection of very large molecules still pre-

Table I. Examples of Matrices Tested for UV-MALDIMS of Peptides and Proteins

	Laser wavelengths[a]	
Matrix	266 nm	337 nm 355 nm
Nicotinic acid	+	–
Pyrazine-2-carboxylic acid	+	–
3-Aminopyrazine-2-carboxylic acid	+	+
Vanillic acid	+	–
Thiourea	+	–
3-Nitrobenzyl alcohol	+	–
Benzoic acid derivatives		
2-Aminobenzoic acid	+	+
2,5-Dihydroxybenzoic acid	+	+
Cinnamic acid derivatives		
Sinapinic acid[b]	+	+
Caffeic acid	+	+
Ferulic acid[b]	+	+
α-Cyano-4-hydroxy cinnamic acid[b]	+	+

[a]266 nm and 355 nm correspond to wavelengths of the frequency-doubled and tripled Nd YAK-laser, respectively; 337 nm corresponds to the wavelength of the N_2-laser.

[b]Matrices preferably used in the author's laboratory.

sents a problem because of their low velocity. The sensitivity is remarkably high. Thus, spectra of as little as 1 fmole of lysozyme have been reported (Karas *et al.*, 1989a). Very little suppression is observed in mixture analysis (Beavis and Chait, 1990). The tolerance to the presence of salts and impurities of various kinds is high, which allows analysis of crude samples of biological origin (Goldberg *et al.*, 1991). This high tolerance has been explained by cocrystallization of the protein with the matrix molecules with concomitant exclusion of impurities from the crystals (Karas *et al.*, 1990). A drawback of MALDI is that the laser irradiance is a very critical parameter. It is important to keep it very close to the desorption threshold to maintain resolution and reduce fragmentation and matrix background (Beavis, 1992b). Calibration is also still not straightforward, and high accuracy requires use of an internal standard similar to the analyte in molecular size and compound class.

2.4. Electrospray Ionization Mass Spectrometry

The three previously described techniques are all based on desorption/ ionization, whereas ESIMS is based on an entirely different ion formation principle. In ESIMS, the analyte is introduced in a solution and the ions formed by desolvation of liquid droplets in a high electrical field in atmospheric pressure. Then the ions are channeled into the high vacuum of the analyzer (Fig. 2) via a series of skimmers and/or a capillary. The principle was first described by Dole *et al.* (1968) and further developed by Fenn and collaborators who in 1988 described that molecular ions with a very high charge state could be obtained with proteins (Fenn *et al.*, 1989). The mechanism of ion formation is not perfectly understood (Mann, 1990). Nevertheless, it is extremely effective and leads to very highly charged states. Thus, application of electrospray ionization to proteins results in a series of multiply charged molecular ions typically in the m/z range 700 to 3000, i.e., one charge per 6 to 30 amino acid residues. In positive ion mass

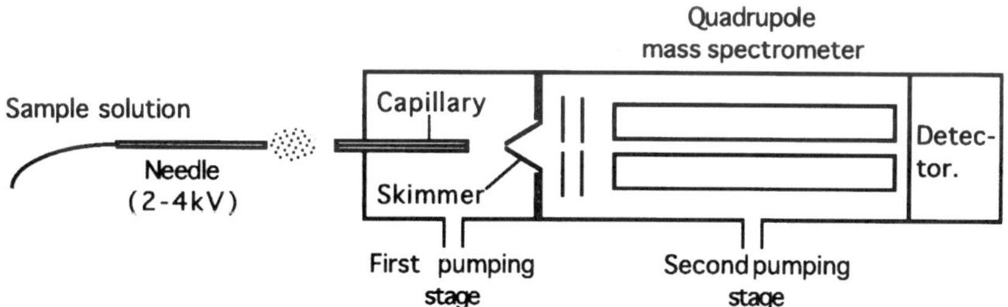

Figure 2. Principle of a single quadrupole electrospray mass spectrometer.

spectrometry, the charged state for a given protein seems to correlate reasonably well with the number of basic residues (Roepstorff *et al.*, 1991).

The formation of a series of highly charged molecular ions directly from a solution of the analyte has several advantages:

1. ESIMS of proteins does not require instruments with high mass range, and the technique consequently can be adapted on standard type instruments. Indeed, most ESIMS studies have been performed on quadrupole instruments, although ESIMS accessories to magnetic sector instruments are also available.
2. The molecular weight of a protein can be determined with high accuracy because it is calculated based on several ions with different charged states. The calibration of the mass spectrometer is also relatively simple because the ions are in an m/z range where standard calibration procedures are applicable.
3. The molecular ions produced have a very low internal energy and fragment ions are rarely observed in ESI-spectra of proteins. However, collision-induced dissociation (CID) is very effective because the highly charged molecular ions fragment due to coulomb repulsive forces even upon a slight energy contribution. CID can be performed either on all ions in the interface between the ESI-source and the mass analyzer or on selected ions in an MS/MS instrument.
4. The ion formation in atmospheric pressure directly from a solution makes ESIMS ideal for coupling with wet separation procedures such as high-performance liquid chromatography (HPLC) or capillary electrophoresis (CE).

The principle of ESIMS requires that the solvents and buffers are volatile. This puts certain restrictions on its compatibility with wet biochemical procedures; it is most suited for analysis of samples that are highly purified in terms of salts and low-molecular-weight contaminants. On the other hand, ESIMS is excellently suited for protein mixture analysis because only little suppression is observed and even minor protein components can be seen in the spectra. Consequently, ESIMS is a viable alternative to traditional procedures to assess the homogeneity of protein preparations.

3. TYPE OF INFORMATION AVAILABLE FROM MASS SPECTRA OF PROTEINS

3.1. Molecular Weight Information

All four mass spectrometric methods described above directly yield molecular ions of peptides and proteins, and thus can be used for molecular weight determination. Their performance in terms of practically achievable mass range, sensitivity, and mass accuracy are listed in Table II. Traditionally, molecular

Table II. Performance of the Mass Spectrometric Methods Used for Protein Studies

Method	Sensitivity limit (pmole)	Practical mass range	Mass accuracy (%)	Commonly used mass analyzer
PD	1–10	20.000	0.05–0.2	Time-of-flight
FAB	1–50	10.000	0.01–0.2	Quadrupole or sector
ESI	0.1–5	>100.000	0.001–0.02	Quadrupole or sector
MALDI	0.001–1	>250.000	0.01–0.2	Time-of-flight

weights of proteins are determined by indirect methods such as gel electrophoresis, gel chromatography, or centrifugation. These methods are, at their best, within 5 and 10%, and much larger deviations are frequently observed due to properties of the protein that are irrelevant to the molecular weight such as hydrophobicity, charge state, and conformation.

The mass spectrometric molecular weight determination depends only on the atomic composition of the molecule. As can be seen from Table II, the molecular weight determination by mass spectrometry is far more precise than that of conventional methods. The sensitivity in the low picomole to femtomole range is adequate and the analysis time short (in the order of minutes). The molecular weight can be obtained for pure compounds as well as for the individual components in peptide or protein mixtures without prior purification. Consequently, the major focus of mass spectrometry in protein studies has been on molecular weight determination.

In principle, all four methods can be used for molecules below 10 kDa. For proteins beyond 10 to 20 kDa, only MALDI and ESI are applicable. In the author's laboratory, all four methods are available and the selection is based on the information required and the ease of sample preparation, instrument operation, and data interpretation. As a consequence, PDMS is the method of choice for analysis of peptides below approximately 5 kDa because PDMS is the easiest to use on routine basis. However, when the sample quantities are below 1–2 pmoles or when unit mass accuracy is needed, the preferred methods are MALDI and ESI, respectively. Purified peptides beyond 5 kDa and proteins are preferably analyzed by ESIMS provided that sufficient amounts (at least a few picomoles) are available. MALDIMS is chosen when sensitivity is a prime demand for analysis of crude protein preparations or for analysis of glycoproteins for which MALDIMS is most likely to yield useful data. FABMS is rarely used for molecular weight determination in our daily work. It may be considered if fragment ion information is needed (see Section 3.2).

Compound-specific suppression may be observed in peptide and protein mixture analysis. The level of suppression is strongly dependent on the mass spectrometric technique used, i.e., FAB > PD > ESI > MALDI. The pronounced suppression in FABMS is due to different surface activity of the components in the liquid matrix (Naylor *et al.*, 1986). In PDMS, the suppression is less pronounced and mainly related to competition for the charge (Nielsen and Roepstorff, 1989) or, if large amounts of sample are applied, caused by competitive binding to the nitrocellulose surface (Wang *et al.*, 1990). ESIMS and MALDI show very little suppression and are both well suited to be used for mixture analysis (Chowdhury *et al.*, 1990; Beavis and Chait, 1990). MALDI is excellent for analysis of very complex mixtures, whereas ESIMS, due to its high dynamic range, can readily disclose the presence of a limited number of minor components.

3.2. Structural Information Based on Fragment Ions

Structure determination based on the presence of fragment ions is traditionally an important application of mass spectrometry. However, the ionization methods used for analysis of underivatized peptides and proteins need to be relatively soft to create sufficient molecular ion abundance. Consequently, too few fragment ions are created to allow complete structure determination. It is possible to overcome this limitation by adding excess energy to the molecular ions to cause them to fragment. Several methods are available for excitation such as bombardment with photons or collision with neutral atoms or surfaces. The most widespread is collision-induced dissociation–mass spectrometry/mass spectrometry (CID-MS/MS) in which the ion of interest is first selected in MS 1 of a multisector instrument, then excited by collision with a neutral gas, and finally the fragment ions are analyzed in MS 2 (Fig. 3).

CID of peptides primarily results in fragmentation of the peptide backbone but also in a number of side chain specific fragments, as listed in Fig 4. The former can be used to assign the sequence, whereas the latter allows distinction between the isomeric amino acid residues Leu and Ile or residues with identical integral residue weight, e.g., Gln and Lys. These latter ions are only observed when CID is performed under high-energy conditions, which normally requires that MS 1 is a magnetic sector mass spectrometer. Protein sequencing using ionization by FAB and CID-MS/MS on a tandem arrangement of two double-focusing mass spectrometers (four-sector mass spectrometers) has been pioneered by Biemann and collaborators (reviewed by Biemann, 1990) and is now pursued in many groups (e.g., Kaur *et al.*, 1990; Carr *et al.*, 1989; Vestling *et al.*, 1990; Stults *et al.*, 1990). The method is applicable to peptides up to a molecular weight of approximately 4000. Above this molecular weight level the energy is not sufficient to break up the molecules.

Peptide sequencing by low-energy CID-MS/MS using triple quadrupole

Configurations:			
triple quadrupole	Q	Q	Q
four sector	EB or BE	field free	EB or BE
hybrid	EB or BE	deceleration (+Q)	Q

Figure 3. Principle of a MS/MS instrument. Different common configurations combining electrostatic (E), magnetic (B), and quadrupole (Q) sectors are shown.

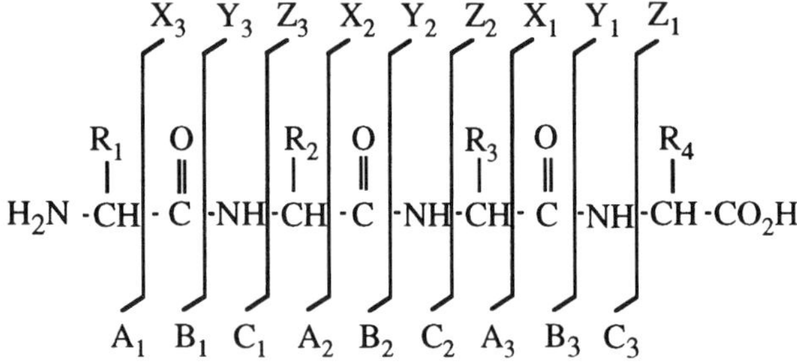

Name		Type of Cleavage	Terminal contained	Assumed structure
A_n^1	a_n^2	Backbone	N	$H\,(NHCHROCO)_{n-1}\ \overset{+}{N}H{=}CHR$
B_n	b_n	Backbone	N	$H\,(NHCHROCO)_{n-1}\ NH\text{-}CHR\text{-}CHR{=}\overset{+}{O}$
C_n''	c_{n+2}	Backbone	N	$[H(NHCHRCO)_n NH_2]H+$
X_n	x_n	Backbone	C	$\overset{+}{O}{=}C\,(NHCHRCO)_n OH$
Y_n''	y_{n+2}	Backbone	C	$[H(NHCHRCO)_n OH]H+$
Z_n''	z_{n+2}	Backbone	C	$[CHR\text{-}CO(NHCHRCO)_{n-1}\ OH]H+$
	d_n	Sidechain	N	$[H(NHCHRCO)_{n-1}\ NH\text{-}CH{=}CHR']H+$
	v_n	Sidechain	C	$[HN{=}CH\text{-}CO\,(NHCHRCO)_{n-1}\ OH]H+$
	w_n	Sidechain	C	$[R'CH{=}CH\text{-}CO\,(NHCHRCO)_{n-1}\ OH]H+$

Figure 4. Common types of fragment ions observed by CID of linear peptides. Nomenclature according to Roepstorff and Fohlmann (1984)[1] and Biemann (1988)[2]. R and R′ represent a side chain and the β-substituent of a side chain, respectively.

mass spectrometers in which the first quadrupole, Q1, is used for ion selection, Q2 is the collision cell, and Q3 used for analysis of fragment ions has been pioneered by Hunt and collaborators (reviewed by Hunt *et al.*, 1990). The spectra obtained with this method are simpler and easier to interpret than those obtained with high-energy CID, but also contain less information. For instance, distinction between Ile and Leu, as mentioned above, is not possible. Recently, it has been demonstrated that electrospray ionization of peptides followed by low-energy CID results in effective fragmentation. Moreover, due to the tension created by the highly charged state of the ions these fragment ions can also be obtained for large peptides and proteins (Smith *et al.*, 1990). A special feature of ESIMS is that CID can be generated in the electrospray interface simply by changing the interface voltages and the fragment ion spectra recorded using a single quadrupole instrument (Smith *et al.*, 1990; Katta *et al.*, 1991). The practical use of this feature requires that only one compound is present at any given time because the interpretation of CID spectra of mixtures would be too difficult.

3.3. Information on Noncovalent Structure and Interaction

Mass spectrometry is not considered to be a method that can yield information about the noncovalent structure of proteins. This is also true for PDMS and FABMS. Surprisingly, ESI and MALDI under certain conditions have been demonstrated to yield information about noncovalently interacting molecules and the conformational state of the molecules.

One of the first MALDI spectra of proteins was that of glucose isomerase containing four noncovalently linked identical subunits. The spectra were unexpectedly dominated by the molecular ion of the tetramer (Karas *et al.*, 1989b). Later the same group investigated the interaction of a peptide with two different lipases. The spectra clearly showed that the peptide interacted with one but not the other lipase (Karas *et al.*, 1990). An attempt to observe the molecular ion of the hemoglobin molecule, which consists of two α- and two β-chains, was not successful (Hillenkamp *et al.*, 1990). In all these examples nicotinic acid was used as matrix. With the now more commonly used matrices, sinnapinic acid and dihydroxybenzoic acid, it has not been possible to observe ions for noncovalently interacting molecules. Little is known about the denaturing properties of the compounds used as matrices in MALDI, but it seems conceivable that a precondition for observation of a noncovalent complex is that it is stable in the matrix solution.

The conformational state of the molecule also affects the charged state of the ions observed in ESIMS. Thus, the ions obtained for proteins in the native state in general contain less charges than molecules in their denatured state (Loo *et al.*, 1991; Katta and Chait, 1991a). Using ESIMS, Katta and Chait (1991a) also demonstrated different hydrogen exchange rates in the denatured and the native state by dissolving the protein in D_2O prior to introduction into the mass spectrometer. Both these observations reflect the state of the molecule in the solution prior to the formation of gaseous ions and do not implicate that the conformation is retained in the gas phase. However, recent observations strongly support that this is the case. McLafferty (1992) has observed that hydrogen exchange experiments can be performed in the gas phase and that here the exchange rate also shows a dependency on the conformation. Several recent studies show that ions can be observed for complexes containing noncovalently linked molecules, the first example being the observation of the intact myoglobin molecule including the noncovalently bound heme group (Katta and Chait, 1991b). These observations only seem possible if the conformation does not change in the transition from solvated molecules to gaseous ions.

The concept of using mass spectrometry to obtain conformational information is still young and future studies will show if it is generally applicable.

4. EXAMPLES OF APPLICATIONS OF MASS SPECTROMETRY TO PROTEIN STUDIES

The examples described in this section do not pretend to comprehensively review the vast literature on the application of mass spectrometry to protein studies. They merely attempt to illustrate possible ways to include mass spectrometry in protein research. The majority of the examples are from the author's laboratory where PDMS now has been used for a decade. This method therefore tends to dominate the review, but, for most of the molecular weight determinations, any of the mass spectrometric techniques described previously might have been used.

4.1. Molecular Weight Determination of Intact Proteins

Determination of the molecular weight of an intact protein primarily serves to specifically characterize the protein. The presently achievable mass accuracy is in most cases sufficient for unique identification of a protein, including its differently posttranslationally modified variants. Molecular weight determination of proteins above 20–30 kDa is only possible with MALDIMS or ESIMS, whereas, in principle, all methods described above can be used for small proteins.

In a comparative study, a number of parvalbumins (11–13 kDa) isolated from different species of fish were analyzed by the different methods. It was found that the molecular weight could be determined with the samples as received using MALDIMS and ESIMS and after further purification also by PDMS, whereas FABMS failed. MALDIMS and ESIMS were preferable in terms of sensitivity and mass accuracy, and ESIMS had the additional advantage that minor components including disulfide-linked dimers were readily detected (Roepstorff *et al.*, 1991). Most species contain several different variants of parvalbumins and often the identification of the specific variants is not well defined because different criteria are used in the literature. Moreover, most parvalbumins are N-terminally blocked. The blocking group is generally presumed to be an acetyl group, but for many of the sequenced parvalbumins this had never been proved. Based on the measured molecular weights it was possible to identify and unambiguously characterize the variants in a given species. The already-sequenced parvalbumins could be identified and the presumed N-terminal acetylation could be verified. In one case (the major component from pike) the sequence was corrected.

Characterization of recombinant proteins is another important field of application. In studies of the acyl-CoA-binding protein (ACBP) (Mikkelsen *et al.*, 1987), the N-terminally acetylated protein from a number of species has been sequenced using the strategy described below. The bovine gene was cloned and expressed in *Escherichia coli* and yeast. Mass spectrometric analysis by ESIMS of the recombinant products confirmed their identity and showed that the N-terminal acetylation did not take place when expressed in these organisms. ^{15}N-labeled recombinant ACBP was prepared for nuclear magnetic resonance (NMR) studies. In this case the mass spectrometric data confirmed complete ^{15}N-labeling and also showed that the recombinant protein surprisingly became N-terminally acetylated under the growth conditions used for labeling, i.e., [^{15}N]ammonium acetate minimal medium. A number of mutant proteins produced by site-directed mutagenesis was also produced and characterized by ESIMS (Table III). The mass accuracy obtained was found to be sufficient to confirm the identity of all the mutants, even for those deviating with only 1 Da from wild-type ACBP, e.g., a Lys to Glu exchange (Roepstorff *et al.*, 1995).

Table III. Molecular Weight Determinations by ESMS for Confirmation of the Identity of Mutants of Recombinant Bovine Acyl-CoA-Binding Protein[a]

Mutation	Mol. wt. measured	Meas. Δu[b]	Cal. Δu[b]	Deviation (% of mol. wt.)
$Asp^{11}\rightarrow His$	9935.4	22.2	22.05	0.001
$Tyr^{28}\rightarrow Asn$	9864.5	−48.7	−49.07	0.003
$Tyr^{31}\rightarrow Asn$	9864.6	−48.6	−49.07	0.004
$Lys^{52,54}\rightarrow Met$	9919.9	6.7	6.04	0.006
$Tyr^{28}\rightarrow Phe$	9897.0	−16.2	−16.00	0.003
$Lys^{32}\rightarrow Arg$	9941.2	28.0	28.01	0.001
$Lys^{32}\rightarrow Glu$	9914.2	1.0	0.94	0.001

[a]Recombinant acyl-CoA-binding protein calculated molecular weight 9913.2 was used for calibration.

[b]Indicates the mass difference between recombinant acyl-CoA-binding protein and the mutated acyl-CoA-binding protein.

4.2. The Use of Mass Spectrometry in Combination with Automatic Edman Degradation in Protein Sequencing

In our laboratory, the strategy for protein sequence determination has been based for a number of years on a combination of molecular weight determination by PDMS and sequencing by automatic Edman degradation (Roepstorff *et al.*, 1989). The sequence determinations of a number of proteins (e.g., Højrup *et al.*, 1986, 1991; Mikkelsen *et al.*, 1987; Klarskov *et al.*, 1989; Knudsen *et al.*, 1989) have demonstrated the strengths of the strategy for determination of the primary structure of proteins, including N-terminally blocked and modified proteins (e.g., Talbo *et al.*, 1990) and simultaneous sequencing of protein isoforms (e.g., Andreasen *et al.*, 1992). PDMS was chosen as mass spectrometric method in the strategy because of the simplicity in sample preparation and the ease of instrument operation and data interpretation; but it also has shortcomings, especially limited practical mass range and occasionally insufficient mass accuracy and sensitivity. Therefore, the strategy has now been extended to include MALDIMS and ESIMS (Fig. 5) to take advantage of the better performance of these methods (Roepstorff and Højrup, 1992). The first example of the use of the extended strategy is illustrated by the sequence determination of a structural protein (Lm-76) from the locust cuticle (Andersen *et al.*, 1993). It has been further developed and tested in the primary structure determination of a number of different proteins several of which were posttranslationally modified (Egorov *et al.*, 1994; Jespersen *et al.*, 1994; Andersen *et al.*, 1995, Haebel *et al.*, 1995; Krogh *et al.*, 1995).

The sequence determination closely followed the strategy outlined in Fig. 5 with the exception that MALDI at that time was not available in our laboratory. First, the molecular weight was determined to be 13,694.8 by ESIMS (Fig. 6) and the intact protein submitted to 46 cycles of Edman degradation, resulting in unambiguous identification of the sequence of the first 44 amino acid residues. A small sample of the protein was submitted to digestion with trypsin. Aliquots taken out 5, 12, 30, and 60 min after the start of the digestion were separated by HPLC and the fractions analyzed by PDMS (Table IV). In the most recent strategy, we have eliminated the separation step in this procedure and analyzed the mixtures directly by MALDIMS. The high sensitivity and excellent ability for mixture analysis of MALDIMS allow a complete time course analysis to be carried out on as little as 1 pmole of protein in nearly real time.

Based on the results from the time course experiment, it was decided to perform a preparative digest and a larger portion of the protein was digested for 30 min. All peptides were isolated and analyzed by PDMS (Fig. 7A). By comparing the molecular weights of the peptides isolated from the preparative digest with that of the intact protein, it became clear that a small peptide (T4) was missing. Its molecular weight was calculated to be approximately 475 and its existence was

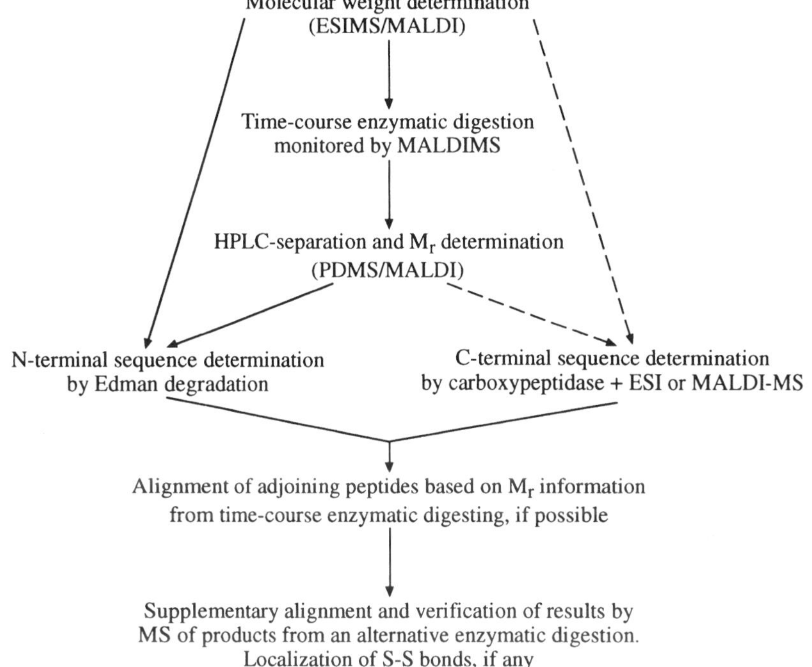

Figure 5. The general strategy for sequence determination of proteins by the combined use of mass spectrometry and Edman degradation (Roepstorff and Højrup, 1992).

confirmed from the molecular weights of three of the peptides identified during time course analysis (Table IV).

Based on their molecular weights two peptides, T1 and T2, could immediately be accounted for in the sequence of the first 44 residues. The order of T3, T4, and T5 could be established from the time course experiment, leaving only the order of the two C-terminal peptides T6 and T7 to be established. According to the amino acid analysis T7 did not contain arginine or lysine. Thus, it was the most probable candidate for being the C-terminal peptide. The sequences of T3, T5, T6, and the first 21 residues of T7 were established by automatic Edman degradation. Digestion of T7 with carboxypeptidase A followed by analysis of the resulting mixture by PDMS identified the sequence of 5 residues from the C-terminus (Fig. 7B). Subdigestion of T7 with chymotrypsin followed by PDMS of the mixture (Fig. 7C) allowed direct identification of the peptides relative to the known part of the sequence. Based on this information the missing part of T7 was isolated and sequenced from a chymotryptic digest of the complete protein. Molecular weight

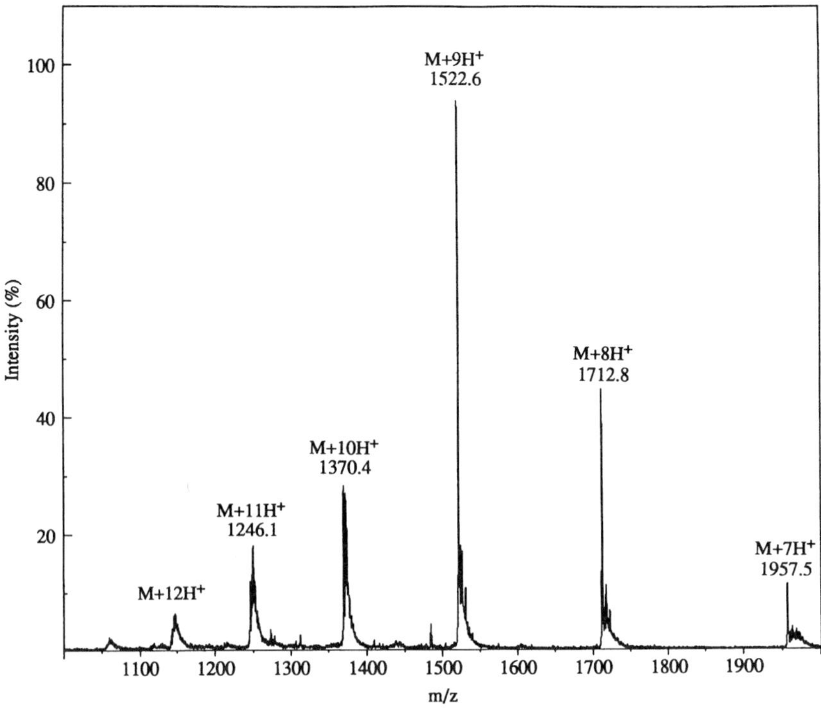

Figure 6. Electrospray spectrum of protein Lm-76. The molecular weight is determined to be 13694.8 based on the series of $(M + nH)^{n+}$ ions. (From Andersen *et al.*, 1992, with permission.)

Table IV. Molecular Weight of Protein Lm-76 as Determined by Electrospray Ionization Mass Spectrometry and Its Tryptic Peptides as Determined by Plasma Desorption Mass Spectrometry[a]

Limit digest tryptic peptides			Time course tryptic peptides		
Peptide	Cal. mol. wt.[b]	Meas. mol. wt.	Peptide	Cal. mol. wt.[b]	Meas. mol. wt.
T1	3107.5	3108.7	T1+T2	4493.0	4495.6
T2	1403.5	1404.1	T2+T3	2568.8	2569.0
T3	1183.3	1184.3	T2+T3+T4	3024.3	3025.0
T4	473.5	ND	T3+T4+T5	4460.0	4457.8
T5	2839.2	2840.2	T4+T5	3294.7	3296.4
T6	1174.4	1173.6	T6+T7	4778.5	4779.6
T7	3622.1	3621.7			
Sum of T1, T2, T3, T4, T5, T6, and T7 − 6 × H_2O					13698.1
Protein Lm-76 as determined by ESMS				13695.4	13694.8

[a]The peptides obtained in the time experiment were analyzed after 5 and 12 min of digestion. The final digest was after 1 hr digestion time. ND, Not determined.

[b]Calculated after sequence determination.

determination of the remaining chymotryptic peptides served to confirm the determined structure.

This protein did not contain disulfide bonds. However, determination of their positions is usually straightforward if the last confirmative digestion is carried out prior to reduction of the disulfide bridges (e.g., Sørensen *et al.*, 1990).

4.3. Direct Sequencing by Mass Spectrometry

The prerequisite for direct sequencing by mass spectrometry is that sequence ions representing cleavage at all the peptide bonds are present in the spectra. CID spectra recorded by CID of MH^+ ions formed by FAB have been found to fulfill this requirement, and a number of proteins has been sequenced solely by mass spectrometry. The following describes the strategy used by Hopper *et al.* (1989) for sequence determination of the N-terminally blocked protein glutaredoxin from rabbit bone marrow by high-energy CID on a four-sector mass spectrometer equipped with a FAB source.

After reduction and alkylation of the disulfide bonds the protein was enzymatically cleaved to produce peptides of a size suitable for MS/MS, i.e., below approximately 2.5 kDa. In principle, it should be possible to perform the MS/MS analysis directly on the obtained peptide mixture. Due to the suppression effects mentioned above, in practice it has been found necessary to submit the mixture to a separation step by HPLC prior to mass spectrometric analysis. The spectra obtained by CID of the selected ^{12}C-only MH^+ ions resulted in series of sequence ions [mainly y and b ions (Fig. 4)] which for more than half of the peptides allowed complete sequence determination and for the others partial sequencing. To fill the gaps and to establish the alignment of the peptides, MS and MS/MS were performed on peptides derived by digestions with four different enzymes. The presence of side-chain fragmentation resulting in w_n and d_n ions allowed direct assignment of 15 out of 19 Leu/Ile in this protein. The two disulfide bonds could be located as described above.

The strategy based on low-energy CID on a triple quadrupole mass spectrometer used by Hunt *et al.* (1989) for sequencing of calbindin isoforms is very similar to the one described above, but instead of selecting the ^{12}C-isotopomer only, a mass window containing the complete isotopic envelope is selected for CID. Widening of the mass window results in considerably increased sensitivity but also in poorer mass accuracy for the fragment ions and consequently more ambiguous interpretation. To facilitate the interpretation and to identify b- or y-series of ions, spectra frequently are recorded before and after derivatization, e.g., *N*-acetylation or methyl esterification. This also solves the Lys/Gln problem because the ϵ-amino group of lysine is acetylated. Distinction between Leu and Ile

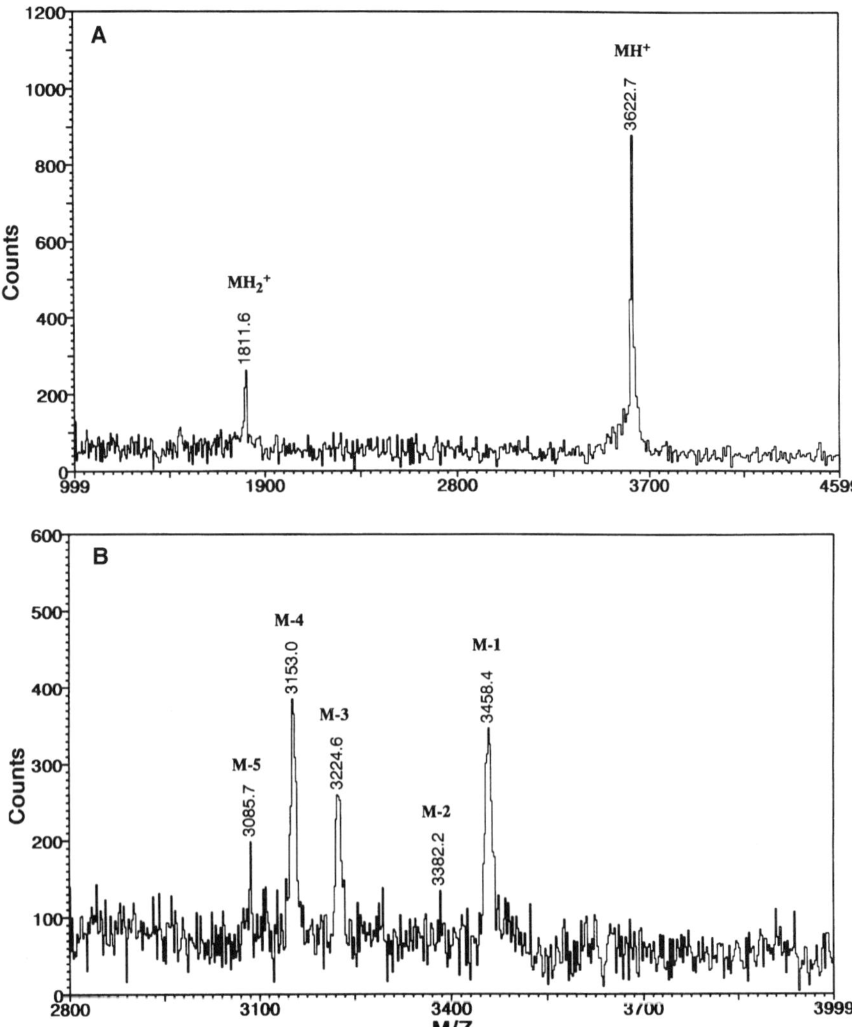

Figure 7. (A) Plasma desorption spectrum of peptide T7 from protein Lm-76. (B) Plasma desorption spectrum of T7 after digestion with carboxypeptidase A. M-1 to M-5 represent the molecular ion after removal of 1 up to 5 amino acid residues from the C-terminus. (C) Plasma desorption spectrum of the peptide mixture obtained by chymotryptic digestion of T7. (From Andersen *et al.*, 1993, with permission.)

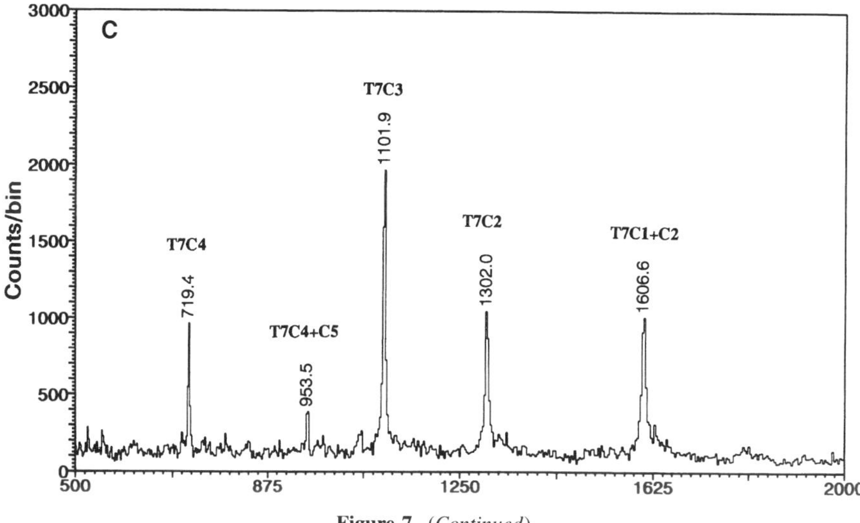

Figure 7. *(Continued)*

is not possible and alternative information such as amino acid analysis data is needed.

4.4. Posttranslationally Modified Proteins

In principle, posttranslational modifications comprise all modifications of the protein after transcription, including processing of the polypeptide chain, formation of disulfide bonds, and formation of a variety of derivatives of the amino acids, e.g., phosphate or sulfate esters, *N*-acyl or -methyl derivatives, and *N*- or *O*-glycosyl derivatives. The two former (processing and disulfide bonds) are considered to be part of the primary structure determination procedure described above. The use of mass spectrometry to elucidate the structure of the latter type of modifications is illustrated by the complete primary structure determination of a locust protein containing three different types of modifications (Talbo *et al.*, 1990).

The strategy for sequence determination essentially followed the early PDMS-based version of the strategy in Fig. 5. PDMS of the intact protein showed that it contained three variants differing in molecular weight by 202 Da. An attempt to perform Edman degradation showed it to be N-terminally blocked. Mass spectrometry of the tryptic peptides located the heterogeneity to one specific peptide (later identified as the C-terminal peptide). By mass spectrometry it was demonstrated that the high-molecular-weight versions of this peptide could be transformed to the low-molecular-weight version by mild acid hydrolysis, indicating that the three peptides contained no, one, and two modifying groups, respectively. Edman degradation of the three peptides showed that one, respectively, two threonyl residues present in the nonmodified peptide could not be identified in the appropriate cycles obtained with the modified peptides. The combined molecular weight and sequencing data indicate that the differences were caused by *O*-glycosylation with an *N*-acetylhexosamine. Carbohydrate analysis after hydrolysis showed the release of *N*-acetyl galactosamine. Another tryptic peptide could be demonstrated to be *O*-glycosylated in all three versions of the protein. The mass spectrometric data further demonstrated the *O*-glycosylation to be sequential, i.e., Thr_{90} was always glycosylated, whereas Thr_{110} was only glycosylated after Thr_{106}.

The mass spectrometry data and failure to perform carboxypeptidase digestion of the C-terminal tryptic peptide indicated that a C-terminal prolyl residue was amidated and not a free acid (1 Da mass difference). This was supported by mass spectrometry of the peptide after methyl esterification and confirmed by HPLC of the isolated proline amide. The combined mass spectrometric and amino acid analysis data suggested that the blocked N-terminus was a pyroglutamyl residue. This was confirmed and the peptide sequenced by PDMS of the permethylated peptide.

4.5. Mapping of Mutants and Isoforms and Interspecies Variation

Characterization of variants of known proteins is a frequent task in our laboratory. The general strategy based on a combination of HPLC and mass spectrometry is outlined in Fig. 8 and exemplified by the determination of hemoglobin mutants (Jensen *et al.*, 1991). The hemoglobin mutants were observed by routine electrophoretic screening at the hematology departments in hospitals and a small blood sample was transferred to our laboratory. At that time we only had access to PDMS which failed to provide information on intact hemoglobin chains. Consequently, at that stage mass spectrometry was omitted. Instead, the site of mutation was located on the α- or β-chain based on retention times of the chains in HPLC and the appropriate chain was isolated. At present the molecular weight of hemoglobin chains can easily be determined by ESIMS (Green *et al.*, 1990) or MALDIMS (Hillenkamp *et al.*, 1990) of intact hemoglobin. The tolerance of MALDIMS even allows analysis of crude hemolysates (Fig. 9). The accuracy achievable by both methods is in most cases sufficient to observe exchange of any single amino acid residue.

The next step is to locate the exact nature and site of mutation. This is achieved by a combined chromatographic and mass spectrometric peptide mapping. Based on the sequence of the natural protein a suitable enzymatic cleavage

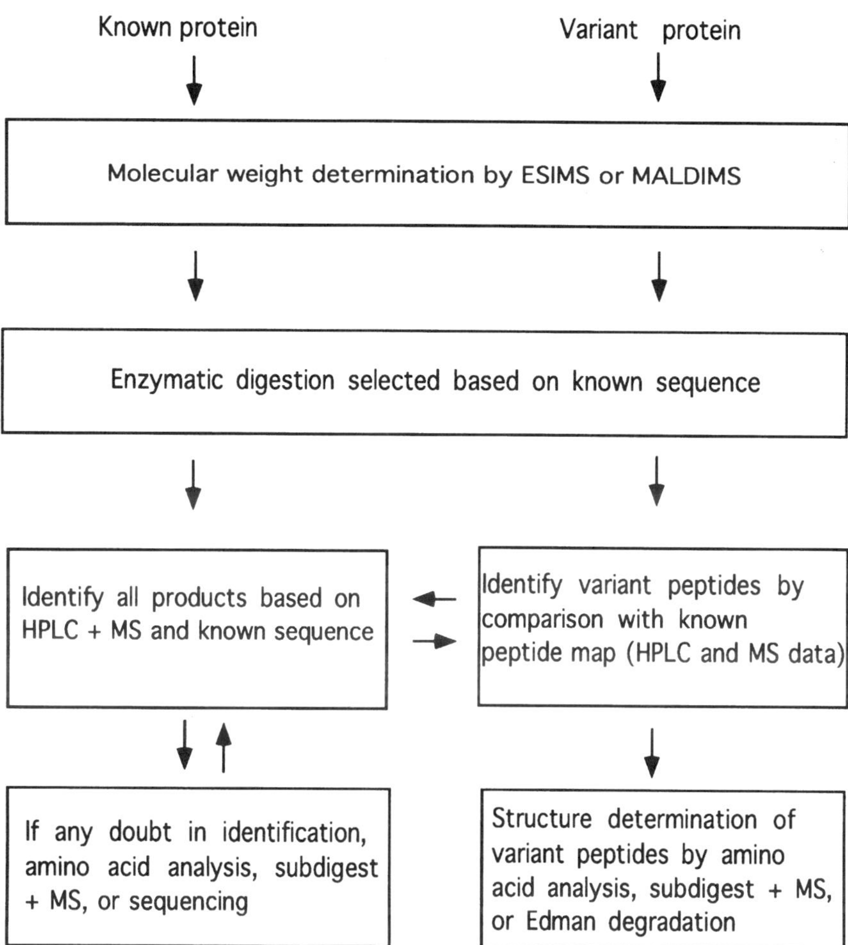

Figure 8. Strategy for analysis of variant proteins using mass spectrometry.

procedure is selected (in the case of hemoglobin, V8-protease for the β-chain and trypsin for the α-chain) and the resulting chromatographic peptide map is recorded. The mutant peptide can be identified and isolated by comparing the maps for the natural and the mutant proteins and its molecular weight as determined by mass spectrometry. The information at this stage is often sufficient to characterize the mutant. If further information is needed, then a subdigest of the mutated peptide followed by mass spectrometry analysis or sequencing either by Edman degradation or MS/MS can resolve possible ambiguities. Once the analytical conditions are established for normal hemoglobin, a complete analysis of a

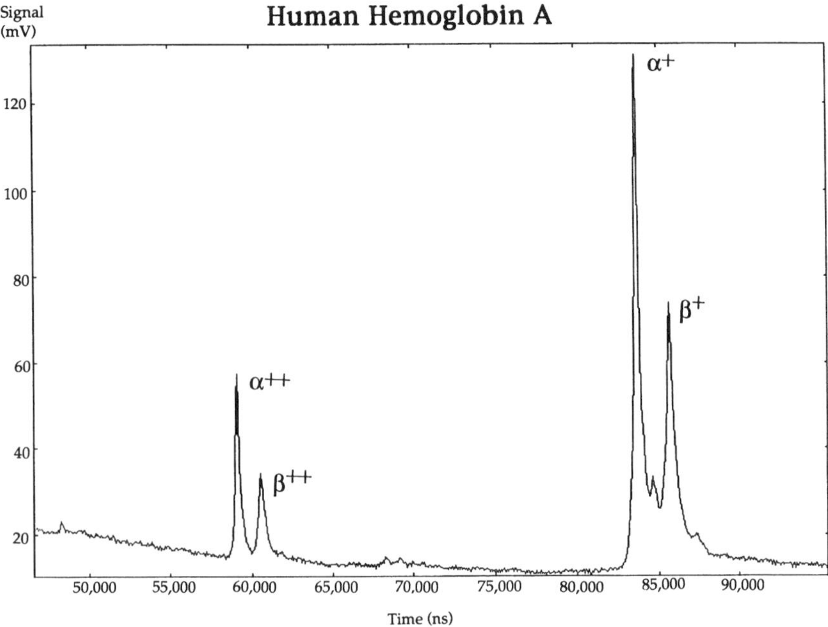

Figure 9. MALDI-spectrum of a blood hemolysate without further purification. The α- and β-chains of hemoglobin A are readily identified.

hemoglobin mutant can be performed in approximately 8 hr using only 15 μl of blood. The advantage of the combined chromatographic and mass spectrometric approach is that mass spectrometrically "silent" mutants, e.g., Leu/Ile, are frequently located in the chromatographic step, whereas chromatographically "silent" mutants can be located in the mass spectrometric procedure.

4.6. Mass Spectrometry Combined with Protein or DNA Sequence Information

The mass spectrometrically determined molecular weight of a protein is often sufficiently accurate to identify the biologically active protein relative to a DNA sequence and also to locate possible posttranslational modifications. This was used for polypeptides isolated in studies of chloroplast gene-coded proteins in photosystem I. The molecular weight, 4066 ± 2, determined by PDMS of an N-terminally blocked polypeptide allowed localization of the polypeptide relative to the sequence of the *psaI* gene and also showed that a methionyl residue inaccessible to CNBr-cleavage was oxidized to the sulfone and that the protein was N-terminally formylated (Scheller *et al.*, 1989). Another protein for which the molecular weight was determined to be 10,821 could be located relative to the *psaE* gene and was found not to be posttranslationally modified (Scheller *et al.*, 1990).

The high mass accuracy presently achievable by mass spectrometry has led us to investigate if it is generally possible to identify a protein relative to protein or DNA sequence data bases. It is our experience that the mass accuracy in most cases is sufficient, provided that the information in the data bases is correct. Unfortunately, this is not always the case. Information about posttranslational processing or modification is often absent and sequence errors occur frequently. A more error-tolerant procedure is to use a mass spectrometric peptide map, i.e., a list of molecular weights of peptides derived by cleavage of the protein with a specific enzyme. It is then possible to search for a limited number of matches between this list and a data base where the protein sequences have been converted into molecular weight lists corresponding to the specific enzymatic cleavage. In our experience, such a procedure is highly tolerant for data base errors and very specific. Sites of sequence errors or posttranslational modification can often be located, and if the investigated protein is not in the data base, it will relate the protein to other known proteins, e.g., the same protein from another species (Mann *et al.*, 1992; Pappin *et al.*, 1993; James *et al.*, 1993). This method is very sensitive and can be used to identify proteins separated by one- and two-dimensional gel electrophoresis after blotting of the protein of interest onto a PVDF membrane and *in situ* digestion (Henzel *et al.*, 1993; Mørtz *et al.*, 1994; Rasmussen *et al.*, 1993).

5. MASS SPECTROMETRY OF PHARMACEUTICAL PROTEINS

Biotechnological and pharmaceutical research groups were some of the first to exploit the potential of mass spectrometry of proteins. The possibility to obtain molecular weight information with hitherto unsurpassed accuracy opened new perspectives for characterization and documentation of proteins irrespective of whether they were extracted from natural sources or prepared by recombinant technology. The first documented pharmaceutically relevant use of mass spectrometry of intact proteins is probably the use of PDMS to follow the preparation of semisynthetic human insulin from porcine insulin by chemical exchange of the C-terminal alanine in the latter residue with a threonine residue. Mass spectrometric documentation was requested from the health authorities prior to release of the product for clinical use. The mass spectra confirmed the expected mass changes between porcine insulin and the intermediate product (B30-des-Ala insulin) and between this product and semisynthetic human insulin and that the molecular weight of the latter was identical with that determined for authentic human insulin (Sundqvist *et al.*, 1984).

Since then, mass spectrometric molecular weight determination of intact proteins combined with peptide mapping have been used in numerous cases to ascertain the identity of recombinant proteins prepared for medical use and their natural counterparts. A typical example is the characterization of natural and recombinant antigen Smp28 from *Schistosoma mansoni*. Schistosomiasis is a parasitic disease affecting 200 million persons worldwide (Cherfas, 1991). A protein Smp28 extracted from the adult worm is a possible candidate for a vaccine against the disease because immunization of rats with this protein demonstrated significant levels of protection against the disease (Balloul *et al.*, 1987). The molecular weight of the natural protein was determined by ESIMS and found to be 56 Da above that predicted from the cDNA sequence. Peptide mapping by LC-ESIMS after digestion with trypsin showed that the N-terminus was acetylated and that one specific methionine residue was oxidized to the sulfoxide. These two modifications accounted for the observed mass difference, and accordingly all other peptide molecular weights were in agreement with those predicted from the cDNA sequence (Bouchon *et al.*, 1994). Peptide mapping of the recombinant protein produced in yeast revealed that the N-terminus in rSmp28 was not acetylated, that five of seven methionines were oxidized, and that one specific cysteine residue was conjugated with mercaptoethanol (Klarskov *et al.*, 1994). The oxidation and conjugation reactions were preparational artifacts and LC-ESIMS must be considered the ideal analytical tool for optimization of the preparative procedures because nearly complete mapping of the protein could be achieved within a few hours using low picomolar quantities.

Many drugs act as antagonists to natural compounds by forming covalent or stable noncovalent complexes with specific proteins. Studies of such complexes to understand the mechanism of action of the drugs are highly relevant in pharmacological research. Mass spectrometry adds new perspectives for studies of covalent complexes and differentiation between covalent and noncovalent complexes, as illustrated, for example, in a recent study of inhibitors of human leucocyte elastase (Knight *et al.*, 1993). ESIMS of the intact enzyme-inhibitor complexes showed that an irreversible inhibitor was covalently bound and the determined molecular weight allowed distinction between two possible inhibition mechanisms, whereas for two reversible inhibitors the inhibitor–enzyme complex was not observed under the conditions used. As mentioned in Section 3.3, it is also possible by carefully choosing the conditions to observe molecular ions of noncovalent complexes between proteins and their ligands. Since ESIMS can be applied to aqueous solutions close to physiological conditions, this method in particular gives promise for studies of structurally specific liquid-phase interactions. Such possibilities open entirely new dimensions for the application of mass spectrometry in pharmaceutical research, but extensive studies are still needed to understand the mechanisms involved and to overcome the many pitfalls (Smith and Light-Wahl, 1993).

6. CONCLUSIVE REMARKS AND FUTURE ASPECTS

In its present state, mass spectrometry is a real alternative to traditional methods in protein chemistry and it opens up the possibility to study certain problems that could not be addressed until now. The mass range covers most known proteins and the mass accuracy far exceeds that of traditional methods for protein molecular weight determination. The sensitivity is in the low picomole to femtomole range, which matches the best sensitivities achieved in traditional procedures. In the author's research group, mass spectrometry has been integrated in protein studies for over a decade and during this period most of the limitations first encountered have been overcome. It is now the first method chosen for routine characterization of any protein or peptide preparation.

What are the future perspectives for mass spectrometry in protein chemistry? I can only express my personal opinion. I feel that protein identification relative to protein or DNA sequence data bases will play an increasingly important role especially with the rapid growth of DNA sequence information as a result of, for example, the Human Genome Project. Mass spectrometric molecular weight determination and peptide mapping perfectly complement the information gained by DNA and cDNA sequencing, and mass spectrometry is likely to become the method of choice for full characterization of proteins sequenced by these techniques. I also believe that the combination of mass spectrometry and Edman degradation for determination of the primary structure of unknown proteins will gain in importance. Until now, direct mass spectrometric sequence determination has contributed very little to sequence determination of complete proteins and probably also in the future will find its main application in the solution of specific problems such as localization of sites of mutation or modification and to sequence determination of N-terminally blocked peptides. The observation of noncovalent interactions by mass spectrometry gives promise for entirely new fields of research and the development in this area needs to be followed closely in the future. Methodologically, I envisage that the sensitivity of the mass spectrometric techniques will be increased by several orders of magnitude within the next few years and also that considerable improvements in mass accuracy are within reach. In the next decade I expect that mass spectrometry will be indispensable in protein research and that access to at least a simple mass spectrometer that permits routine molecular weight determination of proteins will be a prerequisite for all protein research projects.

REFERENCES

Andersen, J. S., Andersen, S. O., Højrup, P., and Roepstorff, P., 1993, Primary structure of a 14 kDa protein (Lm-76) from the cuticle of the migratory locust, *Locusta migratoria*, *Insect Biochem. Molec. Biol.* **23:**391–402.

Andersen, S. O., Rafn, K., Krogh, T. N., Højrup, P., and Roepstorff, P., 1995, Comparison of larval and pupal cuticular protein in *Tenebrio molitor, Insect Biochem. Mol. Biol.* **25:** 177–188.

Andreasen, L., Højrup, P., Andersen, S. O., and Roepstorff, P., 1992, Combined plasma desorption mass spectrometry and Edman degradation applied to simultaneous sequence determination of isoforms of structural proteins from cuticle of *Locusta migratoria*, *Eur. J. Biochem.* **217:**267–273.

Arpino, P. J., and McLafferty, F. W., 1976, Amino acid sequencing of oligopeptides by mass spectrometry, in: *Determination of Organic Structures by Physical Methods*, Vol. 6 (F. C. Nachod, J. J. Zuckerman, and E. W. Randall, eds.), Academic Press, New York, pp. 1–80.

Balloul, J. M., Grzych, J. M., Pierce, R. J., and Capron, A., 1987, A purified 28 kDa protein from *Schistosoma mansoni* adult worms protect rats and mice against experimental schistosomaisis, *J. Immunol.* **138:**3448–3453.

Barber, M., Bordoli, R. S., Sedgwick, R. D., and Tyler, A. N., 1981, Fast atom bombardment of solids (F.A.B.): A new ion source for mass spectrometry, *J. Chem. Soc. Chem. Commun.* **1981:**325–327.

Barber, M., Bordoli, R. S., Elliot, G. J., Sedgwick, R. D., Tyler, A. N., and Green, B. N., 1982, Fast atom bombardment mass spectrometry of bovine insulin and other large peptides, *J. Chem. Soc. Chem. Commun.* **1982:**936–938.

Beavis, R. C., 1992a, Matrix-assisted ultraviolet laser desorption: Evolution and principles, *Org. Mass Spectrom.* **27:**653–659.

Beavis, R. C., 1992b, Phenomenological models for matrix-assisted laser desorption ion yields near the threshold fluence, *Org. Mass Spectrom.* **27:**864–868.

Beavis, R. C., and Chait, B. T., 1990, Rapid sensitive analysis of protein mixtures by mass spectrometry, *Proc. Natl. Acad. Sci. USA* **87:**6873–6877.

Biemann, K., 1988, Contributions of mass spectrometry to peptide and protein structure, *Biomed. Environ. Mass Spectrom.* **16:**99–111.

Biemann, K., 1990, Sequencing of peptides by tandem mass spectrometry and high-energy collision-induced dissociation, in: *Methods in Enzymology*, Vol. 193 (J. A. McCloskey, ed.), Academic Press, San Diego, CA, pp. 455–479.

Boucheron, B., Jacquinod, M., Klarskov, K., Trottein, F., Klein, M., Van Doersselaer, A., Bischoff, R., and Roitsch, C., 1994, Analysis of the primary structure and post-translational modifications of the *Schistosoma mansoni* antigen Smp28, *J. Chromatogr.* **B662:**279–290.

Caprioli, R., and Suter, M. J. F., 1992, Continuous flow fast atom bombardment: Recent advances and applications, *Int. J. Mass Spectrom. Ion Proc.* **118/119:**449–476.

Carr, S. A., Hemling, M. E., Folena-Wasserman, G., Sweet, R. W., Anumula, K., Barr, J. R., Huddleston, M. J., and Taylor, P. J., 1989, Protein and carbohydrate structural analysis of a recombinant soluble CD4 receptor by mass spectrometry, *J. Biol. Chem.* **264:**21286–21295.

Cherfas, F., 1991, New hope for vaccine against schistosomiasis, *Science* **251:**630–631.

Chowdhury, S. K., Katta, V., and Chait, B. T., 1990, An electrospray-ionization with new features, *Rapid Commun. Mass Spectrom.* **4:**81–87.

Dell, A., and Morris, H. R., 1982, Fast atom bombardment-high field magnetic mass spectrometry of 6000 dalton polypeptides, *Biochem. Biophys. Res. Commun.* **106:** 1456–1461.

Desiderio, D. M., 1991, *Mass Spectrometry of Peptides*, CRC Press, Boca Raton, FL.

Dole, M., Mack, L. L., Hines, R. L., Mobley, R. C., Ferguson, L. O., and Alice, M. B., 1968, Molecular beams of macroions, *J. Chem. Phys.* **49:**2240–2249.

Egorov, T. A., Musolyamov, A. Kh., Andersen, J. S., and Roepstorff, P., 1994, The complete amino acid sequence and disulphide bond arrangement of oat alcohol-soluble avenin-3, *Eur. J. Biochem.* **224:**631–638.

Fenn, J. B., Mann, M., Meng, C. K., Wong, S. F., and Whitehouse, C. M., 1989, Electrospray ionization for mass spectrometry of large biomolecules, *Science* **246:**64–71.

Goldberg, D. E., Slater, A. F. G., Beavis, R., Chait, B., Cerami, A., Henderson, G. B., 1991, Hemoglobin degradation in the human malaria pathogen *Plasmodium falciparum*: A catabolic pathway initiated by a specific aspartic protease, *J. Exp. Med.* **173:**961–969.

Green, B. N., Oliver, R. W. A., Falick, A. M., Schackelton, C. M. L., Roitman, E., and Witkanska, H. L., 1990, Electrospray MS, LSIMS and MS/MS for the rapid detection and characterization of variant hemoglonins, in: *Biological Mass Spectrometry* (A. L. Burlingame and J. A. McCloskey, eds.), Elsevier, Amsterdam, pp. 129–146.

Haebel, S., Jensen, C., Andersen, S. O., and Roepstorff, P., 1995, Isoforms of a cuticular protein from larvae of the meal beetle, *Tenebrio moletor*, studied by mass spectrometry in combination with Edman degradation and 2D-PAGE, *Protein Science*, **4:**394–404.

Håkansson, P., Kamensky, I., Sundqvist, B., Fohlman, J., Peterson, P., McNeal, C. J., and Macfarlane, R. D., 1982, 127-I Plasma desorption mass spectrometry of insulin, *J. Am. Chem. Soc.* **104:**2948–2949.

Henzel, W. J., Billeci, T. M., Stults, J. T., Wong, S. C., Grimley, C., and Watanabe, C., 1993, Identifying proteins from two-dimensional gels by molecular mass searching of peptide fragments in protein sequence databases, *Proc. Natl. Acad. Sci. USA* **90:**5011–5015.

Hillenkamp, F., 1989, Laser desorption mass spectrometry: Mechanisms, techniques and applications, in: *Advances in Mass Spectrometry*, Vol. 11A (P. Longevialle, ed.), Heyden and Sons, London, pp. 354–362.

Hillenkamp, F., Karas, M., Ingendoh, A., and Stahl, B., 1990, Matrix assisted UV-laser desorption/ionization: A new approach to mass spectrometry of large biomolecules, in: *Biological Mass Spectrometry* (A. L. Burlingame and J. A. McCloskey, eds.), Elsevier, Amsterdam, pp. 49–58.

Højrup, P., Andersen, S. O., Roepstorff, P., 1986, Primary structure of a structural protein from the cuticle of the migratory locust, *Locusta migratoria*, *Biochem. J.* **236:** 713–720.

Højrup, P., Gerola, P., Mikkelsen, J., Hansen, H. F., El-Shahed, A., Knudsen, J., Roepstorff, P., and Olson, J. M., 1991, The primary structure of the bacteriochlorophyl c binding protein from chlorosomes of *Chlorobium limicola f. thiosulfatophilum*, *Biochem. Biophys. Acta* **1077:**220–224.

Hopper, S., Johnson, R. S., Vath, J. E., and Biemann, K., 1989, Glutaredoxin from rabbit bone marrow, *J. Biol. Chem.* **264:**20438–20447.

Hunt, D. F., Yates III, J. R., Shabanowitz, J., Bruns, M. E., and Bruns, D. E., 1989, Amino acid sequence analysis of two mouse calbindin-D_{9k} isoforms by tandem mass spectrometry, *J. Biol. Chem.* **264:**6580–6586.

Hunt, D. F., Shabanowitz, J., Griffin, P. R., Yates III, J. R., Martino, P. A., and McCormack, A. L., 1990, Protein and oligopeptide sequence analysis by triple quadrupole and quadrupole fourier transform mass spectrometry, in: *Biological Mass Spectrometry* (A. L. Burlingame and J. A. McCloskey, eds.), Elsevier, Amsterdam, pp. 337–362.

James, P., Quadroni, M., Carafoli, E., and Gonnet, G., 1993, Protein identification by mass profile fingerpringing, *Biochem. Biophys. Res. Commun.* **195:**58–64.

Jensen, O. N., Højrup, P., and Roepstorff, P., 1991, Plasma desorption mass spectrometry as a tool in characterization of abnormal proteins. Application to variant human hemoglobins, *Anal. Biochem.* **199:**175–183.

Jespersen, S., Højrup, P., Andersen, S. O., and Roepstorff, P., 1994, The primary structure of an endocuticular protein from two locust species, *Locusta migratoria* and *Schistocerca gregaria*, determined by a combination of mass spectrometry and automatic Edman degradation, *Comp. Biochem. Physiol.* **109B:**125–138.

Jonsson, G. P., Hedin, A. B., Håkansson, P. L., Sundqvist, B. U. R., Säve, G. S., Nielsen, P. F., Roepstorff, P., Johansson, K. E., Kamensky, I., and Lindberg, M. S. L., 1986, Plasma desorption mass spectrometry of peptides and proteins absorbed on nitrocellulose, *Anal. Chem.* **58:**1084–1087.

Karas, M., and Hillenkamp, F., 1988, Laser desorption ionization of proteins with molecular masses exceeding 10,000 daltons, *Anal. Chem.* **60:**2299–2301.

Karas, M., Ingendoh, A., Bahr, U., and Hillenkamp, F., 1989a, Ultraviolet-laser desorption/ionization mass spectrometry of femtomolar amounts of large proteins, *Biomed. Environ. Mass Spectrom.* **18:**841–843.

Karas, M., Bahr, U., Ingendoh, A., and Hillenkamp, F., 1989b, Laser desorption mass spectrometry of mass 100,000 to 250,000, *Angew. Chem. Int.* **28:**760–761.

Karas, M., Bahr, U., Ingendoh, A., Nordhoff, E., Stahl, B., Strupat, K., Hillenkamp, F., 1990, Principles and applications of matrix-assisted UV-laser desorption/ionization mass spectrometry, *Anal. Chim. Acta* **1990:**175–185.

Katta, V., and Chait, B. T., 1991a, Conformational changes in proteins probed by hydrogen-exchange electrospray-ionization mass spectrometry, *Rapid Commun. Mass Spectrom.* **5:**214–217.

Katta, V., and Chait, B. T., 1991b, Observation of the heme-globin complex in native myoglobin by electrospray-ionization mass spectrometry, *J. Am. Chem. Soc.* **113:** 8534–8535.

Katta, V., Chowdhury, S. K., and Chait, B. T., 1991, Use of a single-quadrupole mass spectrometer for collision-induced dissociation studies of multiply charged peptide ions produced by electrospray ionization, *Anal. Chem.* **63:**174–178.

Kaur, S., Medzihradszky, K. F., Yu, Z., Baldwin, M. A., Gillece-Castro, B. L., Walls, F. C., Gibson, B. W., and Burlingame, A. L., 1990, in: *Biological Mass Spectrometry* (A. L. Burlingame and J. A. McCloskey, eds.), Elsevier, Amsterdam, pp. 285–309.

Klarskov, K., Højrup, P., Andersen, S. O., and Roepstorff, P., 1989, Plasma desorption mass spectrometry as an aid in protein sequence determination. Application of the method on a cuticular protein from the migratory locust, *Biochem. J.* **262:**923–930.

Klarskov, K., Roecklin, D., Bouchon, B., Sabatié, J., Van Doersselaer, A., and Bischoff, R., 1994, Analysis of recombinant *Schistosoma mansoni* antigen rSmp28 by on-line liquid chromatography mass spectrometry combined with sodium dodecylsulfate polyacrylamide electrophoresis, *Anal. Biochem.* **216:**127–134.

Knight, W. B., Swiderek, K. M., Sakuma, T., Calaycay, J., Shiveley, J. E., Lee, T. D., Cover, T. R., Shushan, B., Green, B. G., Cabin, R., Shah, S., Mumford, R., Dickinson, T. A., and Griffin, P. R., 1993, Electrospray ionization mass spectrometry as a mechanistic tool: Mass of human leucocyte elastase and a β-lactam-derived E-I complex, *Biochemistry* **32:**2031–2035.

Knudsen, J., Højrup, P., Hansen, H. O., Hansen, F. H., and Roepstorff, P., 1989, Acyl-CoA-binding protein in rat: Purification, binding characteristics, tissue concentrations and amino acid sequence, *Biochem. J.* **262:**513–519.

Krogh, T. N., Skou, L., Roepstorff, P., Andersen, S. O., and Højrup, P., 1995, Primary structure of protein from the wing cuticle of the migratory locust, *Locusta migratoria*, *Insect Biochem. Mol. Biol.*, **25:**319–329.

Loo, J. A., Loo, R. R. O., Udseth, H. R., Edmonds, C. G., and Smith, R. D., 1991, Solvent-induced conformational changes of polypeptides probed by electrospray-ionization mass spectrometry, *Rapid Commun. Mass Spectrom.* **5:**101–105.

Mann, M., 1990, Electrospray: Its potential and limitations as an ionization method for biomolecules, *Org. Mass Spectrom.* **25:**575–587.

Mann, M., Højrup, P., and Roepstorff, P., 1992, Use of mass spectrometric molecular weight information to identify proteins in sequence databases, *Biol. Mass Spectrom.* **22:** 338–345.

McLafferty, F. W., 1992, Presented at the meeting on Biological Mass Spectrometry, Kyoto, September 1992.

Mikkelsen, J., Højrup, P., Nielsen, P. F., Roepstorff, P., and Knudsen, J., 1987, Amino acid sequence of Acyl-CoA-binding protein from cow liver, *Biochem. J.* **245:**857–861.

Mørtz, E., Vorm, O., Mann, M., and Roepstorff, P., 1994, Identification of proteins in polyacrylamide gels by mass spectrometric peptide mapping combined with database search, *Biol. Mass Spectrom.* **23:**249–261.

Naylor, S., Findeis, A. F., Gibson, B. W., and Williams, D. H., 1986, An approach towards the complete FAB analysis of enzymic digests of peptides and proteins, *J. Am. Chem. Soc.* **108:**6359–6363.

Nielsen, P. F., and Roepstorff, P., 1989, Suppression effects in peptide mapping by plasma desorption mass spectrometry, *Biomed. Environm. Mass Spectrom.* **18:**131–137.

Pappin, D. J. C., Højrup, P., and Bleasby, A. J., 1993, Rapid identification of proteins by peptide-mass fingerprinting, *Curr. Biol.* **3:**327–332.

Rasmussen, H., Mørtz, E., Mann, M., Roepstorff, P., and Celis, J. E., 1994, Identification of transformation sensitive proteins recorded in human two dimensional gel protein databases by mass spectrometric peptide mapping alone and in combination with microsequencing, *Electrophoresis* **15:**406–416.

Roepstorff, P., 1990, Sample preparation for plasma desorption mass spectrometry, in: *Methods in Enzymology*, Vol. 193 (J. A. McCloskey, ed.), Academic Press, San Diego, CA, pp. 432–440.

Roepstorff, P., and Fohlman, J., 1984, Proposal for a common nomenclature for sequence ions in mass spectra of peptides, *Biomed. Mass Spectrom.* **11:**601.

Roepstorff, P., and Højrup, P., 1992, A general strategy for the use of molecular weight information in protein purification and sequence determination, in: *Methods in Protein Sequence Analysis 1992* (K. Imahori and F. Sakiyama, eds.), Plenum Press, New York, pp. 149–156.

Roepstorff, P., Klarskov, K., and Højrup, P., 1989, Strategy for the use of plasma desorption mass spectrometry in protein sequence analysis, in: *Methods in Protein Sequence Analysis 1988* (B. Wittmann-Liebold, ed.), Springer Verlag, Berlin, pp. 191–198.

Roepstorff, P., Talbo, G., Klarskov, K., and Højrup, P., 1990, Nitrocellulose, the interface between plasma desorption mass spectrometry and protein chemistry, in: *Biological Mass Spectrometry* (A. L. Burlingame and J. A. McCloskey, eds.), Elsevier, Amsterdam, pp. 25–48.

Roepstorff, P., Klarskov, K., Andersen, J., Mann, M., Vorm, O., Etienne, G., and Parello, J., 1991, Mass spectrometry of proteins: Studies of parvalbumins by plasma desorption, laser desorption and electrospray mass spectrometry, *Int. J. Mass Spectrom. Ion Proc.* **111:**151–172.

Roepstorff, P., Schram, K. H., Andersen, J. S,. Rafn, K., Baldursson, T., Krøll, J., Poulsen, K., Knudsen, J., and Kristiansen, K., 1995, Evaluation of mass spectrometric techniques for characterization of engineered proteins, *Molecular Biotechnology*, in press.

Scheller, H. V., Okkels, J. S., Høj, P. B., Svendsen, I., Roepstorff, P., and Møller, B. L., 1989, The primary structure of a 4.0-kDa photosystem I polypeptide encoded by the chloroplast psaI gene, *J. Biol. Chem.* **264:**18402–18406.

Scheller, H. V,. Okkels, J. S., Roepstorff, P., Jepsen, L. B, and Møller, B. L., 1990, Characterization of a cDNA clone for the PsaE gene from barley and plasma desorption mass spectrometry of the corresponding photosystem I polypeptide PSI-E, in: *Proceedings of the VIIIth International Congress on Photosynthesis, Stockholm, August* (M. Baltscheffsky, ed.), Klüwer Academic Publishers, Dortrecht, pp. 609–612.

Smith, R. D., and Light-Wahl, K. J., 1993, The observation of non-covalent interactions in solution by electrospray ionization mass spectrometry: Promise, pitfalls and prognosis, *Biol. Mass Spectrom.* **22:**493–501.

Smith, R. D., Loo, J. A., Edmonds, C. G., Barinaga, C. J., and Udseth, H. R., 1990, New developments in biochemical mass spectrometry: Electrospray ionization, *Anal. Chem.* **62:**882–899.

Sørensen, H. H., Thomsen, J., Bayne, S., Højrup, P., and Roepstorff, P., 1990, Strategies for determination of disulfide bridges in proteins using plasma desorption mass spectrometry, *Biomed. Environ. Mass Spectrom.* **19:**713–720.

Stults, J. T., Bourell, J. H., Canova-Davis, E., Ling, V. T., Laramee, G. R., Winslow, J. W., Griffin, P. R., Rinderknecht, E., and Vandlen, R. L., 1990, Structural characterization by mass spectrometry of native and recombinant human relaxin, *Biomed. Environ. Mass Spectrom.* **19:**655–664.

Sundqvist, B., Roepstorff, P., Fohlman, J., Hedin, A., Håkansson, P., Kamensky, I., Lindberg, M., Sahlepour, M., and Säve, G., 1984, Molecular weight determination of proteins by Californium plasma desorption mass spectrometry, *Science* **226:**696–698.

Talbo, G., Højrup, P., Rahbek-Nielsen, H., Andersen, S. O., and Roepstorff, P., 1990, Determination of the covalent structure of an N- and C-terminally blocked glycoprotein from endocuticle of *Locusta migratoria*. Combined use of plasma desorption mass spectrometry and Edman degradation to study post-translationally modified proteins, *Eur. J. Biochem.* **195**:495–504.

Torgerson, D. F., Skowronski, R. P., and Macfarlane, R. D., 1974, New approach to the mass spectrometry of non-volatile compounds, *Biochem. Biophys. Res. Commun.* **60:** 616–621.

Vestling, M. M., Murphy, C. M., and Fenselau, C., 1990, Recognition of trypsin autolysis products by high-performance liquid chromatography and mass spectrometry, *Anal. Chem.* **62:**2391–2394.

Wang, R., Chen, L., and Cotter, R. J., 1990, Effects of peptide hydrophobicity and charge state on molecular ion yields in plasma desorption mass spectrometry, *Anal. Chem.* **62:** 1700–1705.

5

Two-, Three-, and Four-Dimensional Nuclear Magnetic Resonance Spectroscopy of Protein Pharmaceuticals

David G. Vander Velde, James Matsuura, and Mark C. Manning

1. INTRODUCTION

In the past decade, nuclear magnetic resonance (NMR) spectroscopy and related computational techniques have evolved to such a degree that it is now possible to completely assign the [^{1}H]-NMR spectrum and, using this information, determine the three-dimensional structure of a protein of more than 100 residues (Clore and Gronenborn, 1991a; Inagaki, 1990; Wagner *et al.*, 1992; Wüthrich, 1986, 1989a,b, 1990). Consequently, it has become as important a tool in the structural determination of proteins as X-ray crystallography. The interest in NMR has been further accelerated by the advent of biotechnology, where it is now possible to produce proteins for use as therapeutic agents.

The field of multidimensional NMR spectroscopy and associated emerging methods have been widely described and reviewed (Braun, 1987; Brünger and Karplus, 1991; Clore and Gronenborn, 1991a; Crippen, 1981; Crippen and Havel, 1988; Croasmun and Carlson, 1987; Fesik, 1988, 1993; Havel, 1991; Havel and

David G. Vander Velde • NMR Laboratory, University of Kansas, Lawrence, Kansas 66045. *James Matsuura and Mark C. Manning* • School of Pharmacy, University of Colorado Health Sciences Center, Denver, Colorado 80262.

Physical Methods to Characterize Pharmaceutical Proteins, edited by James N. Herron *et al.*, Plenum Press, New York, 1995.

Wüthrich, 1984, 1985; Inagaki, 1990; Kaptein *et al.*, 1988; Kessler *et al.*, 1988; Kessler and Steuernagel, 1991; Kuntz *et al.*, 1989; LeMaster, 1989; Nilges *et al.*, 1988a,b; Olejniczak *et al.*, 1984; Oschkinat *et al.*, 1988; Scheek *et al.*, 1989; Wagner *et al.*, 1992; Wider *et al.*, 1984; Wüthrich, 1986, 1989a,b, 1990; Wüthrich *et al.*, 1984). We wish to present an overview of the more common two-, three-, and four-dimensional NMR experiments. Implementation of NMR data in structure determination requires the use of a variety of computational techniques such as distance geometry, simulated annealing, and restrained molecular dynamics. These will be described as well. Finally, we will summarize the work done to date on proteins of pharmaceutical interest. These will include current products as well as those being considered for commercialization.

2. NMR METHODS

The proton NMR spectrum of any of the naturally occurring amino acids, taken in isolation, is relatively easy to understand. In proteins, the structural variety of the amino acids leads to signals that are dispersed over the ~10 ppm ^{1}H-labeled chemical shift range, and the close packing of various anisotropic groups (particularly aromatic rings and carbonyl groups) in a folded protein disperses them even more (Wüthrich, 1986). However, the sheer number of signals in the spectrum of a protein creates such extensive overlap that straightforward analysis of the spectrum is a daunting problem. These problems quickly become intractable in the normal "one-dimensional" NMR spectrum, except for a few favorable cases where a residue of interest occurring a limited number of times in the protein is found in a comparatively uncrowded area of the spectrum (e.g., histidine). The tools of structural and conformational analysis used on small organic molecules, based on one-dimensional NMR techniques for detecting scalar (through bond or J) and dipolar (through space) couplings, namely, homonuclear decoupling and nuclear overhauser enhancement (NOE) difference measurements, will also fail in heavily overlapped spectra. Any position in the spectrum will likely have several signals present, so any attempt to irradiate one peak to locate coupled sites will produce effects at multiple places in the spectrum. Even if feasible, experiments such as these that detect just one correlation at a time would be extremely inefficient on very large biomolecules.

Multidimensional experiments can solve these difficulties by spreading out the heavily overlapped spectrum into multiple dimensions where it is more tractable. Because they simultaneously measure all of the connections between all nuclei of a given type, they are highly efficient. Use of nuclei other than protons to provide one or more of these dimensions, carbon and nitrogen in particular, gives additional dispersion because of their wider chemical shift ranges and also permits complete tracing of all atoms in the peptide backbone. The basic building blocks of the multidimensional experiments, transfers of coherence through various types of through-bond or through-space couplings, can be assembled in different and complementary ways to yield all the connectivities needed to assign each signal to a specific atom in the protein and determine enough of the interatomic distances and torsion angles to determine its structure.

2.1. Description of Multidimensional NMR

Numerous treatments of multidimensional NMR that are rigorous and thorough have appeared (see Section 1), so a similar treatment of the topic will not be presented here. For the less technically oriented reader, a brief nonmathematical description will be provided. The important methods will be surveyed and their utility and applicability to proteins of different sizes described, followed by information on practical experimental considerations, data processing, and spectral analysis. The limitations of the various experiments and the spectrometer hardware needed to perform them will also be noted.

It is useful to begin consideration of multidimensional experiments with a brief description of the one-dimensional experiment. A short (microsecond), high-power radio frequency pulse in the center of the spectral region of interest is applied to the sample and the resulting time domain signal is received, periodically sampled, digitized, and stored in the computer memory. This signal, the free induction decay (FID), consists of a set of decaying exponentials, each with a characteristic frequency and amplitude. The number of points used and the total length of the detection period must be sufficient to give a digital resolution that is smaller than the features of interest (e.g., spin couplings) in the spectrum. A Fourier transform (or other, frequently more computationally intensive methods such as linear prediction) converts this information into the desired frequency domain spectrum of Lorentzian peaks with characteristic frequencies, amplitudes, widths, and phases.

Application of more than one pulse separated by times that are short compared with the NMR relaxation times (typically seconds for the spin $\frac{1}{2}$-nuclei used in studies of biomolecules) creates additional coherences (or energy transfer pathways between coupled spins) that are not visible in the one-dimensional spectrum. The coherences that are generated will evolve as a function of time, as manifested by oscillations of the amplitude or phase of each peak at the frequency of its coupling partners. In order to determine what those frequencies are, a set of FIDs are acquired by incrementally increasing the length of the delay time between the pulses in successive experiments. The number of increments and the total length of the delay that is sampled must be sufficiently long to define the oscillation frequencies adequately (typically a few hundred increments are taken). Defining the frequency of these oscillations by incrementally sampling their evolution is analogous to defining the frequencies present in the free induction decay by periodic sampling and digitization. An evolution period is required for each dimension to be added (i.e., one for a two-dimensional experiment, two in a three-dimensional experiment, and so on). The resulting set of FIDs is Fourier transformed along each of the time axes to convert them to frequency axes, and the frequency of oscillations corresponding to the desired coherences can now be displayed.

In addition to the pulses that generate the coherences, the evolution delay, and the detection of the resulting FIDs, a complete multidimensional experiment has one or two additional elements: (1) a preparation period, during which the spins return to equilibrium between iterations of the pulse sequence, and (2) a mixing period of fixed duration to allow noninstantaneous interactions (particularly NOEs) to build up for observation. This general scheme is shown in Fig. 1. By convention, the first (incremented) delay in a two-dimensional sequence is called t_1 and the detection period for collection of the FID is called t_2. (The additional delays in a higher-dimensional experiment are numbered consecutively.) The frequency axes derived from the delays are numbered f_1, f_2, and so on, in the same order, which is generally the reverse of the order in which they appear in processing (i.e., f_2 appears in the first Fourier transform of a two-dimensional data set).

Specific multidimensional experiments differ in the number and spacing of pulses used to generate the desired coherences. Also, any sequence of pulses will typically generate multiple coherence transfer pathways; all but one of these pathways must be suppressed in order to obtain interpretable data. In the past, this suppression was accomplished by systematically varying the phases of the radio

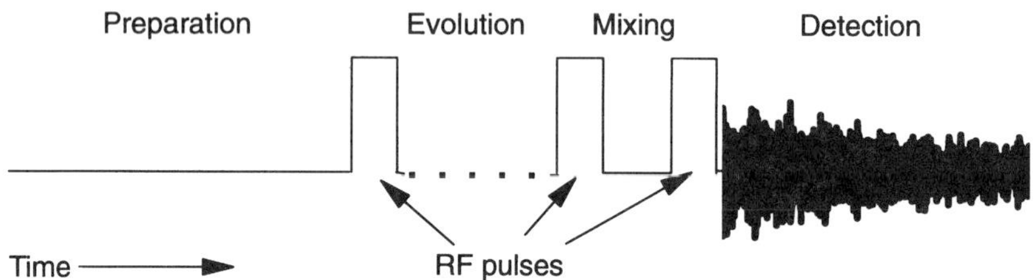

Figure 1. Schematic drawing of the radiofrequency pulses and delays used in a 2D experiment (shown for a NOESY sequence). The time axis is not to scale; typical values are 2 seconds for the preparation period, radiofrequency pulses 5 μseconds, 512 increments of 100 μseconds each in the evolution period, mixing period 50–200 milliseconds, and detection period (2048 point FID collection) 240 milliseconds. Depending on the number of scans required at each value of the evolution delay to give adequate signal intensity, the entire experiment would last 12–36 hours. The resulting 2048 × 512 point data matrix would be expanded to 2048 × 2048 points during processing.

frequency pulses through a defined cycle, which also could be used to suppress various kinds of spurious signals. Even if this procedure works ideally (and because it puts stringent demands on the stability of the spectrometer and the reproducibility of pulse phases and amplitudes, it usually doesn't), it is quite time inefficient. In many cases, the total length of an experiment is fixed by the length of the phase cycles to be much longer than what would be dictated by the number of scans needed solely to obtain satisfactory signal-to-noise ratios. (Typical phase cycles contain 8–16 steps; in the limit, only one scan may provide satisfactory signal intensity.) As the number of dimensions increases, so does the severity of this problem. Experiments using pulsed magnetic field gradient pulses in addition to radio frequency pulses can circumvent this problem (Bax *et al.*, 1980). Interest in these experiments has greatly increased recently because of improved gradient hardware developed for magnetic resonance imaging (Hurd, 1990; Hurd and John, 1991a,b; Vuister *et al.*, 1992). In these sequences, only a single type of coherence is present in the FIDs collected, so the number of scans can be set to the minimum needed for sensitivity. On concentrated samples, two-dimensional spectra can be acquired in just a few minutes using the gradient-enhanced techniques. New NMR spectrometers are routinely equipped with the necessary hardware for these experiments, but many older ones are not.

2.2. Survey of Key Multidimensional Methods

As it has evolved, the nomenclature of multidimensional experiments is strikingly unhelpful to novice users. The commonly used acronyms for many of these experiments, while often clever, convey little or no information about the purpose of the experiment. Attempts to remedy this situation with a systematic nomenclature have met with little success in the case of the older experiments. (However, a hopeful trend is that some of the newer experiments have been given names based on the nuclei that are being correlated and the order in which the correlations are obtained.) Additionally, this subject is complicated by the many variants of some experiments that have been described in the literature; in this chapter, only the basic experimental schemes or a few representative examples will be described. Other reviews or recent papers may be consulted for further details.

2.2.1. HOMONUCLEAR J-COUPLING EXPERIMENTS

The first two-dimensional experiment to be proposed (Jeener, 1971; Aue *et al.*, 1976), the (shift) *co*rrelation *s*pectroscop*y* (or COSY) experiment, shows J-coupling between like (homonuclear) spins. The COSY experiment has been applied to a variety of high natural abundance nuclei; for proteins this basically means protons. Although this experiment is currently little used for proteins or even larger peptides, because of the difficulty of analyzing the spectrum correctly when the peaks overlap extensively, it is the simplest experiment of its type to perform, consisting of only two pulses separated by an evolution time, and it shows all the characteristic features of homonuclear two-dimensional spectra. The two frequency axes on a COSY spectrum are identical. Where each peak intersects itself, a peak (most conveniently viewed on a contour plot) appears. These self- or autocorrelations fall on a diagonal line through the spectrum, conventionally running from the lower left to the upper right corner. Off-diagonal or cross-peaks connect J-coupled spins, the four peaks being found at the corners of a square. Early COSY spectra were often taken at comparatively low resolution, in part because of the limitations of older computers and disk drives and in part because of the inability of data processing software to record the phases of the two-dimensional peaks. "Phase-sensitive" (Marion and Wüthrich, 1983; Rance *et al.*, 1983) experiments that preserve this information, especially as implemented on modern data systems that can efficiently process much larger data sets, give much higher resolution data that show not only the presence of coupling but allow the coupling constant(s) to be approximately determined. Another enhancement was provided by double-quantum filtering (Wokaun and Ernst, 1977; Shaka and Freeman, 1983), which removes uncoupled peaks (e.g., methyl singlets) from the spectrum; these signals provide no useful information but often obscured nearby multiplets of interest.

In a protein spectrum, COSY peaks will only occur inside amino acid residues, because there is no proton J-coupling across amide bonds. Also, couplings do not extend completely through some side chains, e.g., between protons and aromatic protons in Phe, Tyr, Trp, or His. Of the 20 naturally occurring amino acids, 12 generate sets of COSY peaks that are unique, or nearly so; however, there are eight amino acids that give rise to more or less indistinguishable α-β-β' patterns (formally an AMX spin system) that COSY cannot distinguish (Wüthrich, 1986). (For further discussion of these points, see Section 2.4.3.)

The fundamental limitation of the COSY experiment for proteins, however, is the fact that each correlation only extends through a maximum of three bonds (or from one carbon to the next), whereas identification of the complete set of signals from a residue requires multiple jumps through the spectrum (amide NH- $\rightarrow$ α- $\rightarrow$ β,β'- $\rightarrow$ ...). With each jump, signal overlap may introduce ambiguities about the correct connections. For instance, even if a given amide proton is clearly resolved, it may be connected to two or more overlapping α protons. If each α proton is connected to two β protons, even if the two pairs are easily distinguished from each other, which set of β protons is in the same residue as the amide? The COSY spectra of even small peptides may contain a few such ambiguities, and a larger protein will inevitably contain many. A related experiment, which is more complex to perform but provides much more useful data for polypeptides, is known in the literature under two different acronyms: TOCSY (*to*tal *c*orrelation *s*pectroscop*y*) and HOHAHA [*ho*monuclear *Ha*rtmann-*Ha*hn experiment (Bax and Davis, 1985b)]. In this experiment, a mixing period is employed that, if sufficiently long, leads to cross-peaks connecting all of the protons in an extended spin system (which may be all or part of the amino acid residue). A single row or column out of the resulting matrix, starting, for example, from the amide region, will contain all the correlations out to the end of the spin system. This greatly reduces the ambiguities in the spectrum for peptides and small proteins.

For larger proteins, multiple overlap and spectral crowding, particularly for the region between 1.5 to 4.5 δ where the majority of side-chain protons are found, will eventually become serious problems. TOCSY (and COSY) spectra of larger proteins will also show greatly reduced cross-peak intensities because of the increased line width and correspondingly increased relaxation rates of the proton signals. When the proton line widths exceed the coupling constants (4–10 Hz), these experiments will not give complete sets of correlations, although more mobile regions of the proteins having narrower lines may still give some connectivities. In this case, the alternatives are: (1) increase the temperature, which increases molecular motion and narrows the lines (assuming the protein is robust enough to remain stable and nonaggregating in a concentrated solution for the extended periods of time which may be needed for data collection); and (2) use a three-dimensional experiment that detects larger heteronuclear coupling constants that are less sensitive to line broadening (see Section 2.3).

2.2.2. HETERONUCLEAR J-COUPLING EXPERIMENTS

The important heteronuclear couplings in proteins are between protons and ^{13}C or protons and ^{15}N, both of which have low natural abundance (1.1 and 0.3%, respectively). These couplings can be traced in a two-dimensional experiment either by observing the heteronucleus directly in f_2 and the protons indirectly in f_1, or vice versa. The former scheme [often called HETCOR or H, C COSY (Bax and Morris, 1981)] was the first to be developed, is experimentally the simplest, and is the most commonly used for routine work on small molecules, but the latter is almost exclusively used for proteins. Two one-bond experiments of this type are called HMQC [*h*eteronuclear *m*ultiple *q*uantum *c*oherence (Bax and Subramanian, 1986)] and HSQC (*h*eteronuclear *s*ingle *q*uantum *c*oherence). They have the advantages of much higher sensitivity and speed overall and of putting the proton spectrum (which is the most crowded) on the f_2 axis, which has the best digital resolution. In practice, these experiments may be unavailable or only partly implemented on older spectrometers. For natural abundance samples, they require that 99% of the proton signal, which originates from the protons attached to ^{12}C (or the 99.7% from ^{14}N), be efficiently canceled out, effectively by subtraction of alternate scans. Also, for maximum sensitivity gains, the ^{1}H signal must be ^{13}C- or ^{15}N-decoupled, which requires special hardware and substantially higher decoupling power than the corresponding ^{1}H-decoupling of the heteronuclear spectrum (owing to the wider chemical shift range). In this case also, pulsed gradient methods are of considerable benefit, because they permit only the desired $^1H \rightarrow {}^{13}C$ coherence to be selected in a single transient, eliminating spectral artifacts resulting from failure to cancel the ^{1}H on ^{12}C signal; the other advantages of gradient methods described above are also realized.

Data from these experiments are of course useful for making the heteronuclear assignments, but also can assist in making proton assignments by spreading the proton spectrum out and indicating the environment of the attached heteronucleus. The heteronuclear shifts are themselves useful indicators of secondary structure.

The HMBC [or *h*eteronuclear *m*ultiple *b*ond *c*orrelation (Bax and Summers, 1986)] experiment provides information on two- and three-bond couplings while suppressing the larger one-bond couplings. It is an invaluable method for structure assignment and to some extent conformational analysis of small molecules, and it may also be applied to proteins.

Only relatively small proteins are likely to be soluble at high enough concentrations to make these experiments practical at the natural abundance of ^{13}C or ^{15}N. As is the case for virtually all of the higher-dimensional heteronuclear experiments, samples of protein enriched in one or both isotopes are required. Cloning and overexpression of the protein in bacteria or yeast make it possible to do the enrichment efficiently by adding appropriately labeled nutrients [e.g., $^{15}NH_4Cl$, U-^{13}C glucose (Clore *et al.*, 1990d)] to the growth medium as the sole source of nitrogen or carbon or to a medium derived on algae grown on $^{13}CO_2$ (Sørensen and Poulsen, 1992). Protein samples with isotopic labels only in some residues have also been used to give simplified spectra in order to facilitate assignment; this approach has also been used successfully to make some assignments in antibodies and their proteolytic fragments (Kato *et al.*, 1991)

2.2.3. DIPOLAR COUPLING EXPERIMENTS

The basic two-dimensional experiment for measuring through-space or dipolar couplings is NOESY [or NOE *s*pectroscop*y* (Bodenhausen *et al.*, 1984)], a three-pulse sequence that produces a COSY-like spectrum but with the cross-peaks signifying through-space rather than through-bond interactions. As in the case of COSY, the earliest spectra were magnitude mode only, but phase-sensitive presentation of the data gives superior resolution.

The key parameter determining the appearance of the NOESY is the mixing time, t_m, during which the NOEs build in intensity, starting from zero at the beginning of t_m. This delay is of fixed length (or in some experiments, randomly varied around a fixed value, which strongly suppresses COSY-type peaks that would otherwise be present in the spectrum as artifacts). NOE intensities are strongly dependent on distance, scaling with the inverse sixth power of the distance between the nuclei; this necessarily means that the NOE falls to a very small value at relatively short distance (few being observed $\geqslant 5$ Å) compared with the dimensions of a protein. The actual distance is not necessarily related to the intensity of the cross-peak at any one mixing time, but rather to the rate at which the intensity increases as a function of the mixing time. At short mixing times, the NOE buildup is linear, but naturally the absolute intensities will be very small under such conditions, making quantitation difficult. Typically the NOESY experiment will be run at a series of mixing times in order to extract initial slopes from a buildup curve for each cross-peak. Fortunately, highly accurate quantitation is not required in order to obtain reasonable structural data; many investigators only classify NOEs as strong, medium, or weak, corresponding to fixed distance ranges. Several such schemes are in use, with little dependence of the final structure on the exact assumptions made. The overall quality of the structure seems to be more dependent on having a large number of NOEs (correctly assigned) than on the accuracy of the NOEs.

An alternative approach to NOE quantitation has been suggested by Mirau (1988), with impressive results on a cyclic peptide. In this approach, NOESY cross-peak and diagonal peak intensities at a single mixing time are scaled via selective relaxation rate measurements.

Especially in larger proteins, two problems complicate NOESY spectra, particularly at longer mixing times: (1) severe overlap, as each signal can have several NOE coupling partners inside and outside the residue it belongs to, leading to a very large number of cross-peaks; and (2) the phenomenon of spin diffusion, in which NOEs appear not only between nearby spins but to spins close to the nearby spins and so on, eventually leading to a complete loss of useful distance information. When spin diffusion is operating, only the NOEs appearing at very short mixing times actually indicate nearby spins. The overlap may be relieved by increasing the dimensionality of the experiment, spreading the cross-peaks out into a third or even a fourth dimension with homonuclear or heteronuclear axes. Although quantitation of the cross-peak intensities appearing after the long and complex four-dimensional pulse sequences becomes problematic, the approach does work and is the only feasible way to separate and assign the signals in the largest proteins amenable to study by NMR.

The NOE experiment may also fail for small peptides. In an intermediate molecular size region, usually assumed at 500 MHz to be between 500 and 2000 daltons, the NOE changes sign from positive to negative and will be quite small—in the worst case, effectively zero. In some cases both positive and negative NOEs are observed in the same molecule (more mobile regions with segmental motions giving positive NOEs). Changing the temperature and/or solvent, thereby alternating rates of molecular reorientation, may be helpful in pushing the molecule one way or the other; dimethyl sulfoxide (DMSO)–water mixtures at very low temperatures are especially effective at making small molecules act "big," although the relevance of biomolecular conformation determined under such conditions must be established. A more general solution is an alternative pulse sequence called ROESY [*ro*tating frame NOE *s*pectroscop*y* (Bax and Davis, 1985a); or sometimes CAMELSPIN (Bothner-By, 1984)]. In this experiment, the NOEs are invariably positive (although for technical reasons, the cross-peaks appear negative) for molecules of any reasonable molecular weight. Although it can be applied to molecules of any size, ROESY is most often used only when NOESY will not work; the ROESY experiment is somewhat more difficult to do, additional types of artifacts and spurious peaks are possible, and quantitation requires more assumptions compared with NOESY. However, it has also been suggested (Clore *et al.*, 1991c) that for large proteins, ROESY may provide some relief from spin diffusion, because the sign of the effect (and therefore the phase of the cross-peak) changes with each spin diffusion step. Since multiple spin diffusion pathways typically connect any given pair of distant protons, some involving an odd and some an even number of intervening spins, the tendency will be for spin diffusion peaks to cancel out.

NOESY may also be applied to determine the bound-state conformation of small molecules on macromolecules, either as tightly bound stoichiometric complexes or in the presence of exchange. In the former case, isotopic labeling can be used to distinguish the signals of the small molecule from those of the macromolecule, which would otherwise be difficult to do (Fesik *et al.*, 1988; Fesik and Zuiderweg, 1989). In the latter case, the transferred NOE effect can be utilized (Clore and Gronenborn, 1982, 1983). In this experiment, an excess of the small molecule is present. In the bound state, large negative NOEs with rapid buildup times are observed; the molecules free in solution will have positive, slowly changing, typically smaller NOEs. Exchange between the free and bound state will transfer the NOEs originating in the bound state into the free state, where they are readily observed. This technique is experimentally fairly undemanding and has been applied to a wide variety of systems. However, quantitation of the distances is difficult when the exchange becomes slow, owing to the inefficient transfer of the NOE under such conditions (London *et al.*, 1992).

2.3. Three- and Four-Dimensional Experiments

Proteins of 100 residues (10,000–12,000 daltons) or greater typically give such complex and heavily overlapped two-dimensional spectra that assignment is greatly complicated. By extension of the principles described above, two-dimensional spectra can be extended into additional dimensions by inclusion of additional pulses and evolution periods in data collection and additional Fourier transforms of the resulting data. From the standpoint of reducing overlap, more dimensions are always better. Dispersion of the spectrum is obtained much more cheaply by adding dimensions to the data than by buying bigger magnets. The negative consequences of increasing the number of dimensions are increased experiment times, the need for higher sample concentrations (because less signal averaging can be done when so many individual blocks of data must be collected), the need for more complex spectrometer hardware and probes, reduced data intensity at the end of long and complex pulse sequences, and reduced digital resolution in some dimensions. A typical one-dimensional spectrum might contain 32 K points; a typical two-dimensional matrix might be 2 K by 512 points; and in a four-dimensional experiment, as few as 32 points may be collected along some axes.

Three different types of three-dimensional strategies may be distinguished: (1) combining two ^{1}H two-dimensional experiments to reduce overlap by spreading the information out into the third dimension; (2) combining a ^{1}H two-dimensional experiment with a ^{1}H-X nucleus two-dimensional experiment such as HMQC, also to reduce overlap (This approach takes advantage of the wider X-nucleus chemical shift range and also provides the X-nucleus shifts.); and (3) three- or four-dimensional shift correlation experiments that provide either sequential connectivities along the amide backbone or side-chain connectivities.

2.3.1. HOMONUCLEAR THREE-DIMENSIONAL EXPERIMENTS

Significant increases in data dispersion can be achieved by combining two two-dimensional experiments such as TOCSY and NOESY into a three-dimensional experiment (Vuister *et al.*, 1988; Oschkinat *et al.*, 1988) or NOESY and NOESY (Boelens *et al.*, 1989; Breg *et al.*, 1990; Vuister *et al.*, 1991). These were the first three-dimensional experiments to be widely used. The three-dimensional data set can be conveniently analyzed as a set of two-dimensional slices, each of which contains the cross-peaks originating from one or only a few protons whose chemical shift are found at the position of the slice. An advantage of these experiments is that no isotopic labeling is required to perform them, as is typically the case for the heteronuclear experiments described in Section 2.3.2, and the sensitivity is relatively high as only protons are involved. However, the greatest gains in the size of proteins amenable to NMR analysis have resulted from the application of experiments involving a heteronucleus.

2.3.2. HETERONUCLEAR THREE- AND FOUR-DIMENSIONAL EXPERIMENTS

When isotopically labeled protein is available, addition of a heteronuclear axis by combining a ^{1}H two-dimensional experiment with an HMQC or HSQC experiment results in very useful signal dispersion. Many such experiments are possible, some examples being ^{15}N-separated TOCSY and ^{13}C-separated ROESY (Clore *et al.*, 1991c). Data from several complementary experiments of this type may be needed to make full assignments of large proteins. The distance constraints for structure determination and refinement of large proteins generally come from three- and four-dimensional NOESY spectra with ^{13}C and/or ^{15}N axes to provide sufficient signal dispersion.

2.3.3. PEPTIDE BACKBONE ASSIGNMENTS AND SEQUENTIAL CONNECTIVITIES

Sequential connectivities needed to make sequence-specific assignments were initially derived from NOESY spectra. A better alternative, particularly for larger proteins, is to use shift correlation experiments that directly detect the J-couplings of all the atoms in the peptide backbone. A large number of such experiments have been described in the literature (Ikura *et al.*, 1990; Kay *et al.*, 1990a,b, 1991, 1992; Archer *et al.*, 1992; Grzesiek and Bax, 1992; Boucher *et al.*, 1992; Olejniczak *et al.*, 1992b; Richardson *et al.*, 1993; Syperski *et al.*, 1993; Weisemann *et al.*, 1993; Wittekind and Mueller, 1993). They have the following general feature—the need to correlate three different nuclei (^{1}H, ^{13}C, ^{15}N) creates a requirement for doubly labeled protein samples and more sophisticated NMR hardware, such as probes with three radio frequency channels and spectrometers with as many as five separate transmitters. It is desirable in these experiments to simplify the spectra by selectively exciting only α-carbons or only carbonyl carbons at prescribed points in the pulse sequence. Fortunately, the resonances are well separated in the ^{13}C spectrum and the necessary selectivity can be achieved with shaped (typically Gaussian, but many other possibilities exist) low-power pulses. Some of the couplings, particularly to ^{15}N, are small even though they are through one bond. The efficiency of the overall experiment depends substantially on the order in which the correlations are made and the amount of time the magnetization spends on nuclei with unfavorable relaxation properties. Last but not least, a more or less systematic nomenclature has been used to name the experiments! The nuclei correlated are listed in the order the correlations are made; nuclei that the chain of connectivities passes through, but which are not observed, are put in parentheses. For example, an HN(CO)CA experiment starts with an amide proton, goes through the amide nitrogen to the carbonyl carbon (which is not observed), to the α-carbon of the preceding residue. The HNCA experiment (Kay *et al.*, 1990a,b) shows the expected correlations between an amide proton, amide nitrogen, and α-carbon reliably within each residue and in many cases between residues. The HN(CA)HA experiment goes from an amide proton to an α-proton via the amide nitrogen and α-carbon, both within a residue and to the preceding residue. Additional new experiments and refinements of existing experiments will doubtless continue to appear in the literature. The complementary information from several experiments is beneficial for full assignments of large proteins.

2.3.4. SIDE-CHAIN ASSIGNMENT EXPERIMENTS

In large proteins with broad lines, α-β side-chain assignments involving H-C-C-H linkages are much more efficiently made using one-bond couplings involving an X-nucleus rather than three-bond proton–proton couplings. The proton couplings are typically 4 to 10 Hz; when the proton line widths exceed the size of the couplings, COSY or TOCSY experiments will fail. By contrast, the one-bond carbon–proton (125–170 Hz) and carbon–carbon (35–40 Hz) couplings are still usable. Two different approaches, HCCH COSY (Bax *et al.*, 1990; Ikura *et al.*, 1991) and HCCH TOCSY (Fesik *et al.*, 1990; Olejniczak, 1992a) have been employed giving the same information.

In addition to distance constraints from NOESY data, side chain and backbone torsional constraints derived from the value of three-bond heteronuclear coupling constants are also extremely valuable. Several experiments used to extract these have been described (Montelione *et al.*, 1989; Schmieder *et al.*, 1991; Chary *et al.*, 1991; Grzesiek *et al.*, 1992b; Emerson and Montelione, 1992).

2.4. Experimental Considerations

2.4.1. SAMPLE PREPARATION AND EXPERIMENTAL SETUP

In order to be amenable for the types of NMR experiments described here, a protein must be soluble, stable, and nonaggregating at relatively high concentrations ($\geqslant$ 1 mM), often at elevated temperatures to increase molecular motion. Buffers that do not give any interfering resonances are most desirable. In order to limit the exchange rate of amide protons with bulk water, a somewhat acidic pH is beneficial (the exchange rate for backbone amides reaches its minimum about pH 3). Considerable effort may be required to find the optimal conditions of pH, temperature, buffer strength, and so on for the most useful spectra.

A critical part of any experiment on a protein is suppression of the ~55 M water signal (D_2O cannot be used because deuterium would be exchanged into the amide positions, which are of critical importance in determining secondary and tertiary structure; typically, about 10% D_2O is added in order to provide field/frequency lock for stabilization and adjustment of the spectrometer). Unfortunately, the water signal falls near the center of the proton spectrum and toward the edge of the a proton region. Many methods for suppressing this signal have been described; however, they also entail the loss of at least some information from the spectrum. The earliest and most widely employed method (at least until recently), presaturation of the waterline prior to the application of pulses with low-power continuous wave irradiation, typically "bleaches" an area 100- to 150-Hz (0.2–0.3 δ at 500 MHz) wide out of the spectrum. Minimizing the extent of this area is a stringent test of the magnetic field homogeneity and the skill of the operator; recent improvements in the design of the homogeneity correction coils and NMR probes have led to consistently better performance of newer instruments (and, it is hoped, some decrease in the laborious nature of the adjustment process). Since the chemical shift of water is temperature dependent, running the spectrum at several temperatures may uncover most or all of the peaks lost at a single temperature. Other more sophisticated approaches for recovering at least some of the intensity of the peaks in this region have also been developed. Pulsed gradient-based methods offer great promise in this regard; in this case the discrimination is made not on the basis of a peak's frequency, but on the diffusion coefficients of the molecules giving the signals. These "WATERGATE" methods (Piotto *et al.*, 1992) can yield spectra essentially free of artifacts related to the water signal and with signals underneath the waterline completely visible.

2.4.2. DATA PROCESSING

Increases in speed and storage capacity and decreases in cost of computers have greatly aided data processing; on a UNIX-based graphics workstation, typical two-dimensional data sets can be doubly Fourier transformed, phase corrected in both dimensions, and baseplane flattened in less than 5 min (tasks that 10 years ago might take a half day on greatly inferior data). In addition to the data processing software supplied by the instrument manufacturers, some vendors of commercial molecular modeling software support NMR data processing and analysis closely linked to structure calculations such as distance geometry and molecular dynamics. Such programs incorporate features such as integral spreadsheets to assist in the prodigious bookkeeping tasks inherent in assigning and correlating the large numbers of cross-peaks present in multiple NMR data sets.

2.4.3. ASSIGNMENT STRATEGIES

Correct assignment of the peaks in the spectrum is central to structure determination. NOEs observed for peaks that cannot be assigned have no value; peaks that are misassigned will lead to erroneous structures.

Early assignments of NMR spectra were based on a sequential assignment strategy (Wüthrich, 1986) that involved sorting of the signals from each individual residue; identifying each amino acid by type; searching for connections from one residue to the next in the NOESY; identifying di-, tri-, or oligopeptide sequences that are unique in the primary structure; and extending assignments away from each unique site identified until the process had included all the residues in the protein. Even for small proteins, this process was often the slow step in structure determination. It is based on the assignment and recognition of side chains, which involves interpreting the most crowded regions of the spectrum first. Recognition of the patterns from individual amino acids is hampered by the ambiguities resulting from the eight amino acids giving more or less indistinguishable α-β-β only patterns and the fact that restrictions in side-chain motion and packing of interior residues often lead to patterns where expected correlations are missing and chemical shifts fall outside of the expected region.

A second strategy, the main-chain-directed strategy, was proposed by Englander and Wand (Englander and Wand, 1987). In this case, the amide–amide ("main chain") connectivities from the peptide backbone are used to identify regions of secondary structure, starting with α-helix, proceeding through antiparallel β-sheets and parallel β-sheets, and finally, after the residues in these areas are accounted for, to regions without regular secondary structure. The characteristic patterns generated by the regular secondary structures are illustrated in Fig. 2. In this fashion the complex side-chain region of the spectrum can be treated last rather than first, and the secondary structural elements are immediately identified. Most practicing spectroscopists use some combination of the two strategies. The peak assignment and structure determination process may proceed to some extent in parallel, since preliminary structures may assist in making additional assignments or testing those that went into the structural model.

The quality of the final structure is greatly improved if "stereospecific" assignments of groups such as the β and β′ protons on many side chains, the two methyl groups on valine, and so on can be made such that the unique distances from each group to near neighbors can be identified and used (Clore *et al.*, 1991c). Otherwise, a pseudoatom that is assumed to reside at an average position between the real atoms must be defined for structure calculations. Stereospecific assignments also permit identification of the rotameric preferences of closely packed side chains in the protein interior (surface residues are less likely to reside in a single rotamer).

Given that the spectral assignment process is highly time consuming and laborious, requiring in many cases data from numerous multidimensional experiments to be collated, automated assignment by computer is a highly desirable goal. It would seem that a computer could exhaustively test various possible assignments and reject incorrect ones much more quickly than a person could. However, progress using only two-dimensional spectra was limited by spectral overlap, the variability of the patterns observed, and the more or less inevitable presence of artifacts in experimental data. The human mind's pattern recognition capabilities and the judgment of an experienced spectroscopist in recognizing partial patterns and discarding noise and artifacts are extremely difficult to repro-

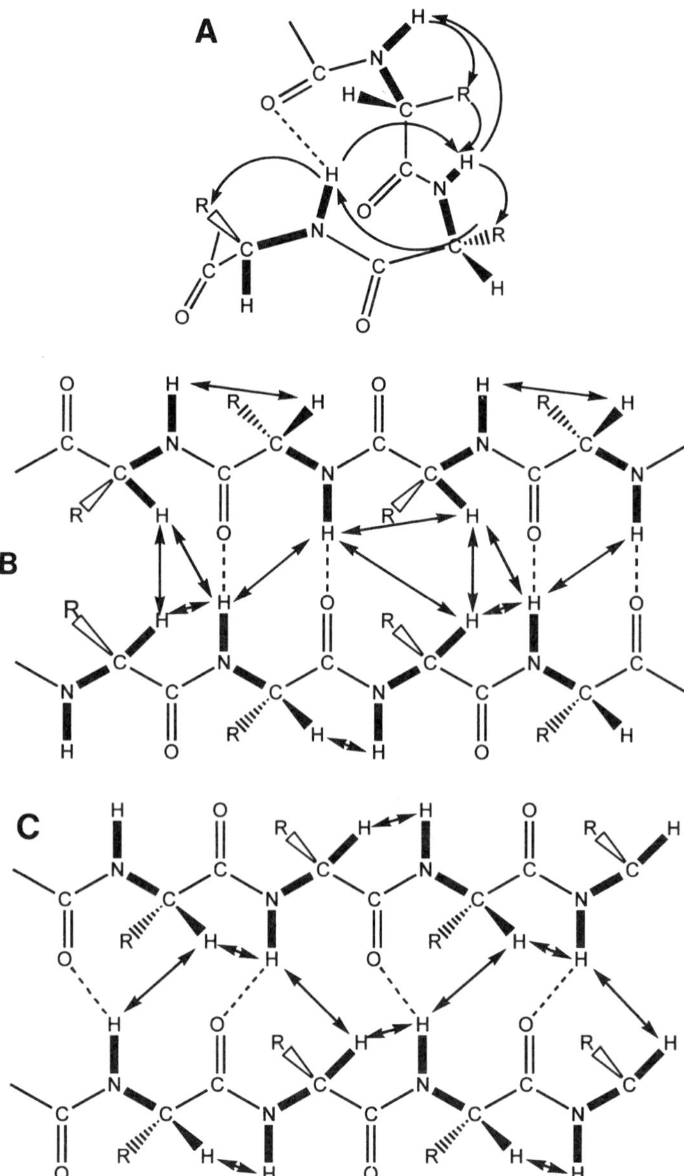

Figure 2. Characteristic cyclic NOE (arrows) and J-coupling (thick lines) connectivities observed for the common secondary structural elements: (A) helix; (B) antiparallel sheet; (C) parallel sheet. Tracing these cyclic paths facilitates peak assignments and identification of secondary structure. (Adapted from Englander and Wand, 1987.)

duce in computer software. Three- and four-dimensional data, which have much reduced overlap, appear to be more tractable (Olejniczak *et al.*, 1992a; Bernstein *et al.*, 1993). Although totally automated assignments may not yet be feasible, the transition from "human-aided" to "computer-aided" assignment has probably been made.

2.4.4. FUTURE PROGRESS

Protein NMR has been a fertile area for continuing advances in methodology, and the size of proteins that have proven to be tractable has steadily risen (Wagner, 1993). The growth in the number of groups using these techniques virtually guarantees continuing increases in the number of protein structures determined by NMR. However, future studies of proteins as large or larger than those that have been studied to date depend on identifying conditions for each protein under which it is soluble and stable enough to perform the experiments. Such conditions cannot always be found for proteins of interest, and searching for the conditions can be time consuming. There are no proven general solutions to this problem or universally applicable strategies for finding a solution for a particular protein, although intriguing recent work has shown that addition of osmolytes such as (deuterated) glycine to protein solutions can substantially increase the thermal stability of at least some proteins (Matthews and Leatherbarrow, 1993). Future progress in the area depends as much on the efforts of protein chemists and biochemists as on spectroscopists.

3. COMPUTATIONAL TECHNIQUES

While the impact of new techniques has been essential in allowing solution structure determination by NMR to occur, the ability to couple that data with computational models has been just as important. Listed below are descriptions of computational methods that permit construction of three-dimensional structures of a protein using NMR data. Together, these computational approaches, coupled with the expanded capabilities of NMR data collection, allow structural determinations that rival those of X-ray crystallography in their precision and quality. In fact, many proteins have now been structurally characterized by both methods.

3.1. Distance Geometry

Distance geometry is one of the most widely used techniques for protein structural determination (Braun, 1987; Braun and Go, 1985; Crippen, 1981; Crippen and Havel, 1988; Havel, 1991; Havel and Wüthrich, 1984, 1985; Inagaki, 1990; Kuntz *et al.*, 1989). It is a mathematical technique for building structures based on a set of constraints on distances between parts of the structure. It has the advantage of generating viable structures relatively quickly (vastly faster than any systematic search). Since many structures can be generated from random starting points using the same constraints, the degree to which the data defines a unique conformation can be evaluated. The disadvantage of distance geometry is that the structures generated are not necessarily energetically favorable. Using structures generated from distance geometry as the starting point for computationally more intensive procedures such as molecular dynamics, where a reasonable starting conformation is necessary because it is impractical to do a large number of simulations, is a natural choice.

From the NMR data on a large protein, hundreds or even thousands of constraints on distances and angles can be identified using the experiments described above; however, each represents only a local part of the conformation (distances $\leqslant 5$ Å, torsion angles around three bonds). Even when the known primary sequence, the covalent geometry and chirality of each residue, and the idealized secondary structure are included as constraints, the vast majority of distances between pairs of atoms in the protein are not determinable from NMR experiments. (Additional distance estimates from other spectroscopic techniques or from theory can in principle also be included in the input to the distance geometry program; these might be particularly valuable if they are longer-range distances than NOEs, but in practice this is infrequently done.) The desired information, a set of *xyz* coordinates for the atoms in the protein, is equivalent to the complete matrix of all such pairwise distances. The function of the distance geometry algorithm is to generate all the unknown distances from the subset of known distances with the fewest number of internal inconsistencies or violations of the constraints.

The experimentally determined distances are always subject to uncertainties, usually entered in the constraints in the form of upper and lower bounds on the distances. From the constraints, a distance matrix of $(N*N1)/2$ distances, where N is the number of atoms in the protein, is generated, with the upper bounds in one half of the matrix and the lower bounds in the other. Most of the distances are initially unknown; these are set to the sum of the van der Waals radii as a lower bound and some very large value as an upper bound. More reasonable estimates are generated by propagating a series of "triangle inequalities" involving each set of three atoms, in which two distance estimates (i–j, i–k) are used to estimate the third (j–k), which is fixed randomly in the range allowed by the bounds. Because this procedure moves the estimated upper and lower values closer together, it is known as *bounds smoothing*. The estimated distances are then converted into *xyz* coordinates by a procedure called *embedding*. Because of the uncertainties in the input data and the use of random distances within the upper and lower bounds, repeated iterations of bounds smoothing and embedding will result in somewhat different structures. These can be evaluated in terms of the number and extent of bounds violations or internal inconsistencies.

As one might expect, the resulting trial structure will be distorted, since no allowance has been made in the distance geometry algorithm for conformational energy. Energy minimization can then be applied to generate a more reasonable structure; annealing (see Section 3.3) or small random displacements of the atoms, followed by further minimization, may be used to sample additional regions of conformational space. Repeated runs must give similar structures if the NMR data are to be believed; however, the reproducibility of the calculated structure does not in itself prove the structure is correct. A necessary and more stringent test is comparison of the experimental NOESY spectrum with one that has been calculated from the structure. In theory, each conformation will eventually converge to a family of closely related structures, although some errors are possible, especially if the number of constraints is limited (Metzler *et al.*, 1989; Oshiro *et al.*, 1991). For example, if the number of distance constraints is less than the number of degrees of freedom of the peptide backbone, inaccurate structures may result (Oshiro *et al.*, 1991).

In addition to distance constraints obtained from NOESY spectra, torsional restraints can be included as well. This is especially important for defining side-chain conformation. Often, distance geometry is the method used to generate a preliminary set of structures that can be further refined by other methods, such as restrained molecular dynamics or simulated annealing.

3.2. Restrained Molecular Dynamics

While most computational techniques provide a static description of a protein structure, molecular dynamics offers the ability to model the motion of a protein in real time. Therefore, it is a powerful alternative to static descriptions generated by molecular mechanics. The field of molecular dynamics is well established and has been widely described (McCammon, 1987; McCammon and Harvey, 1987). The approach of NMR experimentalists has been to modify the force fields to accommodate distance constraints determined from NOE measurements (Scheek *et al.*, 1989; Gronenborn and Clore, 1989; Williamson and Madison, 1990).

3.3. Simulated Annealing

In order to remove the possibility that a minimized structure is trapped in a local energy minimum rather than the global minimum, it is possible to expose the protein computationally to an elevated temperature that will allow it to move across the energy surface and sample a greater amount of conformational space, thereby allowing it to find the global minimum. This approach is called *simulated annealing* (Habazettl *et al.*, 1990; Nilges *et al.*, 1988a–c) and can be used separately or in conjunction with other computational techniques (such as distance geometry) to determine the lowest energy structure from NMR data.

4. CASE HISTORIES

A number of proteins have been characterized using NMR spectroscopy. These studies have ranged from preliminary analyses using simpler one-dimensional techniques to complete determinations of the overall structure in solution. It is important to remember that NMR can be used to evaluate the stability as well as the structure of a peptide or protein. The ever increasing capabilities of NMR spectroscopy have often been used to investigate the behavior of proteins of pharmaceutical interest.

While a description of NMR methods is important, it is equally instructive to examine the application of these techniques to specific systems. The following section describes the work done to date on proteins of pharmaceutical interest. In addition to proteins that have already been approved for use in humans, proteins that are under consideration as potential therapeutic agents are also included in this discussion.

4.1. Interleukin-1β

Interleukin-1β (IL-1β) has been one of the proteins most widely studied by NMR spectroscopy (Clore *et al.*, 1990b–e, 1991a,b; Clore and Gronenborn, 1991c; Driscoll *et al.*, 1990a,b; Gronenborn *et al.*, 1986; Marion *et al.*, 1989; Wingfield *et al.*, 1987, 1989). It is a 153-residue cytokine and is a primary mediator of inflammation. Consequently, it has been an important target molecule for the pharmaceutical industry.

Initial experiments focused on assignment of resonances from aromatic residues (Gronenborn *et al.*, 1986; Wingfield *et al.*, 1989), and using one-dimensional techniques found little differences between six mutants. At the time, complete structural determinations on proteins of this size were not possible. However, in less than two years, the ability to perform multidimensional techniques (three- and four-dimensional) on isotopically labeled proteins increased the capabilities of NMR significantly.

Enrichment of IL-1β in ^{13}C and ^{15}N permitted Clore *et al.* (1990d) to make assignments of the side-chain protons, in addition to those of the backbone.

Assignments were made for all side chains except for 4 (of the 15) lysine side chains. Improved dispersion over ^{15}N-^{1}H three-dimensional spectra (Driscoll *et al.*, 1990a,b) was observed and made nearly complete ^{1}H assignments possible for all 153 residues. The work of Driscoll *et al.* (1990a) on ^{15}N-enriched IL-1β had provided enough information to allow the examination of the backbone protons only for the wild-type protein as well as five mutants (A1C, K27C, K93A, K94A, and C71S) in much more detail than was previously possible. In addition, Driscoll *et al.* (1990b) were able to make assignments for all of the backbone protons, allowing them to determine the positions of secondary structural elements and found them to be consistent with the crystallographic structure (Priestle *et al.*, 1988, 1989).

A low-resolution structure of IL-1β was generated from the previous three-dimensional NMR data using 446 NOESY-derived distance constraints, 45 hydrogen bond distance constraints, and 79 torsion angle constraints (Clore *et al.*, 1990b). A final structure with 12 β-strands in pseudo-threefold symmetry was obtained, similar to the structure determined by X-ray crystallography. While the β-sheet region was relatively well defined, the adjoining loops displayed very high root mean square deviations. The difference in the two structures was calculated to have a 1.5 Å root mean square deviation.

Doubly labeled IL-1β (^{15}N- and ^{13}C-enriched) was the subject of one of the first triple-resonance four-dimensional NMR experiments (Kay *et al.*, 1990a). The increase in resolution of resonances by the use of a fourth dimension led the authors to speculate that the entire NMR spectrum of proteins as large as 40 kDa (300–350 residues) could be assigned. In addition, it appeared that the increased dispersion could allow lower field strengths (<600 MHz) to be used.

Shortly thereafter, Clore *et al.* (1991a) reported four-dimensional ^{13}C–^{13}C-edited NOESY spectra of IL-1β. The authors state that high-quality spectra could be obtained in as little as three days. This work provided NMR spectroscopists with one more method for investigation of proteins in the size range of 100 to 200 residues.

Once four-dimensional NMR data were available, a high-resolution structure of IL-1β was generated, based on over 2700 distance constraints and nearly 400 torsional angle constraints (Clore *et al.*, 1991b). A family of 32 structures was generated using simulated annealing. They displayed a root mean square deviation of 0.82 Å for all atoms and only half of that for the peptide backbone. The structure obtained in this fashion could be considered comparable to an X-ray structure of 2–2.5 Å resolution. Whereas earlier work provided a description of the overall topology of IL-1β, this structure delineates side-chain positions as well, permitting detailed analysis of the structural basis of IL-1β biological activity.

A detailed comparison of the solution conformation versus the X-ray structure has been presented by Clore and Gronenborn (1991c). The root mean square deviation for the two structures is 1.5 Å for all atoms and ~0.9 Å for the backbone atoms. Most of the difference arises from increased flexibility in the loop regions of the NMR structure compared to IL-1β in a single crystal. Positional errors from the two methods seem to be of equal magnitude.

While NMR can be employed to obtain the conformation of a structure in solution, those models are only static pictures of its overall conformation. On the other hand, techniques like ^{15}N T_1 and T_2 relaxation measurements can provide a description of the dynamics of a protein structure. Clore *et al.* (1990c) have measured these values for IL-1β and found evidence for motions on three different time scales (20–50 psec, 0.5–4 nsec, and 30–10,000 nsec). In addition, there appears to be a slow interconversion between the major isomer of IL-1β and a minor form involving 19 residues on part of one face of the molecule.

One of the most exciting aspects of modern NMR spectroscopic studies is the ability to determine the positions of bound water molecules. This provides additional insight into the role of solvation in governing protein structure and stability. Using three-dimensional ROESY techniques, Clore *et al.* (1990e) have identified 11 water molecules that are involved in hydrogen bonding with the backbone and appear to be important on stabilizing the pseudo-threefold structure of IL-1β.

4.2. Interleukin-1α

Unlike IL-1β, IL-1α has received relatively little attention (Gronenborn *et al.*, 1988). Despite a significant lack of homology with IL-1β or with IL-1 receptor antagonist, the two proteins bind to the same receptor. Gronenborn *et al.* (1988) made sequence-specific assignments of the histidine and tryptophan residues and examined some mutants, but no comprehensive structural determination of IL-1α has been reported.

4.3. Interleukin-1 Receptor Antagonist

In addition to the two major forms of IL-1, there is a protein that binds to IL-1 receptors but exhibits antagonistic properties that is termed interleukin-1 receptor antagonist or IL-1ra. The naturally occuring mature form has 152 residues, while the recombinant version produced in *Escherichia coli* possesses 153 residues. Stockman *et al.* (1992) have reported a multidimensional heteronuclear NMR study of IL-1ra using either ^{15}N-labeled or ^{15}N–^{13}C doubly labeled protein. Despite significant differences in homology with IL-1β, the overall structure of IL-1ra was found to be similar. IL-1ra possesses 12 strands of β-sheet structure and two short α-helixes. The β-sheet segments do occur at different residues than in IL-1β, but the size and number of strands are similar. Thus, the activity of these proteins must be controlled by specific side-chain interactions with the receptor, a fact supported by site-directed mutagenesis studies and comparisons with the rat protein.

4.4. Interleukin-4

Interleukin-4 (IL-4) is a 133-residue cytokine responsible for the production of cytotoxic T cells and expression of immunoglobulin E (IgE) receptors (Vercelli *et al.*, 1989). The major structural feature of IL-4 is the central four helix bundle, which adopts a left-handed up–up–down–down folding pattern. The conformation in aqueous solution has been determined in two separate studies, one conducted on recombinant IL-4 (rIL-4) expressed in *E. coli* (Redfield *et al.*, 1991; Smith *et al.*, 1992) and the other a modified IL-4 expressed in the yeast *Saccharomyces cerevisiae* (Garrett *et al.*, 1992; Powers *et al.*, 1992a,b).

The protein produced in yeast was labeled uniformly with both ^{15}N and ^{15}N–^{13}C. The two potential N-linked glycosylation sites were mutated to Asp to prevent any posttranslational modification. Both double- and triple-resonance experiments were employed to determine complete assignments of all resonances (Powers *et al.*, 1992a). Once those were determined, the organization of the secondary structure and overall global fold of the protein could be constructed. Four distinct helixes, packed in a left-handed bundle, were detected, with each helix being from 13 to 19 residues in length (Powers *et al.*, 1992b; Garrett *et al.*, 1992). The tertiary structure of IL-4 was found to be similar to that of other protein hormones.

Similar results were obtained for *E. coli*-produced IL-4. Double-resonance three-dimensional NMR techniques were used on material that was labeled with ^{15}N. Complete assignments were determined for all but three residues (Redfield *et al.*, 1991) and evidence was found for the four-helix bundle. This was further corroborated using doubly labeled ^{13}C–^{15}N protein (Smith *et al.*, 1992).

The four-helix bundle motif is common among protein hormones with pharmaceutical significance. For example, it occurs in human growth hormone (Abdel-Meguid *et al.*, 1987), complement C5a (Zuiderweg *et al.*, 1989), and in granulocyte-macrophage-colony-stimulating factor (Walter *et al.*, 1992). The left-handed up-up-down-down packing of the helixes found in IL-4 is the most common arrangement of helixes in a bundle (Chothia *et al.*, 1981) and appears to be the most stable (Chou *et al.*, 1988; Gilson and Honig, 1989)

Mobility of various segments of IL-4 has been evaluated using [^{15}N]-NMR (Redfield *et al.*, 1992). The central four-helix core of IL-4 appears to be relatively rigid, while the connecting loop regions are fairly mobile.

4.5. Interleukin-6

Interleukin-6 (IL-6) is a multifunctional 185-residue cytokine responsible for the differentiation and growth of stem cells, B cells, T cells, and nerve cells (Ikebuchi *et al.*, 1987; Kishimoto and Hirano, 1988; Satoh *et al.*, 1988). Recently, Nishimura *et al.* (1990) reported partial assignment of the aromatic resonances of IL-6. Using deuterium exchange, the assignment of the imidazole resonances of His^{16} and His^{165} were determined. Iodination of tyrosine residues permitted the characterization of Tyr^{32} resonances and was found not to affect the activity of IL-6. Finally, photochemically induced dynamic nuclear polarization (CIDNP) studies indicated that His^{16}, Tyr^{32}, and Trp^{158} were solvent exposed, while His^{165}, Tyr^{98}, and Tyr^{101} were buried inside the protein. Further studies using [^{1}H]-NMR spectroscopy and chemical modification of Met and Trp residues suggest that Trp^{158} and Met^{162} are near the site recognized by the IL-6 receptor (Nishimura *et al.*, 1991).

4.6. Interleukin-8

Interleukin-8 (IL-8) is another member of the cytokine family that displays numerous biological functions. Consequently, it is known by other names, such as neutrophil activation factor, monocyte-derived chemotactic factor, mitogen-stimulated leucocyte protein, and T-cell chemotactic factor. It is a small protein (72 residues) that possesses sequence homology with other peptides involved in inflammation and the immune system, such as platelet factor 4 and human growth-related protein.

Recently, Clore, Gronenborn, and co-workers have reported detailed structural characterization of IL-8 in aqueous solution (Baldwin *et al.*, 1991; Clore and Gronenborn, 1991b; Clore *et al.*, 1989, 1990a). The recombinant protein was expressed in *E. coli* and was examined using both three- and four-dimensional [^{1}H]-NMR techniques (Clore *et al.*, 1989, 1990a).

Initial studies determined the existence and location of major secondary structural features using conventional two-dimensional [^{1}H]-NMR techniques on unlabeled proteins. Clore *et al.* (1989) identified three strands of a β-sheet and a long, C-terminal α-helix (residues 57–72). Evidence for dimerization was observed, with the monomers interacting at the edges of the β-sheet, forming an extended six-strand sheet. The solution conformation was further refined using distance constraints obtained by three-dimensional NMR methods (Clore *et al.*, 1990a). A set of 1880 distance constraints and 362 torsion angle restraints were used to generate initial structures via distance geometry. These structures were refined using simulated annealing. The final set of 30 structures exhibited a root mean square deviation of 0.41 Å for the backbone atoms and 0.90 Å for all atoms, indicating a highly rigid and refined structure. These small root mean square deviation values correspond to 12–14 restraints per residue and produce a structure that is comparable in resolution to that of a 2 to 2.5 Å resolution crystal structure (Clore and Gronenborn, 1991a).

Comparison with the structure obtained from X-ray crystallography indicates there are slight differences between the solution and crystal structures (Clore and Gronenborn, 1990b; Baldwin, 1991). The β-turn at residues 4 through 7 is ill-defined in the solution structure but well-ordered in the crystal structure, plus there are modest differences in the loop comprising residues 31 to 36. Although these differences could be considered minor, it appears that both regions are important for receptor binding and suggest subtle conformational changes that may govern biological function.

4.7. Insulinlike Growth Factor

There are growth factors that display a marked homology with insulin and exhibit weak hypoglycemic activity as well as potent mitogenic effects. Insulinlike growth factor I (IGF I) (also known as somatomedin C) is a 70-residue polypeptide with known sequence. A two-dimensional NMR study of recombinant IGF I from *E. coli* has been reported. Assignments were made for all backbone protons and secondary structures identified (Sato *et al.*, 1992). Three α-helical regions (residues 8–18, 42–48, 54–61) were found in addition to a β-turn at positions 19–22. These features are similar to the structural motifs found in crystalline insulin (Baker *et al.*, 1988).

4.8. Insulin

While the three-dimensional structure of insulin has been known from X-ray crystallography for some time (Baker *et al.*, 1988), the determination of its solution conformation is relatively recent. For some time, the [^{1}H]-NMR spectrum of insulin resisted complete assignment, and work centered on the aromatic residues (Bradbury and Ramesh, 1981, 1985; Ramesh and Bradbury, 1986; Hua *et al.*, 1989). Only recently have complete assignments for the proton resonances of insulin been assigned (Kline and Justice, 1990). Assignments were complicated by the aggregation behavior of insulin (monomer ↔ dimer ↔ hexamer). Therefore, much of the structural work has centered on insulin derivatives, especially those that are known to be monomeric in solution. For example, des-(B26-B30)-insulin (DPI) is monomeric and was studied in aqueous solution using two-dimensional [^{1}H]-NMR (Knegtel *et al.*, 1991). From the NOESY spectrum, 250 distance constraints were identified. Distance geometry was employed to generate an initial set of structures. These structures then formed the basis of restrained molecular dynamics simulations, which were subsequently energy minimized. The resulting structures resembled the monomeric units in porcine 2Zn insulin, except for increased mobility of the termini of each chain.

The (B9 Asp) mutant of insulin has also been characterized by two-dimensional [^{1}H]-NMR (Jorgensen *et al.*, 1992). It is primarily dimeric in solution. More than 450 NOEs were detected and distance geometry in conjunction with restrained molecular dynamics was used to generate final structures. The root mean square deviation for the final structures was 2.26 Å for the backbone and 3.14Å for all atoms. Overall, both the secondary and tertiary structure were similar to that of 2Zn insulin.

The effect of metals on the structure of insulin has been studied by NMR spectroscopy. Palmieri *et al.* (1988) reported that coordination of either zinc or cadmium within insulin hexamers resulted in upfield shifts of the C-2 resonances of His^{B10}. However, it was the binding of metals such as calcium and cadmium to Glu^{B13} that drove hexamer assembly.

4.9. Interferon-γ

Interferon-γ displays antiviral as well as antitumor effects. It has presented a problem to NMR spectroscopists due to its size; it is a homodimer weighing more than 30 kDa. Recently, assignments were made for a truncated form of recombinant interferon-γ that did not contain the last ten residues (Grzesiek *et al.*, 1992a). The protein was labeled with both ^{15}N and ^{13}C, permitting both three- and four-dimensional experiments to be performed. In particular, triple-resonance ^{15}N and ^{15}N–^{13}C-edited-NOESY spectra (Ikura *et al.*, 1990; Kay *et al.*, 1990a,b) were obtained and formed the basis for determination of the secondary structure of interferon-γ in aqueous solution (Grzesiek *et al.*, 1992a). The secondary structure composition was found to be ~70%, arranged in six helixes per monomer. This was consistent with a low-resolution X-ray structure previously reported (Ealick *et al.*, 1991).

The only other NMR work reported on interferon-γ was analyses of the carbohydrate side chains of recombinant material expressed in Chinese hamster ovary cells (Mutsaers *et al.*, 1986) and in a myelomonocyte cell line (Yamamoto *et al.*, 1989).

4.10. Epidermal Growth Factor

One of the most potent mitogenic factors known, epidermal growth factor (EGF) is also one of the proteins most widely studied by NMR (Campbell *et al.*,

1990; Carver *et al.*, 1986; Cooke *et al.*, 1987, 1990; De Marco, 1983, 1986; Dudgeon *et al.*, 1990; Engler *et al.*, 1990; Han *et al.*, 1988, 1990; Hommel *et al.*, 1991, 1992; Kohda and Inagaki, 1988, 1992a,b; Kohda *et al.*, 1988a,b, 1989, 1991; Koide *et al.*, 1992; Makino *et al.*, 1987; Matsunami *et al.*, 1991; Mayo, 1984, 1985; Mayo *et al.*, 1986a,b; Mayo and Burke, 1987; Montelione *et al.*, 1986, 1987, 1988, 1992; Moy *et al.*, 1989). Due to the numerous works in this area, many investigations will be treated only briefly.

Work on EGF has centered on either the murine (mouse), the rat, or the human form, either in aqueous solution or in the presence of micelles. Each system will be discussed separately. While much work has been done using photo-CIDNP and one-dimensional methods (De Marco *et al.*, 1983, 1986; Mayo, 1984; Mayo *et al.*, 1986a,b), the discussion will emphasize work that has employed multidimensional techniques.

4.10.1. MURINE EGF

Initial assignments and identification of a significant β-sheet were reported by Mayo (1985), using both one- and two-dimensional methods. The secondary structure was further defined by Montelione *et al.* (1986). Both Montelione *et al.* (1986, 1987, 1988) and Kohda and Inagaki (1988) reported complete assignments for all the protons in murine EGF (mEGF). Complete descriptions of the solution conformation have been developed using simulated annealing (Kohda and Inagaki, 1992b) and distance geometry (Kohda *et al.*, 1988a; Montelione *et al.*, 1992). In the work of Kohda and Inagaki (1992b), the effects of both high and low pH (6.8 vs. 2.0) were examined. Little structural difference was observed, except in the region of the C-terminus. In particular, the region surrounding the essential Leu^{47} was disrupted, presumably due to protonation of the side chain of Asp^{46}. Protonation of the N-terminal amino group also caused changes, particularly in the β-sheet core (Kohda *et al.*, 1991).

Isolation of mEGF by high-pressure liquid chromatography (HPLC) produces four distinct forms. The α form has been identified as intact EGF(1-53). The β form has been found to be des-Arg^1-EGF, and the γ form is des-Arg^1, Ser^2-EGF. Mayo and Burke (1987) report that the truncated forms (β and γ) appear to adopt more open or denatured structures. This is in general agreement with the findings of Kohda *et al.* (1988b) on the α and β forms of mEGF. Trypsin digestion of mEGF produced both the 1-48 and the 1-45 fragment. After purification by HPLC, these fragments were studied by [^{1}H]-NMR (Menegatti *et al.*, 1989).

A number of mutants of EGF have been prepared and characterized by multidimensional NMR techniques. For example, Moy *et al.* (1989) have shown that the Ser^{47} mutant of mEGF displays a structure nearly identical to the wild type, suggesting that the inherent differences in the chemical nature of the position-47 side chains that responsible for binding, and not differences in conformation.

4.10.2. RAT EGF

Mayo *et al.* (1989) have reported spectral assignments and secondary structure for rat EGF. In general, the rat protein seemed very similar to the murine and human EGFs.

4.10.3. HUMAN EGF

Carver *et al.* (1986) and Makino *et al.* (1987) reported crude structures of human EGF (hEGF) early on, using two-dimensional NMR spectroscopy. However, the work by Carver *et al.* (1986) was on the 1-48 fragment. Later, Cooke *et al.* (1987) reported a more detailed structure of the 1-48 fragment. The work by Cooke and co-workers employed both distance geometry and restrained molecular dynamics approaches. Other workers have attempted to improve on the structures in various ways. Han *et al.* (1990) have used restrained molecular dynamics to produce a structure of intact hEGF.

Site-directed mutagenesis, in concert with NMR studies, has formed the basis for delineating the structure–function relationships for this hormone. Recently, Matsunami *et al.* (1991) reported NMR studies on seven mutants of hEGF, all replacing Leu^{47}, the residue necessary for strong binding to the receptor. In no case did any mutant have a significantly different structure than the wild type. Other workers have examined mutations at Arg^{41} (Hommel *et al.*, 1991; Engler *et al.*, 1990), Tyr^{13} (Hommel *et al.*, 1991), Tyr^{37} (Engler *et al.*, 1990), Ala^{30} and Asn^{32} (Koide *et al.*, 1992), and Leu^{47} (Dudgeon *et al.*, 1990; Matsunami *et al.*, 1991).

4.10.4. EGF IN MICELLES

The study of EGF conformation and behavior in the presence of micelles has been wide-ranging as well. Mayo *et al.* (1987) found that the solvent accessibility of EGF residues Tyr^{37}, Trp^{49}, and Trp^{50} is markedly reduced in both sodium dodecyl sulfate and dodecylphosphocholine vesicles. A more detailed investigation by Kohda and Inagaki (1992a) indicates that in dodecylphosphocholine micelles, EGF exhibits little perturbation of the core 1-45 residues, but large chemical shift effects on the C-terminal tail (residue 46-53). Finally, a three-dimensional structure was obtained using simulated annealing approaches in combination with the NMR data.

5. SUMMARY

Advances in NMR spectroscopy and related computational methods continue at a rapid pace. In the past three years, the capability to make complete assignments of protein spectra has expanded from a limit of approximately 100 residues to a limit of possibly 400 residues via isotope-edited three- and four-dimensional methods.

REFERENCES

Abdel-Meguid, S. S., Shieh, H.-S., Smith, W. W., Dayringer, H. E., Violand, B. N., and Bentle, L. A., 1987, Three-dimensional structure of a genetically engineered variant of porcine growth hormone, *Proc. Natl. Acad. Sci. USA* **84:**6434–6437.

Archer, S. J., Ikura, M., Torchia, D., and Bax, A., 1992, An alternative 3D NMR technique for correlating backbone ^{15}N with side chain Hβ resonances in larger proteins, *J. Magn. Reson.* **95:**636–641.

Aue, W. P., Bartholdi, E., and Ernst, R. R., 1976, Two-dimensional spectroscopy. Application to nuclear magnetic resonance, *J. Chem. Phys.* **64:**2229–2246.

Baker, E. N., Blundell, T. L., Cutfield, J. F., Cutfield, S. M., Dodson, E. J., Dodson, G. G., Crowfoot Hodgkin, D. M., Hubbard, R. E., Isaacs, N. W., Reynolds, C. D., Sakabe, K., Sakabe, N., and Vijayan, N. M., 1988, The structure of 2Zn pig insulin crystals at 1.5 Å resolution, *Philos. Trans. R. Soc. London B* **319:**369–456.

Baldwin, E. T., Weber, I. T., St. Charles, R., Xuan, J. C., Appella, E., Yamada, M., Matsushima, K., Edwards, B. F., Clore, G. M., and Gronenborn, A. M., 1991, Crystal structure of interleukin 8: Symbiosis of NMR and crystallography, *Proc. Natl. Acad. Sci. USA* **88:**502–506.

Bax, A., and Davis, D. G., 1985a, Practical aspects of two-dimensional transverse NOE spectroscopy, *J. Magn. Reson.* **63:**207–213.

Bax, A., and Davis, D. G., 1985b, MLEV-17 based two-dimensional homonuclear magnetization transfer spectroscopy, *J. Magn. Reson.* **65:**355–360.

Bax, A., and Morris, G., 1981, An improved method for heteronuclear chemical shift correlation by two-dimensional NMR, *J. Magn. Reson.* **42:**501–505.

Bax, A., and Subramanian, S., 1986, Sensitivity-enhanced two-dimensional heteronuclear shift correlation NMR spectroscopy, *J. Magn. Reson.* **67:**565–569.

Bax, A., and Summers, M., 1986, ^{1}H and ^{13}C assignments from sensitivity-enhanced detection of heteronuclear multiple-bond connectivity by 2D multuple quantum NMR, *J. Am. Chem. Soc.* **108:**2093–2094.

Bax, A., DeJong, P. G., Mehlkopf, A. F., and Smidt, J., 1980, Separation of the different orders of NMR multiple-quantum transitions by the use of pulsed field gradients, *Chem. Phys. Lett.* **69:**567–570.

Bax, A., Clore, G. M., and Gronenborn, A. M., 1990, ^{1}H–^{1}H correlation via isotropic mixing of ^{13}C magnetization, a new three-dimensional approach for assigning ^{1}H and ^{13}C spectra of ^{13}C-enriched proteins, *J. Magn. Reson.* **88:**425–431.

Bernstein, R., Cieslar, C., Ross, A., Oschkinat, H., Freund, J., and Holak, T., 1993, Computer-assisted assignment of multidimensional NMR spectra of proteins: Application to 3D NOESY-HMQC and TOCSY-HMQC spectra, *J. Biomol. NMR* **3:**245–251.

Bodenhausen, G., Kogler, H., and Ernst, R. R., 1984, Selection of coherence-transfer pathways in NMR pulse experiments, *J. Magn. Reson.* **58:**370–388.

Boelens, R., Vuister, G. W., Koning, T. M. G., and Kaptein, R., 1989, Observation of spin diffusion in biomolecules by three-dimensional NOE–NOE spectroscopy, *J. Am. Chem. Soc.* **111:**8525–8526.

Bothner-By, A. A., 1984, Structure determination of a tetrasaccharide: Transient nuclear overhauser effects in the rotating frame, *J. Am. Chem. Soc.* **106:**811–813.

Boucher, W., Laue, E. D., Campbell-Burk, S. L., and Domaille, P. J., 1992, Improved 4D NMR experiments for the assignment of backbone nuclei in $^{13}C/^{15}N$ labelled proteins, *J. Biomol. NMR* **2**:631–637.

Bradbury, J. H., and Ramesh, V., 1981, ^{1}H NMR study of the histidine residues of insulin, *J. Mol. Biol.* **150:**609–613.

Bradbury, J. H., and Ramesh, V., 1985, ^{1}H NMR studies of insulin. Assignment of resonances and properties of tyrosine residues, *Biochem. J.* **229:**731–737.

Braun, W., 1987, Distance geometry and related methods for protein structure determination from NMR data, *Quart. Rev. Biophys.* **19:**115–157.

Braun, W., and Go, N., 1985, Calculation of protein conformation by proton–proton distance constraints: A new efficient algorithm. *J. Mol. Biol.* **186:**611–626.

Breg, J., Boelens, R., Vuister, G. W., and Kaptein, R., 1990, 3D NOE–NOE spectroscopy of proteins. Observation of sequential 3D NOE cross peaks in arc repressor, *J. Magn. Reson.* **87:**646–651.

Brünger, A., and Karplus, M., 1991, Molecular dynamics simulations with experimental results, *Acc. Chem. Res.* **24:**54–61.

Campbell, I. D., Baron, M., Cooke, R. M., Dudgeon, T. J., Fallon, A., Harvey, T. S., and Tappin, M. J., 1990, Structure–function relationships in epidermal growth factor (EGF) and transforming growth factor-alpha (TGF-α), *Biochem. Pharmacol.* **40:** 35–40.

Carver, J. A., Cooke, R. M., Esposito, G., Campbell, I. D., Gregory, H., and Sheard, B., 1986, A high-resolution ^{1}H NMR study of the solution structure of human epidermal growth factor, *FEBS Lett.* **205:**77–81.

Chary, K. V. R., Otting, G., and Wüthrich, K., 1991, Measurement of small heteronuclear ^{1}H-^{15}N coupling constants in ^{15}N-labeled proteins by 3D HNNHAB-COSY, *J. Magn. Reson.* **93:**218–224.

Chothia, C., Levitt, M., and Richardson, D., 1981, Helix to helix packing in proteins, *J. Mol. Biol.* **145:**215–250.

Chou, K.-C., Maggiora, G. M., Nemethy, G., and Scheraga, H. A., 1988, Energetics of the structure of the four alpha-helix bundle in proteins, *Proc. Natl. Acad. Sci. USA* **85:** 4295–4299.

Clore, G. M., and Gronenborn, A. M., 1982, Theory and applications of the transferred nuclear overhauser effect to the study of the conformations of small ligands bound to proteins, *J. Magn. Reson.* **48:**402–417.

Clore, G. M., and Gronenborn, A. M., 1983, Theory of the time-dependent transferred nuclear overhauser effect: Applications to structural analysis of ligand–protein complexes in solution, *J. Magn. Reson.* **53:**423–442.

Clore, G. M., and Gronenborn, A. M., 1991a, Structure of larger proteins in solution: Three- and four-dimensional heteronuclear NMR spectroscopy, *Science* **252:**1390–1399.

Clore, G. M., and Gronenborn, A. M., 1991b, Comparison of the solution nuclear magnetic resonance and crystal structures of interleukin-8. Possible implications for the mechanism of receptor binding, *J. Mol. Biol.* **217:**611–620.

Clore, G. M., and Gronenborn, A. M., 1991c, Comparison of the solution nuclear magnetic resonance and X-ray crystal structures of human recombinant interleukin-1β, *J. Mol. Biol.* **221:**47–53.

Clore, G. M., Appella, E., Yamada, M., Matsushima, K., and Gronenborn, A. M., 1989, Determination of the secondary structure of interleukin-8 by nuclear magnetic resonance spectroscopy, *J. Biol. Chem.* **264:**18907–18911.

Clore, G. M., Appella, E., Yamada, M., Matsushima, K., and Gronenborn, A. M., 1990a, Three-dimensional structure of interleukin 8 in solution, *Biochemistry* **29:**1689–1696.

Clore, G. M., Driscoll, P. C., Wingfield, P. T., and Gronenborn, A. M., 1990b, Low resolution structure of interleukin-1β in solution derived from ^{1}H-^{15}N heteronuclear three-dimensional nuclear magnetic resonance spectroscopy, *J. Mol. Biol.* **214:** 811–817.

Clore, G. M., Driscoll, P. C., Wingfield, P. T., and Gronenborn, A. M., 1990c, Analysis of the backbone dynamics of interleukin-1β using two-dimensional inverse detected heteronuclear ^{15}N-^{1}H NMR spectroscopy, *Biochemistry* **29:**7387–7401.

Clore, G. M., Bax, A., Driscoll, P. C., Wingfield, P. T., and Gronenborn, A. M., 1990d, Assignment of the side-chain ^{1}H and ^{13}C resonances of interleukin-1β using double- and triple-resonance heteronuclear three-dimensional NMR spectroscopy, *Biochemistry* **29:**8172–8184.

Clore, G. M., Bax, A., Wingfield, P. T., and Gronenborn, A. M., 1990e, Identification and localization of bound internal water in the solution structure of interleukin 1β by heteronuclear three-dimensional ^{1}H rotating frame Overhauser ^{15}N-^{1}H miultiple quantum coherence NMR spectroscopy, *Biochemistry* **29:**5671–5676.

Clore, G. M., Kay, L. E., Bax, A., and Gronenborn, A. M., 1991a, Four-dimensional $^{13}C/^{13}C$-edited nuclear overhauser enhancement spectroscopy of a protein in solution: Application to interleukin 1β, *Biochemistry* **30:**12–18.

Clore, G. M., Wingfield, P. T., and Gronenborn, A. M., 1991b, High-resolution three-dimensional structure of interleukin 1β in solution by three- and four-dimensional nuclear magnetic resonance spectroscopy, *Biochemistry* **30:**2315–2323.

Clore, G. M., Bax, A., and Gronenborn, A. M., 1991c, Stereospecific assignment of β-methylene protons in larger proteins using 3D ^{15}N-separated Hartmann-Hahn and ^{13}C-separated rotating frame overhauser spectroscopy, *J. Biomol. NMR* **1:**13–22.

Cooke, R. M., Wilkinson, A. J., Baron, M., Pastore, A., Tappin, M. J., Campbell, I. D., Gregory, H., and Sheard, B., 1987, The solution structure of human epidermal growth factor, *Nature* **327:**339–341.

Cooke, R. M., Tappin, M. J., Campbell, I. D., Kohda, D., Miyake, T., Fuwa, T., Miyazawa, Y., and Inagaki, F., 1990, Nuclear magnetic resonance studies of human epidermal growth factor, *Eur. J. Biochem.* **197:**807–815.

Crippen, G. M., 1981, *Distance Geometry and Conformational Calculations*, Chemometrics Research Studies Press, Letchworth, UK.

Crippen, G. M., and Havel, T. F., 1988, *Distance Geometry and Molecular Conformation*, Wiley, New York.

Croasmun, W. R., and Carlson, R. M. K. (eds.), 1987, *Two-Dimensional NMR Spectroscopy. Applications for Chemists and Biochemists*, Verlag Chemie, Berlin.

De Marco, A., Menegatti, E., and Guarneri, M., 1983, Epidermal growth factor: Exposure and dynamics of the aromatic side chains as investigated by photo-CIDNP and variable temperature ^{1}H NMR, *FEBS Lett.* **159:**201–206.

De Marco, A., Mayo, K. H., Bartolotti, F., Scalia, S., Menegatti, E., and Kaptein, R., 1986, Proton NMR and photochemically induced dynamic nuclear polarization studies of peptide fragments obtained by controlled proteolysis of mouse epidermal growth factor, *J. Biol. Chem.* **261:**13510–13516.

Driscoll, P. C., Clore, G. M., Marion, D., Wingfield, P. T., and Gronenborn, A. M., 1990a, Complete resonance assignment for the polypeptide backbone of interleukin-1β using three-dimensional heteronuclear NMR spectroscopy, *Biochemistry* **29:**3542–3556.

Driscoll, P. C., Gronenborn, A. M., Wingfield, P. T., and Clore, G. M., 1990b, Determination of the secondary structure and molecular topology of interleukin-1β by use of two- and three-dimensional heteronuclear ^{15}N-^{1}H NMR spectroscopy, *Biochemistry* **29:**4668–4682.

Dudgeon, T. J., Cooke, R. M., Baron, M., Campbell, I. D., Edwards, R. M., and Fallon, A., 1990, Structure–function analysis of epidermal growth factor: Site-directed mutagenesis and nuclear magnetic resonance, *FEBS Lett.* **261:**392–396.

Ealick, S. E., Cook, W. J., Vijay-Kumar, S., Carson, M., Nagabhushan, T. L., Trotta, P. P., and Bugg, C. E., 1991, Three-dimensional structure of recombinant human interferon-γ, *Science* **252:**698–702.

Emerson, S. D., and Montelione, G. T., 1992, 2D and 3D HCCH TOCSY experiments for determining ^{3}J (Hα-Hβ) coupling constants of amino acid residues, *J. Magn. Reson.* **99:**413–420.

Englander, W., and Wand, S. J., 1987, Main-chain-directed strategy for the assignment of ^{1}H NMR spectra of proteins, *Biochemistry* **26:**5953–5958.

Engler, D. A., Montelione, G. T., and Niyogi, S. K., 1990, Human epidermal growth factor. Distinct roles of tyrosine 37 and arginine 41 in receptor binding as determined by site-directed mutagenesis and nuclear magnetic resoance spectroscopy, *FEBS Lett.* **271:**47–50.

Ernst, R. R., Bodenhausen, G., and Wokaun, A., 1987, *Principles of Nuclear Magnetic Resonance in One and Two Dimensions*, Oxford University Press, New York.

Fesik, S. W., 1988, Isotope-edited NMR spectroscopy, *Nature* **332:**865–866.

Fesik, S. W., 1993, NMR structure-based drug design, *J. Biomol. NMR* **3:**261–269.

Fesik, S. W., and Zuiderweg, E. R. P., 1989, An approach for studying the active site of enzyme/inhibitor complexes using deuterated ligands and 2D NOE difference spectroscopy, *J. Am. Chem. Soc.* **111:**5013–5015.

Fesik, S. W., Luly, J. R., Erickson, J. W., and Abad-Zapatero, C., 1988, Isotope-edited proton NMR study on the structure of a pepsin/inhibitor complex, *Biochemistry* **27:**8297–8301.

Fesik, S. W., Eaton, H. L., Olejniczak, E. T., Zuiderweg, E. R. P., McIntosh, L. P., and Dahlquist, F. W., 1990, 2D and 3D NMR spectroscopy employing ^{13}C-^{13}C magnetization transfer by isotropic mixing. Spin system identification in large proteins, *J. Am. Chem. Soc.* **112:**886–888.

Garrett, D. S., Powers, R., March, C. J., Frieden, E. A., Clore, G. M., and Gronenborn, A. M., 1992, Determination of the secondary structure and folding topology of human interleukin-4 using three-dimensional heteronuclear magnetic resonance spectroscopy, *Biochemistry* **31:**4347–4353.

Gilson, M. K., and Honig, B., 1989, Destabilization of an alpha helix bundle protein by helix dipoles, *Proc. Natl. Acad. Sci. USA* **86:**1524–1528.

Gronenborn, A. M., and Clore, G. M., 1989, Analysis of the relative contributions of the nuclear overhauser interproton distance restraints and the empirical energy function in the calculation of oligonucleotide structures using restrained molecular dynamics, *Biochemistry* **28:**5978–5984.

Gronenborn, A. M., Clore, G. M., Schmeissner, U., and Wingfield, P., 1986, A ^{1}H-NMR study of human interleukin-1β. Sequence-specific assignment of aromatic residues using site-directed mutant proteins, *Eur. J. Biochem.* **161:**37–43.

Gronenborn, A. M., Wingfield, P. T., McDonald, H. R., Schmeissner, U., and Clore, G. M., 1988, Site directed mutants of human interleukin-1α: A ^{1}H NMR and receptor binding study, *FEBS Lett.* **231:**135–138.

Grzesiek, S., and Bax, A., 1992, Improved 3D triple-resonance NMR techniques applied to a 31 kDa protein, *J. Magn. Reson.* **96:**432–440.

Grzesiek, S., Döbeli, H., Gentz, R., Garotta, G., Labhardt, A. M., and Bax, A., 1992a, ^{1}H, ^{13}C, and ^{15}N NMR backbone assignments and secondary structure of human interferon-γ, *Biochemistry* **31:**8180–8190.

Grzesiek, S., Ikura, M., Clore, G. M., Gronenborn, A. M., and Bax, A., 1992b, A 3D triple-resonance NMR technique for qualitative measurement of carbonyl-Hβ J couplings in isotopically enriched proteins, *J. Magn. Reson.* **96:**215–221.

Habazettl, J., Cieslar, C., Oschkinat, H., and Holak, T. A., 1990, 1H NMR assignments of side chain conformations in proteins using a high-dimensional potential in the simulated annealing calculations, *FEBS Lett.* **268:**141–145.

Han, K. H., Ferretti, J. A., Niu, C. H., Lokeshwar, V., Clarke, R., and Katz, D., 1988, Conformational and receptor binding properties of human EGF and TGF-alpha second loop fragments, *J. Mol. Recog.* **1:**116–123.

Han, K. H., Syi, J. L., Brooks, B. R., and Ferretti, J. A., 1990, Solution conformations of the B-loop fragments of human transforming growth factor alpha and epidermal growth factor by 1H nuclear magnetic resonance and restrained molecular dynamics, *Proc. Natl. Acad. Sci. USA* **87:**2818–2822.

Havel, T. F., 1991, An evaluation of computational strategies for use in the determination of protein structure from distance constraints obtained by nuclear magnetic resonance, *Prog. Biophys. Mol. Biol.* **56:**43–78.

Havel, T., and Wüthrich, K., 1984, A distance geometry program for determining the structures of small proteins and other macromolecules from nuclear magnetic resonance measurements of intramolecular ^{1}H-^{1}H proximities in solution, *Bull. Math. Biol.* **46:**673–698.

Havel, T. F., and Wüthrich, K., 1985, An evaluation of the combined use of NMR and distance geometry for the determination of protein conformations in solution, *J. Mol. Biol.* **182:**281–294.

Hommel, U., Dudgeon, T. J., Fallon, A., Edwards, R. M., and Campbell, I. D., 1991, Structure–function relationships in human epidermal growth factor studied by site-directed mutagenesis and ^{1}H NMR, *Biochemistry* **30:**8891–8898.

Hommel, U., Harvey, T. S., Driscoll, P. C., and Campbell, I. D., 1992, Human epidermal growth factor. High resolution solution structure and comparison with human transforming growth factor alpha, *J. Mol. Biol.* **227:**271–282.

Hua, Q.-X., Chen, Y.-J., Wang, C.-C., Wang, D.-C., and Roberts, G. C. K., 1989, High resolution ^{1}H NMR studies of des-(B26-B30)-insulin: Assignment of resonances and properties of aromatic residues, *Biochim. Biophys. Acta* **994:**114–120.

Hurd, R. E., 1990, Gradient-enhanced spectroscopy, *J. Magn. Reson.* **87:**422–428.

Hurd, R. E., and John, B. K., 1991a, Gradient-enhanced proton-detected heteronuclear multiple-quantum coherence spectroscopy, *J. Magn. Reson.* **91:**648–653.

Hurd, R. E., and John, B. K., 1991b, Three-dimensional gradient-enhanced relay-edited proton spectroscopy, GREP-HMQC-COSY, *J. Magn. Reson.* **92:**658–688.

Ikebuchi, K., Wong, C. C., Clark, S. C., Ihle, J. N., Hirai, Y., and Ogawa, M., 1987, Interleukin 6 enhancement of interleukin 3-dependent proliferation of multipotential hemopoietic progenitors, *Proc. Natl. Acad. Sci. USA* **84:**9035–9039.

Ikura, M., Kay, L. E., and Bax, A., 1990, A novel approach for sequential assignment of ^{1}H, ^{13}C, and ^{15}N spectra of larger proteins: Heteronuclear triple-resonance three-dimensional NMR spectroscopy. Application to calmodulin, *Biochemistry* **29:**4659–4667.

Ikura, M., Kay, L. E., and Bax, A., 1991, Improved three-dimensional ^{1}H-^{13}C-^{1}H correlation spectroscopy of a ^{13}C-labeled protein using constant-time evolution, *J. Biomol. NMR* **1:**299–304.

Inagaki, F., 1990, Three-dimensional structures of proteins determined by two-dimensional NMR and distance geometry calculations, *Cell. Struct. Funct.* **15:**237–243.

Jeener, J., 1971, Ampere International Summer School, Basko Polje, Yugoslavia.

Jorgensen, A. M. M., Kristensen, S. M., Led, J. J., and Balschmidt, P., 1992, Three-dimensional solution structure of an insulin dimer. A study of the B9(Asp) mutant of human insulin using nuclear magnetic resonance, distance geometry and restrained molecular dynamics, *J. Mol. Biol.* **227:**1146–1163.

Kaptein, R., Boelens, R., Scheek, R. M., and van Gunsteren, W. F., 1988, Protein structures from NMR, *Biochemistry* **27:**5389–5395.

Kato, K., Matsunaga, C., Odaka, A., Yamato, S., Tahaka, W., Shimada, I., and Arata, Y., 1991, Carbon-13 NMR study of switch-variant anti-dansyl antibodies: Antigen binding and domain–domain interactions, *Biochemistry* **30:**6604–6610.

Kay, L. E., Clore, G. M., Bax, A., and Gronenborn, A. M., 1990a, Four-dimensional heteronuclear triple-resonance NMR spectroscopy of interleukin-1β in solution, *Science* **249:**411–414.

Kay, L. E., Ikura, M., Tschudin, R., and Bax, A., 1990b, Three-dimensional triple-resonance NMR spectroscopy of isotopically enriched proteins, *J. Magn. Reson.* **89:**496–514.

Kay, L. E., Ikura, M., Zhu, G., and Bax, A., 1991, Four-dimensional heteronuclear triple-resonance NMR of isotopically enriched proteins for sequential assignment of backbone atoms, *J. Magn. Reson.* **91:**422–428.

Kay, L. E., Wittekind, M., McCoy, M. A., Friedrichs, M. S., and Mueller, L., 1992, 4D NMR triple resonance experiments for assignments of protein backbone nuclei using shared constant-time evolution periods, *J. Magn. Reson.* **98:**443–450.

Kessler, H., and Steuernagel, S., 1991, in: *Peptide Pharmaceuticals, Approaches to the Design of Novel Drugs. Conformational Determination by NMR Spectroscopy* (D. J. Ward, ed.), Elsevier, New York, pp. 18–46.

Kessler, H., Gehrke, M., and Griesinger, C., 1988, Two-dimensional NMR spectroscopy: Background and overview of the experiments, *Angew. Chem. Int. Ed. Engl.* **27:** 490–536.

Kishimoto, T., and Hirano, T., 1988, Molecular regulation of B lymphocyte response, *Annu. Rev. Immunol.* **6:**485–512.

Kline, A. D., and Justice, R. M., Jr., 1990, Complete sequence-specific ^{1}H NMR assignments for human insulin, *Biochemistry* **29:**2906–2913.

Knegtel, R. M. A., Boelens, R., Ganadu, M. L., and Kaptein, R., 1991, The solution structure of a monoemeric insulin. A two-dimensional ^{1}H NMR study of des-(B26-B30)-insulin in combination with distance geometry and restrained molecular dynamics, *Eur. J. Biochem.* **202:**447–458.

Kohda, D., and Inagaki, F., 1988, Complete sequence-specific 1H nuclear magnetic resonance assignments for mouse epidermal growth factor, *J. Biochem. (Tokyo)* **103:** 554–571.

Kohda, D., and Inagaki, F., 1992a, Structure of epidermal growth factor bound to perdeuterated dodecylphosphocholine micelles determined by two-dimensional NMR and simulated annealing calculations, *Biochemistry* **31:**677–685.

Kohda, D., and Inagaki, F., 1992b, Three-dimensional nuclear magnetic resonance structures of mouse epidermal growth factor in acidic and physiological pH solutions, *Biochemistry* **31:**11928–11939.

Kohda, D., Go, N., Hayashi, K., and Inagaki, F., 1988a, Tertiary structure of mouse epidermal growth factor, *J. Biochem. (Tokyo)* **103:**741–743.

Kohda, D., Kodama, C., Kase, R., Nomoto, H., Hayashi, K., and Inagaki, F., 1988b, A comparative ^{1}H NMR study of mouse α(1-53) and β(2-53) epidermal growth factors, *Biochem. Int.* **16:**647–654.

Kohda, D., Shimada, I., Miyake, T., Fuwa, T., and Inagaki, F., 1989, Polypeptide chain fold of human transforming growth factor alpha analogous to those of mouse and human epidermal growth factors as studied by two-dimensional ^{1}H NMR, *Biochemistry* **28:** 953–958.

Kohda, D., Sawada, T., and Inagaki, F., 1991, Characterization of pH titration shifts for all the nonlabile proton resonances in a protein by two-dimensional NMR: The case of mouse epidermal growth factor, *Biochemistry* **30:**4896–4900.

Koide, H., Muto, Y., Kasai, H., Hoshi, K., Takusari, H., Kohri, K., Takahashi, S., Sasaki, T., Tsukomo, K., Miyake, T., Fuwa, T., Miyazawa, T., and Yokoyama, S., 1992, Recognition of an antiparallel β sheet structure of human epidermal growth factor by its receptor. Site-directed mutagenesis studies, *FEBS Lett.* **302:**39–42.

Kuntz, I. D., Thomason, J. F., and Oshiro, C. M., 1989, Distance geometry, *Methods Enzymol.* **177**:159–204.

LeMaster, D. M., Deuteration in protein proton magnetic resonance, *Methods Enzymol.* **177**:23–43.

London, R. E., Perlman, M. E., and Davis, D. G., 1992, Relaxation-matrix analysis of the transferred nuclear Overhauser effect for finite exchange rates, *J. Magn. Reson.* **97**:79–98.

Makino, K., Morimoto, M., Nishi, M., Sakamoto, S., Tamura, A., Inooka, H., and Akasaka, K., 1987, Proton nuclear magnetic resonance study on the solution conformation of human epidermal growth factor, *Proc. Natl. Acad. Sci. USA* **84**:7841–7845.

Marion, D., and Wüthrich, K., 1983, Application of phase sensitive two-dimensional correlated spectroscopy (COSY) for measurements of ^{1}H-^{1}H spin–spin coupling constants in proteins, *Biochem. Biophys. Res. Commun.* **113**:967–974.

Marion, D., Driscoll, P. C., Kay, L. E., Wingfield, P. T., Bax, A., Gronenborn, A. M., and Clore, G. M., 1989, Overcoming the overlap problem in the assignment of ^{1}H NMR spectra of larger proteins by use of three-dimensional heteronuclear ^{1}H-^{15}N Hartmann-Hahn-multiple quantum coherence and nuclear overhauser-multiple quantum coherence spectroscopy: Application to interleukin 1β, *Biochemistry* **28**:6150–6156.

Matsunami, R. K., Yette, M. L., Stevens, A., and Niyogi, S. K., 1991, Mutational analysis of leucine 47 in human epidermal growth factor, *J. Cell. Biochem.* **46**:242–249.

Matthews, S. J., and Leatherbarrow, R. J., 1993, The use of osmolytes to facilitate protein NMR spectroscopy, *J. Biomol. NMR* **3**:597–600.

Mayo, K. H., 1984, Epidermal growth factor from the mouse. Structural characterization by proton nuclear magnetic resonance and nuclear overhauser experiments at 500 MHz, *Biochemistry* **23**:3960–3973.

Mayo, K. H., 1985, Epidermal growth factor from the mouse. Physical evidence for a tiered beta-sheet domain: Two-dimensional NMR correlated spectroscopy and nuclear overhauser experiments on backbone amide protons, *Biochemistry* **24**:3783–3794.

Mayo, K. H., and Burke, C., 1987, Structural and dynamical comparison of alpha-, beta-, and gamma-forms of murine epidermal growth factor, *Eur. J. Biochem.* **169**:201–207.

Mayo, K. H., Schaudies, P., Savage, C. R., De Marco, A., and Kaptein, R., 1986a, Structural characterization and exposure of aromatic residues in epidermal growth factor from rat, *Biochem. J.* **239**:13–18.

Mayo, K. H., DeMarco, A., and Kaptein, R., 1986b, Photo-CIDNP nuclear magnetic resonance as a probe for conformational changes in epidermal growth factor, *Biochim. Biophys. Acta* **874**:181–186.

Mayo, K. H., DeMarco, A., Menegatti, E., and Kaptein, R., 1987, Interaction of epidermal growth factor with micelles monitored by photochemically induced dynamic nuclear polarization-^{1}H NMR spectroscopy, *J. Biol. Chem.* **262**:14899–14904.

Mayo, K. H., Cavalli, R. C., Peters, A. R., Boelens, R., and Kaptein, R., 1989, Sequence-specific ^{1}H NMR assignments and peptide backbone conformation in rat epidermal growth factor, *Biochem. J.* **257**:197–205.

McCammon, J. A., 1987, Computer-aided molecular design, *Science* **238**:486–491.

McCammon, J. A., and Harvey, S. C., 1987, *Dynamics of Proteins and Nucleic Acids*, Cambridge University Press, Cambridge.

Menegatti, E., Scalia, A., Bortolotti, F., Ascenzi, P., and De Marco, A., 1989, Controlled proteolysis of mouse epidermal growth factor. An RP-HPLC and ^{1}H NMR study, *Int. J. Peptide Protein Res.* **34:**161–165.

Metzler, W. J., Hare, D. R., and Pardi, A., 1989, Limited sampling of conformational space by the distance geometry algorithm: Implications for structures generated from NMR data, *Biochemistry* **28:**7045–7052.

Mirau, P., 1988, Quantitative interpretation of a single NOESY spectrum, *J. Magn. Reson.* **80:**439–447.

Montelione, G. T., Wüthrich, K., Nice, E. C., Burgess, A. W., and Scheraga, H. A., 1986, Identification of two anti-parallel beta-sheet conformation in the solution structure of murine epidermal growth factor by proton magnetic resonance, *Proc. Natl. Acad. Sci. USA* **83:**8594–8598.

Montelione, G. T., Wüthrich, K., Nice, E. C., Burgess, A. W., and Scheraga, H. A., 1987, Solution structure of murine epidermal growth factor: Determination of the polypeptide backbone chain-fold by NMR and distance geometry, *Proc. Natl. Acad. Sci. USA* **84:**5226–5230.

Montelione, G. T., Wüthrich, K., and Scheraga, H. A., 1988, Sequence-specific ^{1}H NMR assignments and identification of slowly exchanging amide protons in murine epidermal growth factor, *Biochemistry* **27:**2235–2243.

Montelione, G. T., Winkler, M. E., Rauenbeller, P., and Wagner, G., 1989, Accurate measurements of long-range heteronuclear coupling constants from 2D NMR spectra of isotope-enriched proteins, *J. Magn. Reson.* **82:**198–204.

Montelione, G. T., Wüthrich, K., Burgess, A. W., Nice, E. C., Wagner, G., Gibson, K. D., and Scheraga, H. A., 1992, Solution structure of murine epidermal growth factor determined by NMR spectroscopy and refined by energy minimization with restraints, *Biochemistry* **31:**236–249.

Moy, F. J., Scheraga, H. A., Liu, J.-F., Wu, R., and Montelione, G. T., 1989, Conformational characterization of a single-site mutant of murine epidermal growth factor (EGF) by ^{1}H NMR provides evidence that leucine-47 is involved in the interactions with the EGF receptor, *Proc. Natl. Acad. Sci. USA* **86:**9836–9840.

Mutsaers, J. H. G. M., Kamerling, J. P., Devos, R., Guisez, Y., Fiers, W., and Vliegenthart, J. F. G., 1986, Structural studies of the carbohydrate chains of human γ-interferon, *Eur. J. Biochem.* **156:**651–654.

Nilges, M., Clore, G. M., and Gronenborn, A. M., 1988a, Determination of three-dimensional structures of proteins from interproton distance data by hybrid distance geometry-dynamical simulated annealing calculations, *FEBS Lett.* **228:**317–324.

Nilges, M., Clore, G. M., and Gronenborn, A. M., 1988b, Determination of three-dimensional structures of proteins from interproton distance data by dynamical simulated annealing from a random array of atoms, *FEBS Lett.* **239:**129–136.

Nishimura, C., Hanzawa, H., Itoh, S., Yasukawa, K., Shimada, I., Kishimoto, T., and Arata, Y., 1990, Proton nuclear magnetic resonance study of interleukin-6: Chemical modifications and partial spectral assignments for the aromatic residues, *Biochim. Biophys. Acta* **1041:**243–249.

Nishimura, C., Ekida, T., Masuda, S., Futatsugi, K., Itoh, S., Yasukawa, K., Kishimoto,

T., and Arata, Y., 1991, Chemical modification and ^{1}H NMR studies on the receptor binding region of human interleukin-6, *Eur. J. Biochem.* **196:**377–384.

Olejniczak, E. T., Poulsen, F. M., and Dobson, C. M., 1984, Distance dependence of proton nuclear overhauser effects in proteins, *J. Magn. Reson.* **59:**518–523.

Olejniczak, E. T., Xu, R. X., and Fesik, S. W., 1992a, A 4D HCCH-TOCSY experiment for assigning the side chain ^{1}H and ^{13}C resonances of proteins, *J. Biomol. NMR* **2:**655–659.

Olejniczak, E. T., Xu, R. X., Petros, A. M., and Fesik, S. W., 1992b, Optimized constant-time 4D HNCAHA and HN(CO)CAHA experiments. Applications to the backbone assignments of the FKBP/ascomycin complex, *J. Magn. Reson.* **100:**444–450.

Oschkinat, H., Griesinger, C., Krautlis, P. J., Sørensen, O. W., Ernst, R. R., Gronenborn, A. M., and Clore, G. M., 1988, Three-dimensional NMR spectroscopy of a protein in solution, *Nature* **332:**374–376.

Oshiro, C. M., Thomason, J., and Kuntz, I. D., 1991, Effects of limited input distance constraints upon the distance geometry algorithm, *Biopolymers* **31:**1049–1064.

Palmieri, R., Lee, R. W.-K., and Dunn, M. F., 1988, ^{1}H Fourier transform NMR studies of insulin: Coordination of calcium to the Glu(B13) site drives hexamer assembly and induces a conformational change, *Biochemistry* **27:**3387–3397.

Piotto, M., Saudek, V., and Sklenár, V., 1992, Gradient-tailored excitation for single-quantum NMR spectroscopy of aqueous solutions, *J. Biomol. NMR* **2:**661–665.

Powers, R., Garrett, D. S., March, C. J., Frieden, E. A., Gronenborn, A. M., and Clore, G. M., 1992a, ^{1}H, ^{15}N, ^{13}C, and ^{13}CO assignments of human interleukin-4 using three-dimensional double- and triple-resonance heteronuclear magnetic resonance spectroscopy, *Biochemistry* **31:**4334–4346.

Powers, R., Garrett, D. S., March, C. J., Frieden, E. A., Gronenborn, A. M., and Clore, G. M., 1992b, Three-dimensional solution structure of human interleukin-4 by multi-dimensional heteronuclear magnetic resonance spectroscopy, *Science* **256:**1673–1677.

Priestle, J. P., Schär, H.-P., and Grütter, M. G., 1988, Crystal structure of the cytokine interleukin-1β, *EMBO J.* **7:**339–343.

Priestle, J. P., Schär, H. P., and Grutter, M. G., 1989, Crystallographic refinement of interleukin 1β at 2.0 Å resolution, *Proc. Natl. Acad. Sci. USA* **86:**9667–9671.

Ramhesh, V., and Bradbury, J. H., 1986, ^{1}H NMR studies of insulin, *Int. J. Pept. Protein Res.* **28:**146–153.

Rance, M., Sørensen, O. W., Bodenhausen, G., Wagner, G., Ernst, R. R., and Wüthrich, K., 1983, Improved spectral resolution in COSY ^{1}H NMR spectra of proteins via double quantum filtering, *Biochem. Biophys. Res. Commun.* **117:**479–485.

Redfield, C., Smith, L. J., Boyd, J., Lawrence, G. M. P., Edwards, R. G., Smith, R. A. G., and Dobson, C. M., 1991, Secondary structure and topology of human interleukin-4 in solution, *Biochemistry* **30:**11029–11035.

Redfield, C., Boyd, J., Smith, L. J., Smith, R. A. G., and Dobson, C. M., 1992, Loop mobility in a four-helix bundle protein: ^{15}N NMR relaxation measurements on human interleukin-4, *Biochemistry* **31:**10431–10437.

Richardson, J. M., Clowes, R. T., Boucher, W., Domaille, P. J., Hardman, C. H., Keeler, J., and Laue, E. D., 1993, The use of heteronuclear cross polarization to enhance the sensitivity of triple-resonance NMR experiments. Improved 4D HCCNNH pulse sequences, *J. Magn. Reson. Ser. B* **101:**223–227.

Sato, A., Nishimura, S., Ohkubo, T., Kyogoku, Y., Koyama, S., Kobayashi, M., Yasuda, T., and Kobayashi, Y., 1992, ^{1}H NMR assignment and secondary structure of human insulin-like growth factor-I (IGF-I) in solution, *J. Biochem. (Tokyo)* **111:**529–536.

Satoh, T., Nakamura, S., Taga, T., Matsuda, T., Hirano, T., Kishimoto, T., and Kaziro, Y., 1988, Induction of neuronal differentiation in PC12 cells by B-cell stimulatory factor 2/interleukin 6, *Mol. Cell. Biol.* **8:**3546–3549.

Scheek, R. M., van Gunsteren, W. F., and Kaptein, R., 1989, Molecular dynamics simulation techniques for determination of molecular structures from nuclear magnetic resonance data, *Methods Enzymol.* **177:**204–218.

Schmieder, P., Thanabal, V., McIntosh, L. P., Dahlquist, F. W., and Wagner, G., 1991, Measurements of Ha-HN vicinal coupling constants in a protein with large linewidths in a new 3D ^{1}H-^{15}N-^{13}C quadruple resonance NMR experiment, *J. Am. Chem. Soc.* **113:**6323–6324.

Shaka, A. J., and Freeman, R., 1983, Simplification of NMR spectra by filtration through multiple-quantum coherence, *J. Magn. Reson.* **51:**169–173.

Smith, L. J., Redfield, C., Boyd, J., Lawrence, G. M. P., Edwards, R. G., Smith, R. A. G., and Dobson, C. M., 1992, Human interleukin-4. The solution structure of a four-helix bundle protein, *J. Mol. Biol.* **224:**899–904.

Sørensen, P., and Poulsen, F., 1992, A simple and economical algal culture system for stable isotopic labelling, *J. Biomol. NMR* **2:**99–101.

Stockman, B. J., Scahill, T. A., Roy, M., Ulrich, E. L., Strakalaitis, N. A., Brunner, D. P., Yem, A. W., and Deibel, M. R., Jr., 1992, Secondary structure and topology of interleukin-1 receptor antagonist protein determined by heteronuclear three-dimensional NMR spectroscopy, *Biochemistry* **31:**5237–5245.

Syperski, T., Wider, G., Bushweller, J. H., and Wüthrich, K., 1993, 3D ^{13}C-^{15}N heteronuclear two-spin coherence spectroscopy for polypeptide backbone assignments in ^{13}C-^{15}N double-labeled proteins, *J. Biomol. NMR* **3:**127–132.

Vercelli, D., Jabara, H. H., Arai, K., Yokota, T., and Geha, R. S., 1989, Endogenous interleukin 6 plays an obligatory role in interleukin 4-dependent human Ige synthesis, *Eur. J. Immunol.* **19:** 1419–1424.

Vuister, G. W., Boelens, R., and Kaptein, R., 1988, Nonselective three-dimensional NMR spectroscopy. The 3D NOE-HOHAHA experiment, *J. Magn. Reson.* **80:**176–185.

Vuister, G. W., Boelens, R., Padilla, A., and Kaptein, R., 1991, Statistical analysis of double NOE transfer pathways in proteins as measured in 3D NOE–NOE spectroscopy, *J. Biomol. NMR* **1:**421–438.

Vuister, G. W., Boelens, R., Kaptein, R., Burgering, M., and van Zijl, P. C. M., 1992, Gradient-enhanced 3D NOESY-HMQC spectroscopy, *J. Biomol. NMR* **2:**301–305.

Wagner, G., 1993, Prospects for NMR of large proteins, *J. Biomol. NMR* **3:**375–385.

Wagner, G., Hyberts, S. G., and Havel, T. F., 1992, NMR structure determination in solution: A critique and comparison with X-ray crystallography, *Annu. Rev. Biophys. Biomol. Struct.* **21:**167–198.

Walter, M. R., Cook, W. J., Ealick, S. E., Nagabhushan, T. L., Trotta, P. P., and Bugg, C. E., 1992, Three-dimensional structure of recombinant human granulocyte-macrophage colony-stimulating factor, *J. Mol. Biol.* **224:**1075–1085.

Weisemann, R., Rüterjans, H., and Bermel, B., 1993, 3D triple-resonance NMR techniques for the sequential assignment of NH and ^{15}N resonances in ^{15}N- and ^{13}C-labelled proteins, *J. Biomol. NMR* **3:**113–120.

Weiss, M. A., Nguyen, D. T., Khait, I., Inouye, K., Frank, B. H., Beckage, M., O'Shea, E., Shoelson, S. E., Karplus, M., and Neuringer, L. J., 1989, Two-dimensional NMR and photo-CIDNP studies of the insulin monomer: Assignment of aromatic resonances with application to protein folding, structure, and dynamics, *Biochemistry* **28:**9855–9873.

Wider, G., Macura, S., Kumar, A., Ernst, R. R., and Wüthrich, K., 1984, Homonuclear two-dimensional ^{1}H NMR of proteins. Experimental procedures, *J. Magn. Reson.* **56:** 207–234.

Wilkinson, A. J., Cooke, R. M., Pastore, A., Campbell, I. D., Sheard, B., and Gregory, H., 1987, The determination of the three-dimensional structure of human epidermal growth factor from ^{1}H NMR data, *Protides Biol. Fluids* **35:**453–456.

Williamson, M. P., and Madison, V. S., 1990, Three-dimensional structure of porcine C5a desArg from 1H nuclear magnetic resonance data, *Biochemistry* **29:**2895–2905.

Wingfield, P., Graber, P., Movva, N. R., Gronenborn, A. M., and MacDonald, H. R., 1987, N-terminal-methionylated interleukin-1β has reduced receptor-binding affinity, *FEBS Lett.* **215:**160–164.

Wingfield, P., Graber, P., Shaw, A. R., Gronenborn, A. M., Clore, G. M., and MacDonald, H. R., 1989, Preparation, characterization and application of interleukin-1β mutant proteins with surface-accessible cysteine residues, *Eur. J. Biochem.* **179:**565–571.

Wittekind, M., and Mueller, L., 1993, HNCACB, a high-sensitivity 3D NMR experiment to correlate amide-proton and nitrogen resonances with the alpha- and beta-carbon resonances in proteins, *J. Magn. Reson. Ser. B* **101:**201–205.

Wokaun, A., and Ernst, R. R., 1977, Selective detection of multiple-quantum transitions in NMR by two-dimensional spectroscopy, *Chem. Phys. Lett.* **52:**407–412.

Wüthrich, K., 1986, *NMR of Proteins and Nucleic Acids*, John Wiley and Sons, New York.

Wüthrich, K., 1989a, Protein structure determination in solution by nuclear magnetic resonance spectroscopy, *Science* **243:**45–50.

Wüthrich, K., 1989b, Determination of three-dimensional protein structures in solution by nuclear magnetic resonance: An overview, *Methods Enzymol.* **177:**125–131.

Wüthrich, K., 1990, Protein structure determination in solution by NMR spectroscopy, *J. Biol. Chem.* **265:**22059–22062.

Wüthrich, K., Billeter, M., and Braun, W., 1984, Polypeptide secondary structure determination by NMR observation of short proton–proton distances, *J. Mol. Biol.* **180:**715–740.

Yamamoto, S., Hase, S., Fukuda, S., Sano, O., and Ikenaka, T., 1989, Structures of the sugar chains of interferon-γ produced by human myelomonocyte cell line HBL-38, *J. Biochem. (Tokyo)* **105:**547–555.

Zuiderweg, E. R. P., Nettesheim, D. G., Mollison, K. W., and Carter, G. W., 1989, Tertiary structure of human complement C5a in solution from nuclear magnetic resonance data, *Biochemistry* **28:**172–185.

6

Thermodynamic Strategies for Rational Protein and Drug Design

Kenneth P. Murphy and Ernesto Freire

1. INTRODUCTION

The rapidly increasing availability of high-resolution protein structures from X-ray crystallography and multidimensional nuclear magnetic resonance (NMR) opens the possibility of using structural information in the design of proteins and pharmaceuticals. A prerequisite to structure-based design strategies is understanding which features dictate the energetics of stabilization of that structure. Recently, we have shown that it is possible to predict rather accurately the energetics of folding of a globular protein, if its crystal structure is known, using a semi-empirically derived set of structural energetic parameters (Murphy *et al.*, 1992; Murphy and Freire, 1992). This approach also shows promise for predicting the energetics of binding of peptides and proteins (Murphy *et al.*, 1993, 1995). In this chapter we will review the structural energetic analysis and the implications of this analysis regarding the general energetic features of assembly processes that bear upon the rational design of proteins and drugs. A structural energetic analysis can be envisioned as part of an overall scheme of rational protein or drug design as shown schematically in Fig. 1.

Kenneth P. Murphy • Department of Biochemistry, University of Iowa, Iowa City, Iowa 52242. *Ernesto Freire* • Department of Biology and Biocalorimetry Center, The Johns Hopkins University, Baltimore, Maryland 21218.

Physical Methods to Characterize Pharmaceutical Proteins, edited by James N. Herron *et al.*, Plenum Press, New York, 1995.

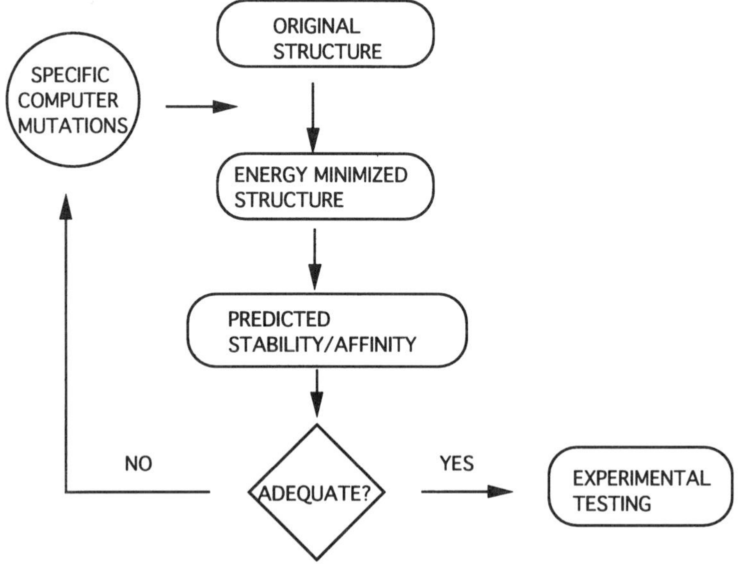

Figure 1. Schematic representation of structure-based approach to drug or protein design.

2. THERMODYNAMIC DESCRIPTION OF PROTEIN STABILITY AND LIGAND BINDING

2.1. Interactions Important to Folding and Binding

The folding of a protein and the binding of a ligand to a protein are analogous processes. They differ primarily in that the former is an intramolecular process, while the latter is an intermolecular one. The primary forces, or interactions, that must be considered in the design of proteins and drugs therefore are the same. They consist primarily of hydrogen bonding, the hydrophobic effect, configurational cntropy, and clcctrostatic and protonation cffccts. In thc binding of a ligand to a protein, a cratic entropy must also be included that reflects the intermolecular character of the process.

In the context in which it is used here, hydrogen bonding refers to the *overall* energetics associated with the formation of a hydrogen bond and the burial of the participating groups into the protein interior or into the protein–ligand interface. Thus, the energetics of the hydrogen bond formation per se are considered together with the van der Waals interactions and dehydration energetics. Likewise, the hydrophobic effect will refer to the *overall* energetics involved with the burial of hydrophobic surface, including van der Waals and dehydration.

The protein backbone and side chains have access to many more conformations (degrees of freedom) in the unfolded state than in the native state. This gives rise to a configurational entropy term that opposes folding. Similarly, the side chains in the binding pocket lose degrees of freedom upon binding of a ligand; in general, the ligand itself loses degrees of freedom, which results in an unfavorable configurational entropy change. Binding can also result in decreased flexibility in the protein, which carries an entropic cost. These unfavorable entropy changes must be offset by the favorable contributions of hydrogen bonding and the hydrophobic effect.

2.2. Formal Description of Stability and Binding

Protein stability and ligand binding are defined using the equations of thermodynamics. The Gibbs free energy change, $\Delta G°$, is composed of two components, the enthalpy change, $\Delta H°$, and the entropy change, $\Delta S°$:

$$\Delta G° = \Delta H° - T\,\Delta S° \qquad (1)$$

where T is the absolute temperature in Kelvin.

The Gibbs energy change is related to an equilibrium constant, K, as:

$$\Delta G° = -R\,T \ln K \qquad (2)$$

where R is the gas constant.

In biological systems, $\Delta H°$ and $\Delta S°$ are also dependent on temperature through the change in heat capacity, ΔC_p:

$$\Delta H° = \Delta H°_R + \Delta C_p\,(T - T_R) \qquad (3)$$

and

$$\Delta S° = \Delta S°_R + \Delta C_p \ln (T/T_R) \qquad (4)$$

where T_R is any convenient reference temperature and $\Delta H°_R$ and $\Delta S°_R$ are the values of $\Delta H°$ and $\Delta S°$ at that reference temperature.

By substituting Eqs. (3) and (4) into (1), we get a description of $\Delta G°$ as a function of temperature:

$$\Delta G° = \Delta H°_R - T\,\Delta S°_R + \Delta C_p[(T - T_R) - T \ln (T/T_R)] \qquad (5)$$

2.3. Group Additivity and Accessible Surface Area

For the purposes of structure-based design of proteins and pharmaceuticals, we would like to be able to relate the thermodynamic parameters in Eqs. (3) to (5) to some structural features. One structural feature that has proven extremely useful in dealing with protein folding–unfolding energetics is accessible surface area (ASA) (Lee and Richards, 1971). Changes in apolar ASA were initially correlated with the hydrophobic effect (Hermann, 1972; Chothia, 1975) and later with the solvation of other constituent groups of proteins (Rashin, 1984; Eisenberg and McLachlan, 1986). For the case of protein unfolding transitions, the ASA of the unfolded chain is taken as the sum of the ASA of each residue as calculated in an extended Ala-X_{aa}-Ala tripeptide plus an additional correction for the carboxy and amino groups at the ends. The change in accessible surface area, ΔASA, is then the difference between this sum and the ASA of the native protein. In the case of ligand binding, ΔASA is the difference between the sum of the ASA of the unligated protein plus the ligand and the ASA of the complex.

Recently, several groups have shown that changes in ASA can be used to rationalize and predict ΔH° and ΔC_p for protein unfolding reactions (Murphy *et al.*, 1992; Murphy and Freire, 1992; Oobatake and Ooi, 1992; Spolar *et al.*, 1992; Makhatadze and Privalov, 1993; Freire, 1993) and for peptide–protein binding (Murphy *et al.*, 1993). The ΔS° can also be predicted reasonably well by inclusion of some additional terms.

While these various treatments differ in the particulars, the basic approach is the same. In all cases it is assumed that the functional groups of which the protein consists contribute independently to the energetics; that is, group additivity is assumed. Consequently, the contribution of apolar surface to, say, the change in heat capacity is the same whether that apolar surface is present in a leucine or an arginine.

Assuming group additivity and assuming that the energetics are proportional to the ASA, any thermodynamic quantity, ΔX, is given as:

$$\Delta X = \Sigma \Delta \mathrm{ASA}_i \; \Delta x_i + \Delta x_{other} \tag{6}$$

where $\Delta \mathrm{ASA}_i$ is the change in ASA upon binding (or folding) for all functional groups of type i, Δx_i is the contribution to ΔX per unit ASA of group i, and Δx_{other} accounts for all other contributions to ΔX that are not proportional to ASA. These other contributions will include configurational entropy changes, protonation events, salt-bridge formation, and so forth, depending on the individual system being considered.

In our own work we have partitioned the change in ASA into apolar and polar surface.* Consequently, the change in heat capacity can be described as:

$$\Delta C_p = \Delta \mathrm{ASA}_{ap} \; \Delta c_{p,ap} + \Delta \mathrm{ASA}_{pol} \; \Delta c_{p,pol} \tag{7}$$

In general, ΔC_p seems to be accounted for almost entirely by solvation effects

*Further subdivisions can be included, such as considering the apolar surface as aliphatic or aromatic (Connelly and Thomson, 1992; Makhatadze and Privalov, 1993).

(Murphy and Gill, 1991; Murphy *et al.*, 1992; Murphy and Freire, 1992; Privalov and Makhatadze, 1992; Spolar *et al.*, 1992) however, contributions due to protonation/deprotonation or other specific effects must be taken into account explicitly when they are present.

As Eqs. (3) to (5) are entirely general, we can define the reference temperatures in any manner that is convenient. For the purpose of relating structure and energetics, it is convenient to define the reference temperatures in Eqs. (3) and (4) as those temperatures at which the change in apolar surface makes no contribution to the energetics. These reference temperatures are designated at T_H^* for the enthalpy and T_S^* for the entropy. Thus we can write:

$$\Delta H^\circ = \Delta \mathrm{ASA}_{pol}\, \Delta h_{pol} + \Delta C_p\, (T - T_H^*) + \Delta h_{other} \tag{8}$$

Note that in Eq. (8) Δh_{pol} is defined as the area-normalized polar contribution at the temperature T_H^*. Similarly, we can write for ΔS°:

$$\Delta S^\circ = \Delta \mathrm{ASA}_{pol}\, \Delta s_{pol} + \Delta C_p \ln (T/T_S^*) + \Delta s_{other} \tag{9}$$

Here, changes in configurational entropy, Δs_{conf}, are expected to be a large contributor to Δs_{other}, and it is worth including this term separately from contributions such as protonation effects.

$$\Delta S^\circ = \Delta \mathrm{ASA}_{pol}\, \Delta s_{pol} + \Delta C_p \ln (T/T_S^*) + \Delta s_{conf} + \Delta s_{other} \tag{9a}$$

It should also be noted that in the case of drug binding to a protein, ΔS° depends on the standard state (i.e., choice of concentration units). This point will be discussed in more detail below.

With Eqs. (7), (8), and (9a), we should be able to describe the ΔG° of a folding–unfolding transition in a protein or of the binding of a ligand (drug) to a protein in the absence of protonation or other specific effects, provided a structure is available to determine ΔASA values and that the other parameters ($\Delta c_{p,ap}$, $\Delta c_{p,pol}$, T_H^*, T_S^*, Δh_{pol}, Δs_{pol}, Δs_{conf}) are known.

2.4. Determination of Empirical Parameters

For the case of protein folding–unfolding transitions, we have determined empirically most of these parameters, either from model compound data or from analysis of the protein thermodynamic data.

Values of $\Delta c_{p,ap}$ and $\Delta c_{p,pol}$ have been derived from crystalline cyclic dipeptide dissolution data (Murphy and Gill, 1990, 1991) that have been normalized to ASA values (Murphy *et al.*, 1992; Murphy and Freire, 1992). Accessible surface areas were calculated using the Lee and Richards algorithm (Lee and Richards, 1971), as implemented by Scott Presnell (University of California, San Diego). It is important to realize that the values of the empirical parameters that are proportional to ASA will vary depending on the algorithm and radii used in determining ASA values (Livingstone *et al.*, 1991; Murphy and Freire, 1992).

The value of $T_H{}^*$ has been taken as 100 °C, which is the enthalpy convergence temperature for proteins (Murphy and Gill, 1991). At this temperature the denaturation ΔH of many globular proteins, normalized to the number of amino acids, is essentially a constant.* For the considered set of proteins, the change in apolar ASA per residue varies, while the change in polar ASA is essentially constant (Murphy *et al.*, 1992; Murphy and Freire, 1992). Thus, at this temperature, ΔH is independent of ΔASA_{ap} and depends only on ΔASA_{pol}. The value of Δh_{pol} therefore can be determined by dividing the ΔH at $T_H{}^*$ by ΔASA_{pol}.

The value of $T_S{}^*$, at which $\Delta S°$ is independent of ΔASA_{ap}, appears to be universal for any process in which apolar groups are exposed to water. This is true for the dissolution of apolar compounds from the gas, liquid, or solid phases and for the denaturation of globular proteins (Murphy *et al.*, 1990). This temperature is 112 °C (Baldwin, 1986; Murphy *et al.*, 1990).

At this temperature, in the absence of protonation effects, the average $\Delta S°$ of unfolding, equal to 4.3 cal K^{-1} (mole-res)$^{-1}$, is the sum of configurational and polar contributions. There is no way of separating these terms from the protein data, since all the proteins have essentially the same ΔASA_{pol} per residue. Theoretical calculations and consideration of model compound data can provide some insight into the relative contributions of these two terms.

There are two rotatable bonds in the peptide backbone of each residue. As a first approximation, the peptide bond can be considered to have two conformations (*cis* and *trans*), while the other two bonds can be considered to have three possible conformations each (t, g^+, and g^-), giving a total of 18 conformations in the unfolded state. Consequently, the configurational entropy change of a backbone with no side chains in going from a unique configuration to the unfolded state is about 5.7 cal K^{-1} mole^{-1} (R ln 18). Schellman (1955) arrived at a comparable value between 3 and 7.2 cal K^{-1} mole^{-1} by similar arguments. The presence of a side chain will decrease the configurational entropy of the backbone (Némethy *et al.*, 1966; Creamer and Rose, 1992). Némethy *et al.* (1966) have estimated the decrease in backbone configurational entropy due to the side chain to be between −2.4 and −4.8 cal K^{-1} mole^{-1} relative to glycine. Thus, the backbone entropy will be maximal for glycine (~5.5 cal K^{-1} mole^{-1}), will drop to ~3 cal K^{-1} mole^{-1} for alanine, and will be as low as 0.7 cal K^{-1} mole^{-1} for amino acids with larger side chains. Amino acid side chains will also contribute to the configurational entropy gain upon unfolding. The contribution of each side chain, however,

*This convergence temperature is obtained by extrapolating the experimental ΔH values for the proteins in the database assuming a constant ΔCp. Within the temperature range of about 0 to 80°C, a linear temperature dependence of the enthalpy accurately accounts for the experimental data. A temperature-dependent ΔCp may need to be considered outside of this temperature range.

whether the side chain is buried in the interior of a protein or exposed to the solvent. Side chains buried in the interior of a protein can be considered to be completely immobilized. Therefore, the entropy change associated with the passage of a side chain from the interior to a solvent-exposed but still folded state can be approximated by the entropy of side chains at protein surfaces or exposed helices. The absolute entropy of side chains in an α-helix has been estimated using Monte Carlo methods to range between 0.2 and 4.7 cal K^{-1} $mole^{-1}$ with an average of 2.8 cal K^{-1} $mole^{-1}$ (Creamer and Rose, 1992). Finally, the additional gain in configurational entropy of the side chain in going from a solvent-exposed folded conformation to a coil has been estimated to be anywhere between 0.5 and 0.75 cal K^{-1} $mole^{-1}$ (Némethy *et al.*, 1966; Creamer and Rose, 1992).

According to the discussion above, the configurational entropy for protein folding–unfolding can be written in terms of a minimum of three different types of contributions (Freire *et al.*, 1993): (1) $\Delta S_{bu\rightarrow ex}$, the entropy change associated with the transfer of a side chain that is buried in the interior of the protein to its surface; (2) $\Delta S_{ex\rightarrow u}$, the entropy change gained by a surface-exposed side chain when the peptide backbone unfolds; and (3) ΔS_{bb}, the entropy change gained by the backbone itself upon unfolding. The total entropy change can be written as:

$$\Delta S_{conf} = \sum_i \Delta S_{ex\rightarrow u,i} + \sum_j \Delta S_{bu\rightarrow ex,j} + \Delta S_{bb} \tag{10}$$

where the summation i runs over all amino acid side chains and the summation j runs only over those amino acids that are buried. ΔS_{bb} is the actual backbone entropy, including the restrictions imposed by the presence of side chains.

If the configurational entropy values discussed above are taken into consideration together with Eq. (10), it follows that the configurational entropy change associated with protein folding should be anywhere between 4 and 5 cal K^{-1} $mole^{-1}$. The presence of disulfide bridges or other covalent links will decrease this value. Also, a decrease from those values will be observed if side chains in the interior of a protein have configurational entropies larger than zero.

The contribution of polar solvation to $\Delta S°$ at $T_S{}^*$ can be evaluated from model compound data. The difference between the $\Delta S°$ of dissolution of linear alkanes and linear alcohols at $T_S{}^*$ is 8.3 cal K^{-1} $mole^{-1}$ with the alcohols having the lower value (Murphy, 1995). While this has been interpreted as indicating a negative $\Delta S°$ of hydration of the polar hydroxyl group (Oobatake and Ooi, 1992; Privalov and Makhatadze, 1993), the same difference is seen in comparing the $\Delta S°$ of vaporization of the alkanes and alcohols, suggesting that this difference may be due to changes in internal degrees of freedom of the model compounds and not to hydration (Murphy, 1995).

The assumption that the hydration $\Delta S°$ of polar groups is significant has led to overestimations of the configurational entropy. As the $\Delta S°$ of denaturation at $T_S{}^*$ is the sum of the configurational and polar contributions, a large negative polar

$\Delta S°$ will lead to a large positive configurational $\Delta S°$. Assuming that all the $\Delta S°$ of dissolution of polar groups is due to hydration (and not to changes in internal degrees of freedom), Privalov and Makhatadze (1993) reached the conclusion that ΔS_{conf} is about 19 cal K^{-1} $mole^{-1}$, a value significantly larger than any reasonable estimate of this term, as discussed above. Recent experimental results (d'Aquino *et al.*, 1995, have shown that the estimates of Privulov and Makhatadze (1993) are incorrect. Thus it would appear that the polar groups do not make a very significant contribution to $\Delta S°$ at the reference temperature and that the value of 4.3 cal K^{-1} $mole^{-1}$ is close to the average ΔS_{conf} per residue. To a first approximation then:

$$\Delta S_{conf} = (4.3 \pm 0.2)\, N_{res} \tag{11}$$

where N_{res} is the number of amino acid residues involved in the transition.

The empirical parameters used for relating structure to energetics are summarized in Table I. The structural information required is the ΔASA_{pol}, ΔASA_{ap}, and the number of residues involved in the transition. With this information and the parameters in Table I, Eqs. (7) to (11) can be used to estimate energetics from structure.

2.5. Limitations of the Model

Before proceeding with illustrations on the use of this structural energetic approach, it is important to consider the limitations of the model. The primary limitation of the model is the assumption of group additivity. While group additivity has found wide application in solution thermodynamics (Gill and Wadsö, 1976; Nichols *et al.*, 1976; Savage and Wood, 1976; Cabani *et al.*, 1981; Makhatadze and Privalov, 1990, 1993; Murphy and Gill, 1990), it is generally most applicable when dealing with homologous systems. In applying group additivity,

Table I. Empirical Parameters for Estimating Energetics from Structure

Parameter	Estimated value
$\Delta c_p°$, ap	0.44 ± 0.02 cal K^{-1} (mole-$Å^2$)$^{-1}$
$\Delta c_p°$, pol	-0.26 ± 0.03 cal K^{-1} (mole-$Å^2$)$^{-1}$
T_H*	100 ± 6 °C
T_S*	112 ± 1 °C
Δh_{pol}	35 ± 3 cal (mol$-Å^2$)$^{-1}$ of buried polar surface

we are assuming that each Å^2 of polar (or apolar) surface makes the same contribution to the energetics as every other Å^2. This clearly cannot be the case since van der Waals interaction, hydrogen bonding, and so on will vary from group to group depending on its specific environment, particularly its packing density. The success of the structural energetic approach results from the fact that we generally consider relatively large changes in ASA so that the average contributions indicated by the empirical parameters can be expected to apply to the systems under study. When very specific changes are being considered, such as point mutations, deviations from the average values are much more likely to be observed. This is an area of current investigation in this laboratory and involves the inclusion of local deviations in packing density that will affect the magnitude of van der Waals interactions. Richards (1977) has demonstrated before that the packing density of proteins is similar to the packing density of amino acid crystals and much larger than that of liquids. Furthermore, the packing density does not exhibit a significant deviation among proteins even though main-chain regions participating in secondary structure have generally a higher density than side-chain regions interacting only through van der Waals contacts. It is expected that, within the near future, the parametrization given by Eqs. (7) to (9) will be further refined by the explicit inclusion of local packing densities into the fundamental energetic parameters.

3. APPLICATION TO PROTEIN STABILITY

The structural energetic analysis described in Section 2 has been applied to 15 proteins for which both good crystallographic and calorimetric data are available. The experimental $\Delta H°$ values have been corrected for protonation effects. The proteins considered are listed in Table II along with their protein data bank (PDB) designation (Bernstein *et al.*, 1977; Abola *et al.*, 1987) and ΔASA_{ap} and ΔASA_{pol}. The comparison of the calculated and experimental thermodynamic quantities is given in Table III. The values of $\Delta H°$ and $\Delta S°$ of denaturation are compared at 60 °C, which is near the median melting temperature of the considered proteins. Comparison at this temperature minimizes extrapolation errors due to experimental uncertainties in ΔC_p (Murphy *et al.*, 1992; Murphy and Freire, 1992).

The comparison in Table III shows that the calculated values agree with the experimental values quite well, nearly within the experimental error. The root mean square error in the calculation is 6.0% for ΔC_p, 5.7% for $\Delta H°$, and 6.9% for $\Delta S°$. This is illustrated in Fig. 2 in which the fractional error in calculating each thermodynamic parameter is shown for each protein. Typical experimental error is indicated by the horizontal lines.

The excellent agreement between the calculated and experimentally deter-

Table II. Globular Proteins for which Both Calorimetric and Crystallographic Data Are Available

Protein	PDB designation	ΔASA_{ap} (Å^2)	ΔASA_{pol} (Å^2)	N_{res}
Cytochrome *c*		5809	3792	104
Carbonic anhydrase	2CAB	15950	10590	256
Chymotrypsin	4CHA	13970	8762	239
α-Lactalbumin	1AL	6773	4814	122
Hen lysozyme	1LYM	6844	5473	129
Myoglobin	4MBN	8873	5927	153
Staph. nuclease	1SNC	8151	5344	135
Papain	9PAP	13070	8692	212
Parvalbumin	5CPV	5770	4027	108
Pepsinogen	1PSG	22396	13498	365
RNase A	7RSA	6004	4963	124
Trypsin	1TLD	12370	8903	220
RNase T1		5000	3797	104
B2 domain of IgG binding protein G	1PGX	2797	2015	57
BPTI	5PTI	2745	2043	58

Table III. Comparison of Experimental and Calculated Values for the Thermodynamics of Folding–Unfolding Transitions of Globular Proteins[a]

	ΔC_p (kcal K^{-1} $mole^{-1}$)		$\Delta H°$ (kcal $mole^{-1}$)		$\Delta S°$ (cal K^{-1} $mole^{-1}$)	
Protein	Exp.	Calc.	Exp.	Calc.	Exp.	Calc.
Cytochrome *c*	1.60	1.67	78.0	66.7	208	206
Carbonic anhydrase	3.80	4.42	182.0	193.7	530	459
Chymotrypsin	3.00	4.01	175.0	146.3	552	446
α-Lactalbumin	1.80	1.80	96.0	96.6	241	264
Hen lysozyme	1.60	1.66	112.0	125.3	310	314
Myoglobin	2.60	2.45	101.0	109.4	282	302
Staph. nuclease	2.40	2.28	92.0	95.9	280	250
Papain	3.30	3.62	152.0	159.4	395	386
Parvalbumin	1.30	1.55	93.0	79.0	241	240
Pepsinogen	6.10	6.57	233.0	209.7	654	617
RNase A	1.40	1.41	111.0	117.2	325	328
Trypsin	2.90	3.25	186.0	181.5	544	474
RNase T1	1.40	1.26	84.5	82.4	274	264
B2 domain of IgG binding protein G	0.69	0.74	43.83	41.16	130	135
BPTI	0.67[b]	0.68	44.73	43.37	141	151

[a]ΔH° and ΔS° values are compared at 60°C as described in the text.
[b]Calculated as the average ΔC_p between 0 and 75°C. From the data of Makhatadze *et al.* (1993).

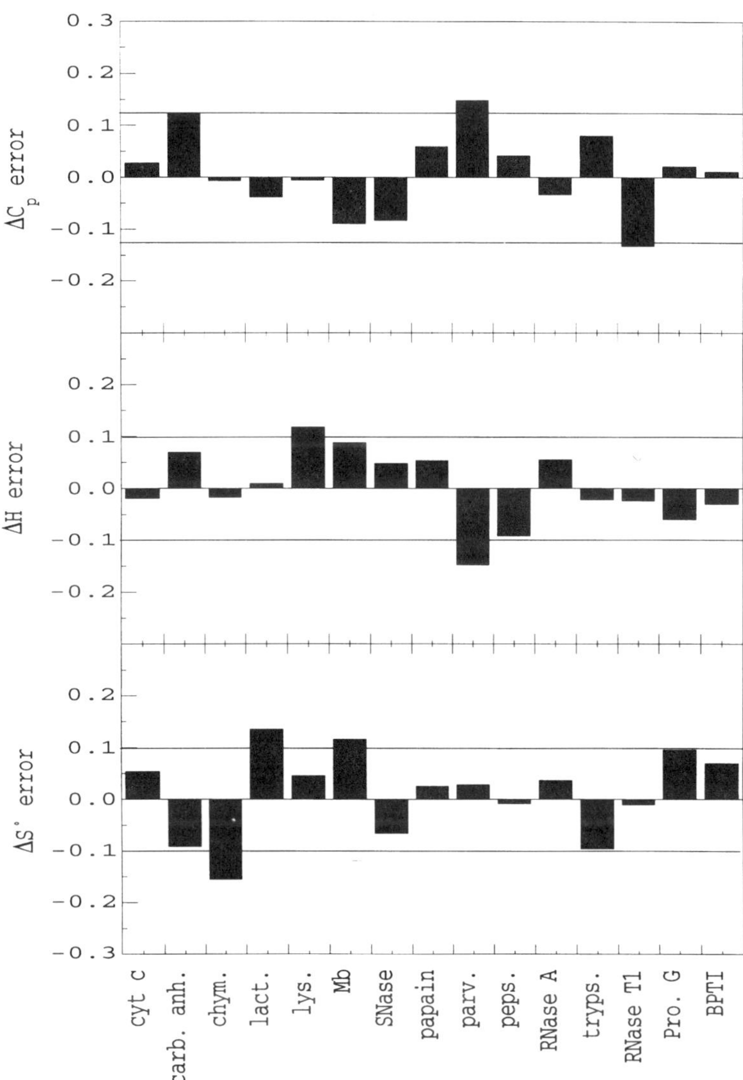

Figure 2. Fractional error in estimating thermodynamics of folding–unfolding transitions for globular proteins. The horizontal line represents typical experimental error.

mined values of ΔH°, ΔS°, ΔC_p of denaturation for the globular proteins illustrates the utility of the structural energetic approach and of the parameters listed in Table I. These parameters suggest a physical picture of the contributions of various interactions to the stability of proteins. It should be reemphasized that the parameters in Table I are largely empirical and therefore independent of theoretical discussions aimed at dissecting the relative contributions of specific atomic interactions. Nevertheless, the physical basis of protein stability suggested here is supported by several lines of evidence, including comparisons to model compound data (Murphy and Gill, 1990, 1991) and the effects of alcohol on the thermodynamics of protein stability (Fu and Freire, 1992). A more detailed discussion of these issues is given elsewhere (Murphy, 1993).

Figure 3 illustrates the breakdown of the stability of a typical protein according to the above model. The burial of apolar and polar surface both stabilize the protein. The burial of apolar surface provides less stability than has typically been supposed (see, for example, Chothia, 1975), presumably because the constraints of the peptide backbone and hydrogen bonding limit the ability of the apolar surface to achieve optimal packing (Creighton, 1991; Murphy and Gill, 1991). The polar contribution to ΔG° is primarily enthalpic and reflects the contributions of both hydrogen bonding and van der Waals interactions (Murphy and Freire, 1992). A large van der Waals contribution from the hydrogen-bonding amide groups is

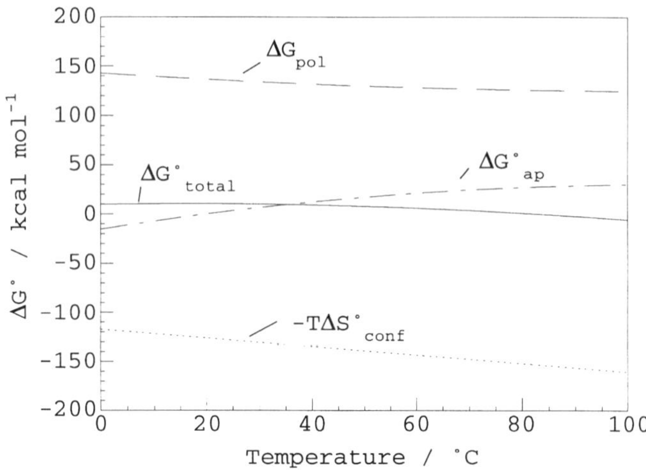

Figure 3. Contributions of polar and apolar surface area burial and configurational entropy to the overall stability of a "typical" globular protein with 100 amino acid residues; $\Delta ASA_{ap} = 5600$ Å^2 and $\Delta ASA_{pol} = 3900$ Å^2.

supported by the observation of exceptionally tight packing of the protein backbone (Richards, 1992). The stabilizing contributions of the apolar and polar surfaces are offset by the configurational entropy as given in Eq. (10). The overall modest stability of globular proteins is achieved primarily by the sum of these three contributions plus other specific electrostatic, protonation, covalent links, or ligand interactions.

4. APPLICATION TO PROTEIN–LIGAND INTERACTIONS

The same interactions responsible for the folding and stability of proteins also dictate the energetics of ligand binding to proteins. In both cases there are hydrogen bonds, hydrophobic interactions, configurational entropy effects, and electrostatic, protonation, and other specific effects. Consequently, Eqs. (7) to (10) can also be applied to protein–ligand interactions.

4.1. Entropic Effects in Binding Interactions

While the general interactions in protein folding and ligand binding are the same, there are important differences with regard to entropic effects. In protein folding there are major restrictions imposed on both the backbone and the side chains. In contrast, when a ligand binds to a protein, the major restrictions occurring in the protein are limited to the side chains, with little or no change in the backbone. These considerations require a further partitioning of the configurational entropy.

4.1.1. CONFIGURATIONAL ENTROPY EFFECTS

The total configurational entropy involved in the formation of a protein–ligand complex includes changes in the protein as well as changes in the ligand. As in the folding–unfolding case, these changes can include changes in backbone entropy as well as changes in side-chain entropies. In most binding cases, the configurational entropy changes are restricted to those side chains located in the binding pocket. If the ligand molecule is a peptide with no structure in solution, ΔS_{conf} will also include contributions from the backbone. Inspection of the structure of the complex should indicate which side chains are involved in the binding, but structures of both the complex and the free protein and ligand are necessary to determine whether backbone interactions are changed.

4.1.2. OTHER ENTROPIC CONSIDERATIONS

In addition to solvation and configurational entropy changes, entropy changes arising from vibrational, rotational, and translational degrees of freedom must also be considered in binding processes (Page and Jencks, 1971; Searle and Williams, 1992; Murphy *et al.*, 1994). The translational contribution to $\Delta S°$ of binding has often been theoretically treated using the Sakur-Tetrode equation, which describes the translational entropy for a particle in a box (see, for example, Bromberg, 1984, p. 713). Application of the Sakur-Tetrode equation to protein–ligand binding predicts a translational entropy change of about -25 cal K^{-1} $mole^{-1}$ (Searle *et al.*, 1992), which corresponds to ~7.5 kcal $mole^{-1}$ in free energy. This is equivalent to a decrease of 10^6-fold in the "intrinsic" binding affinity.

The Sakur-Tetrode equation, based on the quantum states of a particle in a box, requires the volume of the box as a parameter. For a molecule in the gas phase, the volume available for free translation, and hence the volume of the box, essentially is the molar volume of the gas, at least at low pressures. For processes in solution, however, application of the Sakur-Tetrode equation using the volume of the solution apparently leads to overestimates of the entropy changes (Murphy *et al.*, 1994). Use of the cratic entropy seems to provide a better estimate for translational contributions in solution (Murphy *et al.*, 1994).

For the binding of a single ligand to a protein, using a molar standard state, the cratic entropy is ~8 cal K^{-1} $mole^{-1}$ (Kauzmann, 1959), which amounts to 2.4 kcal $mole^{-1}$ in terms of free energy. This corresponds to a decrease of only 50-fold in the "intrinsic" affinity.

4.2. Binding of Angiotensin II to an Antibody

Recently, we have studied the energetics of binding of angiotensin II (AII) to a monoclonal antibody, Ab131, in collaboration with Amzel and co-workers (Murphy *et al.*, 1993). Angiotensin II is a peptide hormone involved in the regulation of blood pressure as part of the renin–angiotensin system and is an octapeptide of sequence Asp-Arg-Val-Tyr-His-Pro-Phe. The anti-idiotypic antibody serves as an analogue of the AII receptor.

From the calorimetric titrations of Ab131 with AII, it was shown that the binding at 30 °C is exothermic, with a $\Delta H°$ of -8.9 ± 0.7 cal K^{-1} $mole^{-1}$, and a ΔC_p of -240 ± 20 cal K^{-1} $mole^{-1}$ (Murphy *et al.*, 1993). Using the published binding constant at 4 °C (Budisavljevic *et al.*, 1988), the $\Delta G°$ at 30 °C is estimated as -11 ± 0.1 kcal $mole^{-1}$ and the $\Delta S°$ as 6.9 ± 2.3 cal K^{-1} $mole^{-1}$. The large negative $\Delta H°$, modest positive $\Delta S°$, and significant negative ΔC_p appear to be typical of antibody–antigen binding, having also been observed for the binding of antibodies to the antigen p185HER2 (Kelley *et al.*, 1992) and to cytochrome *c* (Murphy *et al.*, 1995). By studying the binding of AII to Ab131 in buffers with different heats of ionization, we also determined that approximately one proton is released per AII bound (Murphy *et al.*, 1993).

4.2.1. STRUCTURAL ENERGETICS OF ANGIOTENSIN II BINDING

Analysis of the AII/Ab131 structure (Garcia *et al.*, 1992) indicates that 993 $Å^2$ of apolar surface and 745 $Å^2$ of polar surface are buried on binding. The surface area calculation assumes that the free AII can be approximated as an extended peptide and that there is little or no change in the antibody structure on binding. According to Eq. (7), the calculated ΔC_p of binding is -250 ± 30 cal K^{-1} $mole^{-1}$, in excellent agreement with the experimental value.

The contribution of buried surface area to $\Delta H°$ of binding is calculated from Eqs. (7) and (8) to be -8.6 ± 3.2 kcal $mole^{-1}$. While this number is in good agreement with the experimentally determined value of -8.9 ± 0.7 kcal $mole^{-1}$, it does not include the $\Delta H°$ associated with the proton release. As the source of this proton has not been determined, we cannot estimate this quantity. Depending on the group(s) involved, this quantity can be in the range of 0 to 7 kcal $mole^{-1}$ (Christensen *et al.*, 1976).

The calculation of the $\Delta S°$ of binding involves more approximations than the calculation of either $\Delta G°$ or $\Delta H°$, as might be expected from the discussion above. The solvent contribution to $\Delta S°$ can be estimated as $\Delta C_p \ln (T/T_S^*)$ using the calculated ΔC_p. This yields a value of 61 ± 7 cal K^{-1} $mole^{-1}$ (Murphy *et al.*, 1993). Opposing this favorable change in entropy are the folding of the AII peptide, the freezing of side chains in the binding site, and the cratic entropy change.

The $\Delta S°$ associated with the folding of the AII molecule on binding can be calculated using the average value from folding–unfolding transitions in proteins as given by Eq. (11). This yields a value of -34.4 ± 0.6 cal K^{-1} $mole^{-1}$. Alternatively, one can consider the contribution of the individual residues as suggested by Eq. (10). Values for the contribution of individual residues to ΔS_{conf} for unfolding reactions have been estimated using the data of O'Neil and DeGrado (1990) and an average residue contribution of 4.3 cal K^{-1} $mole^{-1}$ (Murphy *et al.*, 1993). Using the sequence of AII and values for individual amino acids, we calculate a value of -35.9 ± 0.6 cal K^{-1} $mole^{-1}$, not significantly different from the value calculated using the protein average.

The contribution of freezing side chains in the binding site can be estimated by again considering the structure. There are six side chains in the binding site that are expected to become immobilized on binding (Garcia *et al.*, 1992). The average contribution to ΔS_{conf} is -2.8 ± 1.7 cal K^{-1} $mole^{-1}$ (Creamer and Rose, 1992). This suggests a $\Delta S°$ contribution from side-chain freezing of -16.8 ± 4.2 cal K^{-1} $mole^{-1}$.

The value of $\Delta S°$ also includes the cratic term of -8 cal K^{-1} $mole^{-1}$. The sum of the solvent, AII configurational, side-chain freezing, and cratic entropy changes is 1.8 ± 8 cal K^{-1} $mole^{-1}$, which agrees within error with the experimentally determined value of 6.9 kcal K^{-1} $mole^{-1}$.

It is interesting to note that the modest experimental $\Delta S°$ results from compensation between rather large solvent and configurational terms. This entropic compensation seems to be a general feature of biological systems in which the cost of inducing order, either by folding or binding, is offset, at least to some extent, by the release of ordered solvent.

The free energy change, $\Delta G°$, is then calculated to be -9 ± 4 kcal $mole^{-1}$, which corresponds to a binding constant of 3×10^6. These values are somewhat below the experimentally determined quantities, but are still quite reasonable estimates considering the simplicity of the approach used to arrive at them. While more studies must be performed to generalize the approach, this initial application of the structural energetic analysis is encouraging.

4.2.2. BINDING OF ANGIOTENSIN ANALOGUES

The above example illustrates that the structural energetic analysis can be used to estimate binding energetics when a structure of the complex is known, given appropriate assumptions about the structures of the free components. We can also extend this approach to consider complexes whose structures have not been determined, if reasonable models of those structures can be made.

There are several analogues of AII whose inhibition of AII binding to Ab131 has been investigated (Garcia, 1991). Figure 4 lists the analogues that have been studied and their (IC_{50}) scoring. The inhibition studies were performed using an enzyme-linked immunosorbent assay as described elsewhere (Picard *et al.*, 1986).

Estimates of the binding affinity of the various analogues were determined using the structure of the AII/Ab131 complex (Garcia *et al.*, 1992). Analogues were modeled by deleting the appropriate amino acid residue from the coordinate file and then calculating the changes in ASA. Energetics were estimated in the same manner as for AII binding, except that the ligand contribution to ΔS_{conf} was calculated using individual values for each amino acid and scaling the entropy of side-chain freezing to the number of residues in the analogue peptide. The results are summarized in Table IV.

As seen in Table IV, the IC_{50} ratio (given as the ratio of K_{AII} to K_{ana}) is in fairly good agreement with the published ratios for angiotensin III and for the C6 peptide. For C5, however, the calculations predict an IC_{50} lower than that ob-

Analog	Amino Acid Position 1	2	3	4	5	6	7	8	IC50 Score
AII	Asp-	Arg-	Val-	Tyr-	Ile-	His-	Pro-	Phe	++
AIII		Arg-	Val-	Tyr-	Ile-	His-	Pro-	Phe	++
C6			Val-	Tyr-	Ile-	His-	Pro-	Phe	+
C5				Tyr-	Ile-	His-	Pro-	Phe	-

Scoring:	Ratio of [analogue]:[AII] for IC_{50}
++	1-10
+	10-100
-	>100

Figure 4. Sequences and IC_{50} scores for angiotensin II analogues binding to Ab131.

served. The reason for the lack of agreement for the C5 peptide is not clear and cannot be evaluated at this point since actual binding data for these analogues are not available.

4.3. Antibody Binding to Cytochrome *c*

The ideas presented above can also lend insight into ligand-binding systems in the absence of high-resolution structural data. This is illustrated in recent work

Table IV. Estimates of Binding Constants for Angiotensin II Analogues[a]

Analogue	ΔASA_{ap}	ΔASA_{pol}	log K	IC_{50}
Angiotensin II	−993	−745	6.07	1
Angiotensin III	−922	−644	5.06	10
C6	−874	−537	3.53	345
C5	−828	−493	3.98	123

[a]Binding constants and IC_{50} values were calculated as described in the text.

on the binding of two different antibodies, E3 and E8, to cytochrome *c* (cyt *c*). The binding of these antibodies to cyt *c* has been studied extensively by Paterson and co-workers (Cooper *et al.*, 1987; Collawn *et al.*, 1988; Paterson *et al.*, 1990; Mayne *et al.*, 1992).

Two of their observations are of particular interest. Paterson *et al.* (1990) have used two-dimensional NMR to monitor the effect of E8 binding on the hydrogen–deuterium exchange kinetics of amide protons in cyt *c*. They found that amide protons protected from exchange were localized to the epitope as determined by previous epitope mapping studies. The epitope on the cyt *c* surface, defined in this way, was found to consist of approximately 750 $Å^2$ of surface area (Paterson *et al.*, 1990).

In contrast, similar studies on the effect of binding E3 to cyt *c* indicated protection of amide protons both in and beyond the previously mapped epitope (Mayne *et al.*, 1992). This suggested to the authors that the binding of E3 to cyt *c* resulted in greater configurational restriction than did the binding of E8 (Mayne *et al.*, 1992).

In order to test this hypothesis, we undertook a calorimetric study of the binding of these two antibodies to cyt *c* (Murphy *et al.*, 1995). The calorimetrically determined binding energetics are summarized in Table V. At 25 °C, the favorable binding free energy is primarily enthalpic in origin; however, the $\Delta S°$ of binding for both antibodies is small and favorable. The $\Delta S°$ of binding for E3 is slightly more favorable than that for E8, but they are nearly the same within the experimental error.

As discussed above, the observed $\Delta S°$ of binding has several contributions including those from solvent release and configurational changes. According to the above discussion, the solvent contribution to $\Delta S°$ is given as:

$$\Delta S°_{solv} = \Delta C_p \ln (T/T_S{}^*) \tag{13}$$

where $T_S{}^*$ is the entropy reference temperature of 385 K (112 °C). For E3 and E8, the calculated solvent contributions to $\Delta S°$ at 25 °C are 90 ± 15 cal K^{-1} $mole^{-1}$ and 42 ± 10 cal K^{-1} $mole^{-1}$. Subtracting $\Delta S°_{solv}$ and $\Delta S°_{cratic}$ (-8 cal K^{-1} $mole^{-1}$) from the experimental values yields the estimated configurational contributions to $\Delta S°$.

Table V. Thermodynamics of Binding of Antibodies E8 and E3 to Horse Cyt *c* at 25°C[a]

Antibody	Δ (kcal $mole^{-1}$) G°	Δ (kcal $mole^{-1}$) H°	Δ (cal k^{-1} $mole^{-1}$) S°	Δ (cal k^{-1} $mole^{-1}$) C_p
E8	-10.6 ± 0.7	-9.5 ± 0.9	3.7 ± 3.8	-165 ± 38
E3	-9.7 ± 0.6	-7.3 ± 0.4	8.8 ± 2.4	-350 ± 58

[a]Adapted from Murphy *et al.* (1993a).

For E3 this value is -73 ± 15 cal K^{-1} $mole^{-1}$, while for E8 it is -30 ± 11 cal K^{-1} $mole^{-1}$. Thus the configurational $\Delta S°$ associated with E3 binding to cyt *c* is more than twice that for E8 binding, as predicted from the hydrogen–deuterium exchange results.

The relationship between changes in surface areas and energetics indicated by Eqs. (7) and (8) also suggests that the energetics should provide a means for estimating ΔASA values. According to Eq. (8), and in the absence of protonation effects, the value of $\Delta H°$ at $T_H{}^*$ is directly proportional to ΔASA_{pol}. Using the value from Tables I and V, we calculate ΔASA_{pol} as -960 Å^2 for E3 and -625 Å^2 for E8. These values can then be used with the experimental values of ΔC_p and Eq. (7) to calculate ΔASA_{ap} for the binding of the two antibodies as -1330 Å^2 for E3 and -730 for E8.

From these values we estimate that the binding of E3 to cyt *c* buries a total of 2300 Å^2 of surface area, while the binding of E8 buries a total of 1360 Å^2. Inspection of known crystallographic structures of antibody–antigen complexes for protein antigens indicates that the antigen and antibody each contribute about one half of the surface area buried (Davies *et al.*, 1990). Thus we can estimate that the cyt *c* epitope to which E3 binds is approximately 1150 Å^2, while that for E8 is approximately 680 Å^2. These estimates agree rather well with the values estimated from epitope mapping studies of 1200 Å^2 for E3 (Mayne *et al.*, 1992) and 750 Å^2 for E8 (Paterson *et al.*, 1990).

5. CONCLUSIONS

The structural thermodynamic approach provides a set of promising new tools for the development of rational strategies for protein and drug design. The advantage of this approach is that the fundamental energetic parameters used in the calculations are derived from direct calorimetric measurements. In that sense, the resulting energies are effective energies and explicitly or implicitly include all interactions that are found in the system (i.e., protein, ligand, and solvent interactions). As the research progresses and the structural thermodynamic database is expanded, the set of fundamental energetic parameters is expected to be refined to the point that they will allow accurate prediction of the effects of small molecular changes. Two areas are of particular importance: the accurate prediction of configurational entropy changes and the incorporation of local fluctuations in packing densities into the fundamental energy functions.

ACKNOWLEDGMENTS. This work was sponsored by grants from the NIH (RR04328, NS24520, GM37911) to EF. KPM was also supported by the Roy J. Carver Charitable Trust.

REFERENCES

Abola, E. E., Bernstein, F. C., Bryant, S. H., Koetzle, T. F., and Weng, J., 1987, Protein data bank, in: *Crystallographic Databases—Information Content, Software Systems, Scientific Applications* (F. H. Allen, G. Bergerhoff, and R. Sievers, eds.), Data Commission of the International Union of Crystallography, Bonn/Cambridge/Chester, pp. 107–132.

Baldwin, R. L., 1986, Temperature dependence of the hydrophobic interaction in protein folding, *Proc. Natl. Acad. Sci. USA* **83:**8069–8072.

Bernstein, F. C., Koetzle, T. F., Williams, G. J. B., Meyer, E. F., Jr., Brice, M. D., Rodgers, J. R., Kennard, O., Shimanouchi, T., and Tasumi, M., 1977, The protein data bank: A computer-based archival file for macromolecular structures, *J. Mol. Biol.* **112:** 535–542.

Bromberg, J. P., 1984, *Physical Chemistry*, 2nd ed., Allyn and Bacon, Boston.

Budisavljevic, M., Geniteau-Legendre, M., Baudouin, B., Pontillon, F., Verroust, P. J., and Ronco, P. M., 1988, Angiotensin II (AII)-related idiotypic network. II. Heterogeneity and fine specificity of AII internal images analyzed with monoclonal antibodies, *J. Immunol.* **140:**3059–3065.

Cabani, S., Gianni, P., Mollica, V., and Lepori, L., 1981, Group contributions to the thermodynamic properties of non-ionic organic solutes in dilute aqueous solution, *J. Solution Chem.* **10:**563–595.

Chothia, C., 1975, Structural invariants in protein folding, *Nature* **254:**304–308.

Christensen, J. J., Hansen, L. D., and Izatt, R. M., 1976, *Handbook of Proton Ionization Heats and Related Thermodynamic Quantities*, John Wiley and Sons, New York.

Collawn, J. F., Wallace, C. J. A., Proudfoot, A. E. I., and Paterson, Y., 1988, Monoclonal antibodies as probes of conformational changes in protein-engineered cytochrome *c*, *J. Biol. Chem.* **263:**8625–8634.

Connelly, P. R., and Thomson, J. A., 1992, Heat capacity changes and hydrophobic interactions in the binding of FK506 and rapamycin to the FK506 binding protein, *Proc. Natl. Acad. Sci. USA* **89:**4781–4785.

Cooper, H. M., Jemmerson, R., Hunt, D. F., Griffin, P. R., Yates, J. R., Shabanowitz, J., Zhu, N. Z., and Paterson, Y., 1987, Site-directed chemical modification of horse cytochrome *c* results in changes in antigenicity due to local and long-range conformational perturbations, *J. Biol. Chem.* **262:**11591–11597.

Creamer, T. P., and Rose, G. D., 1992, Side-chain entropy opposes α-helix formation but rationalizes experimentally determined helix-forming propensities, *Proc. Natl. Acad. Sci. USA* **89:**5937–5941.

Creighton, T. E., 1991, Stability of folded conformations, *Curr. Opin. Struct. Biol.* **1:**5–16.

Davies, D. R., Padlan, E. A., and Sheriff, S., 1990, Antibody–antigen complexes, *Annu. Rev. Biochem.* **59:**439–473.

Eisenberg, D., and McLachlan, A. D., 1986, Solvation energy in protein folding and binding, *Nature* **319:**199–203.

Freire, E., 1993, Structural thermodynamics: Prediction of protein stability and protein binding affinities, *Arch. Biochim. Biophys.* **303:**181–184.

Freire, E., Haynie, D. T., and Xie, D., 1993, Molecular basis of cooperativity in protein folding. IV. CORE: A general cooperative folding model, *Proteins* **17:**111–123.

Fu, L., and Freire, E., 1992, On the origin of enthalpy and entropy convergence temperatures in protein folding, *Proc. Natl. Acad. Sci. USA* **89:**9335–9338.

Garcia, K. C., 1991, Ph.D. thesis, Structural Studies of Hormone Recognition: Anti-idiotypic Recognition in an Angiotensin II Defined Idiotypic Series. The Johns Hopkins University.

Garcia, K. C., Ronco, P. M., Verroust, P. J., Brünger, A. T., and Amzel, L. M., 1992, Three-dimensional structure of an angiotensin II-Fab complex at 3 Å: Hormone recognition by an anti-idiotypic antibody, *Science* **257:**502–507.

Gill, S. J., and Wadsö, I., 1976, An equation of state describing hydrophobic interactions, *Proc. Natl. Acad. Sci. USA* **73:**2955–2958.

Hermann, R. B., 1972, Theory of hydrophobic bonding. II. The correlation of hydrocarbon solubility in water with solvent cavity surface area, *J. Phys. Chem.* **76:**2754–2759.

Kauzmann, W., 1959, Some factors in the interpretation of protein denaturation, *Adv. Protein Chem.* **14:**1–63.

Kelley, R. F., O'Connell, M. P., Carter, P., Presta, L., Eigenbrot, C., Covarrubias, M., Snedecor, B., Bourell, J. H., and Vetterlein, D., 1992, Antigen binding thermodynamics and antiproliferative effects of chimeric and humanized anti-p185^{HER2} antibody Fab fragments, *Biochemistry* **31:**5434–5441.

Lee, B., and Richards, F. M., 1971, The interpretation of protein structures: Estimation of static accessibility, *J. Mol. Biol.* **55:**379–400.

Livingstone, J. R., Spolar, R. S., and Record, M. T., Jr., 1991, Contribution to the thermodynamics of protein folding from the reduction in water-accessible nonpolar surface area, *Biochemistry* **30:**4237–4244.

Makhatadze, G. I., and Privalov, P. L., 1990, Heat capacity of proteins. I. Partial molar heat capacity of individual amino acid residues in aqueous solution: Hydration effect, *J. Mol. Biol.* **213:**375–384.

Makhatadze, G. I., and Privalov, P. L., 1993, Contribution of hydration to protein folding thermodynamics. I. The enthalpy of hydration, *J. Mol. Biol.* **232:**639–659.

Makhatadze, G. I., Kim, K.-S., Woodward, C., and Privalov, P. L., 1993, Thermodynamics of BPTI folding, *Protein Sci.* **2:**2028–2036.

Mayne, L., Paterson, Y., Cerasoli, D., and Englander, S. W., 1992, Effect of antibody binding on protein motions studied by hydrogen-exchange labeling and two-dimensional NMR, *Biochemistry* **31:**10678–10685.

Murphy, K. P., 1993, Hydration and convergence temperatures: On the use and interpretation of correlation plots, *Biophys. Chem.* **51:**311–326.

Murphy, K. P., 1995, Noncovalent forces important to the conformational stability of protein structures, in: *Protein Stability and Folding* (B. A. Shirley, ed.), Humana Press, Clifton, NJ, 1–34

Murphy, K. P., and Freire, E., 1992, Thermodynamics of structural stability and cooperative folding behavior in proteins, *Adv. Protein Chem.* **43:**313–361.

Murphy, K. P., and Gill, S. J., 1990, Group additivity thermodynamics for dissolution of solid cyclic dipeptides into water, *Thermochim. Acta* **172:**11–20.

Murphy, K. P., and Gill, S. J., 1991, Solid model compounds and the thermodynamics of protein unfolding, *J. Mol. Biol.* **222:**699–709.

Murphy, K. P., Privalov, P. L., and Gill, S. J., 1990, Common features of protein unfolding and dissolution of hydrophobic compounds, *Science* **247:**559–561.

Murphy, K. P., Bhakuni, V., Xie, D., and Freire, E., 1992, The molecular basis of cooperativity in protein folding III: Identification of cooperative folding units and folding intermediates, *J. Mol. Biol.* **227:**293–306.

Murphy, K. P., Xie, D., Garcia, K. C., Amzel, L. M., and Freire, E., 1993, Structural energetics of peptide recognition: Angiotensin II/antibody binding, *Proteins* **15:** 113–120.

Murphy, K. P, Xie, D., Thompson, K. S., Amzel, L. M., and Freire, E., 1994, Entropy in biological binding processes: Estimation of translational entropy loss, *Proteins*, **18:**63–67.

Murphy, K. P., Freire, E., and Paterson, Y., 1995, Configurational effects in antibody–antigen interactions determined by micro-calorimetry, *Proteins*, **21:**83–90.

Némethy, G., Leach, S. J., and Scheraga, H. A., 1966, The influence of amino acid side chains on the free energy of helix-coil transitions, *J. Phys. Chem.* **70:**998–1004.

Nichols, N., Sköld, R., Spink, C., and Wadsö, I., 1976, Additivity relations for the heat capacities of non-electrolytes in aqueous solution, *J. Chem. Thermodynamics* **8:**1081–1093.

O'Neil, K., and DeGrado, W., 1990, A thermodynamic scale for the helix-forming tendencies of the commonly occurring amino acids, *Science* **250:**646–651.

Oobatake, M., and Ooi, T., 1992, Hydration and heat stability effects on protein unfolding, *Prog. Biophys. Molec. Biol.* **59:**237–284.

Page, M. I., and Jencks, W. P., 1971, Entropic contributions to rate accelerations in enzymic and intramolecular reactions and the chelate effect, *Proc. Natl. Acad. Sci. USA* **68:** 1678–1683.

Paterson, Y., Englander, S. W., and Roder, H., 1990, An antibody binding site on cytochrome *c* defined by hydrogen exchange and two-dimensional NMR, *Science* **249:** 755–759.

Picard, C., Ronco, P., Moullier, P., Yao, J., Baudouin, B., Geniteau Legendre, M., and Verroust, P., 1986, Epitope diversity of angiotensin II analyzed with monoclonal antibodies, *Immunology* **57:**19–24.

Privalov, P. L., and Makhatadze, G. I., 1992, Contribution of hydration and non-covalent interactions to the heat capacity effect on protein unfolding, *J. Mol. Biol.* **224:** 715–723.

Privalov, P. L., and Makhatadze, G. I., 1993, Contribution of hydration to protein folding thermodynamics. II. The entropy and Gibbs energy of hydration, *J. Mol. Biol.* **232:** 660–679.

Rashin, A. A., 1984, Buried surface area, conformational entropy, and protein stability, *Biopolymers* **23:**1605–1620.

Richards, F. M., 1977, Areas, volumes, packing and protein structure, *Annu. Rev. Biophys. Bioeng.* **6:**151–176.

Richards, F. M., 1992, Folded and unfolded proteins: An introduction, in: *Protein Folding* (T. E. Creighton, ed.), W. H. Freeman and Company, New York, pp. 1–58.

Savage, J. J., and Wood, R. H., 1976, Enthalpy of dilution of aqueous mixtures of amides, sugars, urea, ethylene glycol, and pentaerythritol at 25 °C: Enthalpy of interaction of the hydrocarbon, amide, and hydroxyl functional groups in dilute aqueous solutions, *J. Solution Chem.* **5:**733–750.

Schellman, J. A., 1955, The stability of hydrogen-bonded peptide structures in aqueous solution, *Compt-Rend. Lab. Carlsberg Sér. Chim.* **29:**230–259.

Searle, M. S., and Williams, D. H., 1992, The cost of conformational order: Entropy changes in molecular associations, *J. Am. Chem. Soc.* **114:**10690–10697.

Searle, M. S., Williams, D. H., and Gerhard, U., 1992, Partitioning of free energy contributions in the estimation of binding constants: Residual motions and consequences for amide–amide hydrogen bond strengths, *J. Am. Chem. Soc.* **114:**10697–10704.

Spolar, R. S., Livingstone, J. R., and Record, M. T., Jr., 1992, Use of liquid hydrocarbon and amide transfer data to estimate contributions to thermodynamic functions of protein folding from the removal of nonpolar and polar surface from water, *Biochemistry* **31:** 3947–3955.

7

Chromatographic Techniques for the Characterization of Proteins

Joost J. M. Holthuis and Reinoud J. Driebergen

1. INTRODUCTION

The improvements in the chromatographic analysis of proteins and peptides during the last decade contributed significantly to the development of (recombinant) pharmaceutical proteins. The availability of these analytical methods for the characterization and purity determination of proteins enabled the improvement of the overall manufacturing process of pharmaceutical proteins.

As with all other pharmaceutical preparations, (recombinant) pharmaceutical protein products must be thoroughly characterized. Protein drugs must meet the same standards of safety, purity, and potency as conventional drugs. The characterization of the structure of a protein is complex because of the presence of a primary, secondary, tertiary, and quarternary structure. Small structural changes in a protein can influence the physicochemical properties as well as the activity and potency of the protein. Different purification steps are used during the manufacturing that may change the structural characteristics of the protein. Such changes may range from minor modifications (e.g., deamidation of an Asp) (Patel and Borchardt, 1990) to unfolding of the protein chain yielding biologically

Joost J. M. Holthuis • OctoPlus b.v., P.O. Box 722, 2300 AS Leiden, The Netherlands. *Reinoud J. Driebergen* • Ares Serono, 15bis, Chemin des Mines, CH-1211 Geneva, Switzerland.

Physical Methods to Characterize Pharmaceutical Proteins, edited by James N. Herron *et al.*, Plenum Press, New York, 1995.

inactive molecules. Unfolding may result in the decomposition (such as oxydation and fragmentation) of the protein due to the fact that certain parts of the molecule, normally part of the inner side of the tertiary structure, are exposed to chemicals, water, and so forth.

For the determination of the purity of proteins the detection of impurities is of importance. The impurities are process, host, or product related (Garnick *et al.*, 1988). Product-related impurities can be the result of posttranslational changes or variability but are in most cases decomposition products that are formed during the purification process or upon storage. Table I presents an overview of possible processes that can change the structure of a protein (Geigert, 1989). Recently an extended overview was published on the inactivation of proteins during chromatographic analysis (Sadana, 1992).

The protein can also be changed intentionally. With the introduction of large amounts of recombinant proteins, more and more time is spent on the development of chemical modifications of the proteins in order to change their disposition *in vivo*. Covalent attachment of, e.g., polyethylene glycol (PEG) chains shows promise for increasing the stability and protein solubility. The resulting product contains a mixture of species characterized by a distribution in both the number and position of attachment of the PEG molecules. The situation is even more complicated by the fact that PEG is inherently polydispersal. Characterization of such a pegylated protein is important in order to be able to assure product consistency.

The qualitative and quantitative analysis of a pharmaceutical protein and its related compounds relies on the use of sophisticated analytical methods for demonstrating the structural identity and homogeneity of the proteins to be separated. The application of multiple methods is imperative to ensure that a protein is thoroughly identified and characterized and its purity accurately assessed. For the analysis of pharmaceutical proteins the method must be able to

Table I. Possible Modifications of Proteins

Degradation pathways	Mutation	Posttranslational process	Chemical conjugation
Fragmentation	Site-directed mutagenesis	Glycosylation	Pegylation
Deamidation		Formylation	
Oxidation		Acylation	
Disulfide scrambling		Phosphorylation	
Oligomerization		Sulfatation	
Aggregation			
Cross-linking			
Proteolysis			
Denaturation			

separate compounds that may only differ by one or a few amino acids or functional groups. Considering the low diffusion coefficients of large biopolymers, it is a real challenge to develop analytical chromatographic methods that enable the required separation.

Proteins are ampholytes and therefore the pH and the ionic strength of the mobile phase play an important role in their chromatographic behavior. In addition, temperature, protein concentration, and adsorbent surface characteristics affect the analysis in all modes of interactive chromatography as well as in noninteractive (size-exclusion) chromatography. These parameters are important because the three-dimensional structure of the protein may change when the protein is in a nonphysiological environment and may be sensitive to slight changes in this environment; these parameters also can be used to manipulate the protein structure in order to achieve separation.

This chapter presents an overview of the analytical chromatographic methods used for the characterization of proteins with an emphasis on pharmaceutical proteins. This chapter does not pretend to give a detailed and complete overview on the chromatography of proteins and peptides, but rather presents an overview of approaches used to characterize proteins, including the significant problems that can arise and relevant examples. General information on the chromatography of peptides and proteins, including the column materials used, can be found elsewhere (Mant and Hodges, 1991).

2. REVERSED-PHASE CHROMATOGRAPHY

2.1. General

Reversed-phase chromatography (RPC) is the most frequently applied chromatographic technique for the analysis (especially for the determination of the purity) of peptides and intact proteins as well as for the analysis of peptide fragments obtained after enzymatic or chemical cleavage of the protein (see also section 2.4.3). In RPC the protein generally loses its native structure and denatures (Lau *et al.*, 1984). Denaturation occurs when the hydrophobic and coulombic interactions with the stationary and/or mobile phase are stronger than those that are maintaining the protein native structure. The level of denaturation often depends on the time the sample resides on the column. The elution of a protein is determined by the presence of hydrophobic amino acids as well as by the exposure of hydrophobic regions to the column surface, determined by a sequence of hydrophobic amino acids. The retention of the intact protein is based on the number of hydrophobic residues that interact with the hydrophobic column support and therefore is dependent on whether these hydrophobic areas are present on the outside of the protein when dissolved in the mobile phase. The less polar residues can be shielded by conformational effects, which results in an exposition of the more polar residues at the protein's surface during the analysis (Nugent *et al.*, 1988). For peptides up to 15 amino acids there is a correlation between the hydrophobicity and retention on a C_4, C_8, and C_{18} bonded phase. With larger compounds also the polypeptide chain length must be taken into account because the length of the chain determines the spatial structure. Using the overall hydrophobicity of proteins (based on retention parameters derived from small peptides), a correlation was found with retention behavior, under denaturing conditions, for proteins ranging in molecular weight from 3.5 to 32 kDa (Mant *et al.*, 1989). The conformation of the protein determines the exposure of hydrophobic regions to the stationary phase, and thus regulates the retention of the protein. Resolution of the denatured protein in RPC is achieved on the basis of the overall hydrophobicity of the sample and is mainly dominated by solvent effects.

Conformational changes on reversed-phase and hydrophobic supports can lead to multiple states of the protein, which can result in multiple or distorted peaks. The control of the peak shape and the number of peaks in protein chromatography require the understanding of equilibria and the kinetic processes involved. One can distinguish three kinetic processes where protein structural changes can affect behavior. In the first case, the changes are slow relative to the chromatographic migration time and the process is apparently irreversible. In such cases several chromatographic peaks, representing native and (partially) denatured compounds, may be observed. In the second case, retention and half-lives are comparable and distorted peaks can be observed. Finally, rapid changes occur and the system may act as though it were behaving ideally, while at the same time significant changes have taken place on the chromatographic surface. In principle one would like to prevent such changes in order to enhance the possibility of analyzing the native biologically active protein (Karger and Blanco, 1989); but in RPC this is almost an impossible task. Therefore, in order to prevent conformational changes during the analysis, it is important to choose the circumstances in such a way that the analyte is present in one single conformation, where in most cases the denaturation of the protein is complete. The unfolding of proteins in a chromatographic system can be controlled by the hydrophobicity of the packing material. Sometimes it is possible to chromatograph proteins under reversed-phase conditions in the folded conformation by carefully choosing the column material (Hanson *et al.*, 1992).

Changes in the quaternary structure often result in a significant change in the molecular weight or hydrophobic properties of the molecule because of dissociation or aggregation of the molecule. Oligomers are high-molecular-weight compounds and in most cases more hydrophylic when compared with the monomeric proteins. Oligomers are therefore easily separated from the corresponding monomers/subunits by RPC.

Peptide mapping is often used in the quality control of pharmaceutical proteins and is then used to detect lot-to-lot variations. Isolation of peptides following the chromatographic analysis and the determination of the amino acid sequence of these peptides are tools that are used to determine the full sequence of a protein. Peptide mapping of proteins is used to generate information on the primary structure of proteins. Peptides produced from proteins by two or more different chain cleavage methods and analyzed by RPC give information on the identity of the protein. Peptide mapping consists of an enzymatic (e.g., trypsine) or chemical [e.g., cyanobromide (CNBr)] digestion often preceded by reduction and carboxymethylation of free sulfhydryl groups with iodoacetic acid. With peptide mapping it is possible to detect small changes in the protein, e.g., replacement of a single amino acid. The replacement of one amino acid in a peptide has in general a dramatic influence on the retention behavior on RPC. In general, RPC gives a high-resolution separation of tryptic digests. A typical run takes between 30 and 60 min in order to obtain optimum separation of the peptide mixture. The use of short reversed-phase columns packed with 2-μm pellicular particles can reduce the run time of a peptide to 15 min or less at elevated temperatures, as was shown by Kalghatgi and Horváth (1988) for recombinant tissue plasminogen activator and recombinant human growth hormone (rhGH). For the determination of impurities the method is less suitable. In general, it is difficult to identify protein impurities below 10% by peptide mapping using RPC with UV detection, as was shown by Clogston *et al.* (1992) for recombinant human granulocyte colony stimulating factor.

2.2. Stationary Phase

The stationary phase used in RPC contains a hydrophobic surface and often consists of coated silica, polymeric bonded silica, and cross-linked polymers (Mant and Hodges, 1991; Kennedy *et al.*, 1989). The matrices are in most cases coated with hydrophobic ligands such as (in increasing order of hydrophobicity), C_2, C_4, C_8, C_{18}, phenyl, and cyanopropyl groups. In general, the retention time increases with increasing hydrophobicity of the ligands. Decreasing particle size results in an increasing surface area, and thus in a higher resolution.

The most frequently used column materials are silica based with a pore size of 100 to 300 Å. These phases are only stable at pH 2 to 8. The phases consisting of a polymer coating on silica are more inert to chemical degradation when compared with the conventional silica bound packings. These stationary phases are stable in the pH region of 2 to 10.

Replacement of the silica as base of the reversed-phase material by a polymer increases the stability further. An example is the C_4, C_8, and C_{18} anchored to the polymer skeleton, e.g., polyvinylalcohol. Other examples are cross-linked hydrophobic packings like hydroxylated polyether (e.g., TSK), polystyrenedivinylbenzene (e.g., PLRP-S), polystyrene (Maa and Horváth, 1988), polymethacrylate, and polymeric fluorocarbon. These stationary phases are stable in the pH region of 1 to 13 (Kato *et al.*, 1990). The pore size (50–4000 Å) of the column material is of importance and must match the molecular size of the sample in order to eliminate undesired exclusion. Polymeric (polystyrene) packings with particle diameters of 5 μm and pores of 300 Å are in general superior in terms of resolution and sample recovery (Burton *et al.*, 1988).

Reversed-phase material based on silica and polymer-based material often show a different selectivity, probably due to unwanted interactions between silanol groups and the protein, e.g., for proinsulin and insulin (Linde and Welinder, 1991a,b).

2.3. Mobile Phase

The mobile phase often consists of mixtures of organic solvents to which organic or anorganic acids are added. Isocratic elution of the proteins is possible; however, retention of the protein is strongly affected by the percentage of the organic modifier. It is very difficult to prepare the mobile phase to be so reproducible that consistent data are obtained. Therefore, a gradient elution system is usually used that results in higher resolution and is most effective in the control of the retention time. Elution is started with aqueous acid solutions of, e.g., trifluoroacetic acid (TFA), phosphoric acid, perchloric acid, and heptafluorobutyric acid. Compounds such as phosphoric acid, perfluorinated carboxylic acids, and trialkylammonium phosphate act as ion pairing agents. The pH of the aqueous solutions is in the range of 2 to 5. The use of hydrochloric acid (HCl) instead of TFA has some advantages, such as a greater optical clarity that enables the detection at low wavelengths.

A recent development is the use of formic acid instead of phosphoric acid or TFA in the mobile phase for the chromatography of proteins. Formic acid enables the detection of nonaromatic peptides at low wavelengths, and because of its volatility it is easily removed. Detection at 214 nm is possible with good resolution. Formic acid is a mild acid that does not damage the column (Poll and Harding, 1989, 1991).

The starting eluent is followed by a displacing eluent that is an organic solvent such as methanol, acetonitrile, 1-propanol, or 2-propanol. The mobile phase causes denaturation by disrupting the intramolecular hydrophobic and coulombic forces. The organic solvents can be ranked according to decreasing elutropic strength as compared to water which results in the following order:

propanol > acetonitrile > methanol > water. Experimental variables that affect the elution behavior in RPC are temperature, pH, solvents, buffers, additives, flow rate, gradient program, and sample pretreatment (Nugent *et al.*, 1988).

RPC is very suitable for the analysis of hydrophilic proteins and peptides. More hydrophobic proteins, e.g., membrane proteins, often give poor recovery and multiple peaks and may not elute or may be considerably retarded. Protein aggregates can result in plugging of the stationary phase and poor recovery. The fact that many proteins are not fully recovered from the analytical column often makes RPC not suitable for quantitative analysis. These problems are sometimes solved by optimization of the elution conditions and stationary phase and/or a correct sample preparation. Aggregation can sometimes be prevented by modifying the mobile phase by the addition of sodium dodecyl sulfate (SDS), 3-[(3-cholamidopropyl)dimethylammonio]-1-propanesulfonate (CHAPS). Solubilization of the protein can be obtained by using acids, detergents, and/or chaotropic agents, e.g., SDS, CHAPS, guanidine, Triton-X-100, octyl glucoside, or urea. Also, sodium cholate or Emulgen 991 have been used, in order to solubilize membrane proteins prior to chromatographic analysis (Kato *et al.*, 1987). The addition of such compounds results in dissolution of the protein, but very often it also results in loss of the secondary and tertiary structure. Because of the denaturation of the proteins in RPC, it is not always suitable to study small changes in the protein conformation.

2.4. Examples

2.4.1. CONFORMATIONAL CHANGES DURING CHROMATOGRAPHY AND RETENTION BEHAVIOR

Thévenon and Regnier (1989) studied the influence of formic acid concentration on retention and protein structure. The conformation of the proteins is strongly dependent on the acid concentration. At concentrations below 10% the acid acts as an amine-pairing agent and the protein is partially denaturated. Between 10 and 30% the protein is further denaturated and the hydrophobic residues become exposed to the solvent. The hydrophobicity of the species is increased and hence the retention increases. Above 30% the solution becomes so polar that the hydrophobic residues begin to migrate into the interior of some highly denatured structure and they are finally shielded from the aqueous phase. The decrease in retention at concentrations above 30% is not only due to the higher solvophilic power of formic acid relative to water but also to various conformational states of the proteins (Thévenon and Regnier, 1989).

The temperature may also influence the retention behavior and resolution

(Cohen *et al.*, 1984). The effect of the temperature can be explained by different levels of unfolding of the protein in the mobile phase. The effect of the temperature was shown in chromatographic experiments with ribonuclease A (RNase) on an *n*-butyl column. RNase elutes in two peaks at 25 °C and as one peak at 37 °C. The first peak proved to be the native peak while the second peak is the denaturated compound (Fig. 1).

The chromatographic behavior of RNase A on RPC is not ideal (Lu *et al.*, 1986). Cohen *et al.* (1985) showed that the peak shape can be distorted due to refolding of the protein during elution. The binding of the proteins to the column material was also determined by the components of the mobile phase. This indicated that broad peak shapes can be due to reversible conformational changes during the elution of a protein (Cohen *et al.*, 1985).

The choice of organic solvent can influence the conformation of the protein in solution and influences the binding of proteins to C_8 alkyl-bonded silica. It was shown for several proteins that 1-propanol induces conformational changes of the surface-associated protein (Katzenstein *et al.*, 1986). At low pH, a lower concen-

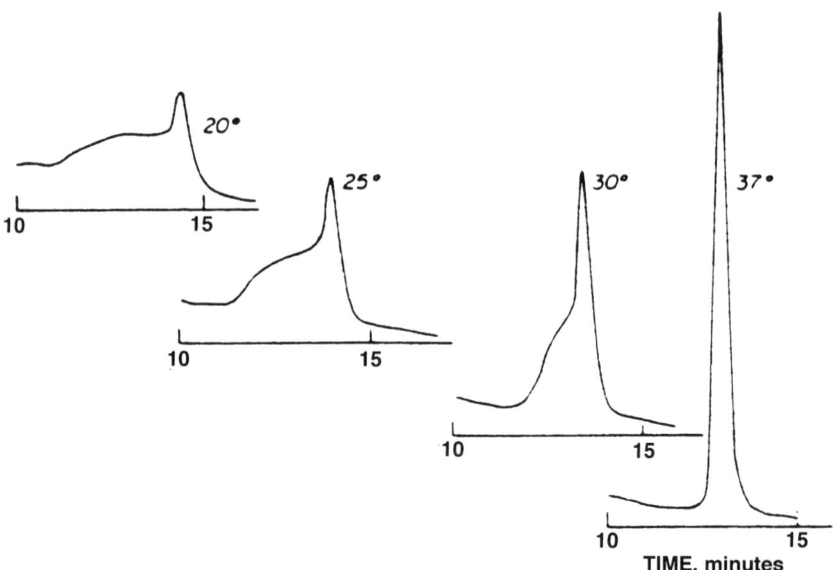

Figure 1. RPLC behavior of pancreatic ribonuclease A (RNase) as a function of temperature at pH 2.2. Conditions—solvent A: 10 mM H_3PO_4, pH 2.2; solvent B: 55/45 (v/v)% $H_{H3}PO_4$ concentration is 10 mM; gradient: 5–85% B in 30 min, linear; flow rate: 1 ml/min; column: 10-μm C4 LiChrospher SI 500, 100 × 4.6 mm inner diameter; sample size: 140 μg RNase; detection: 254-nm 0.1 AUFS (Cohen *et al.*, 1984).

tration of 1-propanol is required to induce conformational changes as compared to neutral conditions (Sadler *et al.*, 1984).

Retention parameter, Z [slope of log capacity factor (k') versus log molar concentration of organic modifier 1-propanol in the mobile phase] and log I (the value of log k' at 1 M 1-propanol) can be used to describe the relation between the retention and the state of unfolding of a protein (Lin and Karger, 1990). The Z and I values increased from folded, surface unfolded, urea unfolded, to reduced unfolded for a series of small globular proteins. These experiments proved that proteins with their disulfide bridges cleaved have the largest degree of unfolding. Each protein was present in four conformational states. This relation between the parameters and the state of unfolding opens possibilities for the measurement of the rate or state of unfolding. The differences between strongly related compounds with regard to the change in conformation on the column during elution can be used to obtain an optimal separation, as was shown by Oroszlan *et al.* (1992), who studied the retention of rhGH and the retention of methionine–rhGH (Met-rhGH). Retention of rhGH decreases with increasing column temperature when 1-propanol is used as an organic modifier, while the retention increases with temperature when acetonitrile is employed. Desorption and elution with 1-propanol was correlated with a solvent-induced conformational change. The retention increase obtained with acetonitrile correlated with the temperature rise and was caused by a partial structural change yielding a more hydrophobic species. In this case a surface-driven process was suggested. In general, lower protein yields from reversed chromatography are obtained with acetonitrile when compared with 1-propanol. The change in the β-sheet structure of bovine α-chymotrypsinogen on alkyl- and phenyl-bonded phases when eluted at different pH and concentration (30–40%) of 1-propanol proved to be independent of the alkyl chain length. The structure was changed to the same extent after leaving the protein interaction with the column material and releasing it back into solution (Drake *et al.*, 1989), suggesting a solvent-driven rather than a surface-driven process.

Depending on the type of column used it is sometimes possible to perform RPC without an organic solvent. A polymeric stationary phase, divinylbenzene based, in combination with acetic acid gradients in water without organic modifiers was used to determine rhGH in a crude acetic acid extract of *Escherichia coli* in which the protein was expressed. Other proteins such as interleukin-1β, glucagon, and insulin are eluted as sharp symmetrical peaks with acetic acid gradients in water, acetonitrile, or isopropanol (Welinder and Sørensen, 1991).

2.4.2. SEPARATION OF (INTACT) PROTEINS

For the development and quality control of pharmaceutical proteins it is often necessary to have analytical methods available that separate proteins that differ only in one or a few amino acids. One example is the separation of muteins, proteins that differ by only a few amino acids caused by a mutation in the DNA sequence. Other examples are the separation of oxidized proteins and the determination of the number and the position of Cys residues.

Under carefully controlled conditions it is possible to separate mixtures of insulins that have minimal differences in structure, e.g., a single uncharged residue (McLeod and Wood, 1984). From this it is concluded that retention of insulin proteins on the reversed-phase column is determined by the entire exposed surface of the molecule rather than a specific region.

RPC can also be used for the separation of related recombinant proteins in the fermentation broth after a selective on-line sample clean up (Hayashi *et al.*, 1987). Human recombinant epidermal growth factors (rhEGF) were determined quantitatively in cultured media. The analytes were isolated on a C_{18} protein-coated precolumn and separated on an analytical TSK ODS 120T column. The rhEGF[1-53] was separated from rhEGF[1-47] and [1-51]. The isolation of the compounds on the precolumn was superior when compared with off-line protein precipitation with trichloroacetic acid (TCA).

The replacement of one amino acid can change the conformation of a protein molecule. Such conformational changes can often be monitored by RPC. Brems *et al.* (1988) studied the effect of replacing Lys 112 with Leu in bovine growth hormone (bGH) (approx. 190 amino acids) on the folding process. The retention time of the Leu mutant was increased from 16.5 to 25 min when compared with the wild type under denaturing conditions. The conformation of the mutant in its native state was indistinguishable from that of the wild type. These results indicate that the mutant form is more sensitive for the denaturing agent guanidine–HCl (3–6 M) than the wild type.

Kunitani *et al.* (1986) studied the retention behavior of several recombinant human interleukin-2 (IL-2) muteins. These muteins differ in the presence of Ala at position 1 and contain different amino acids at positions 58, 104, 105, and 125 (Table II). The replacement of one single amino acid resulted in most cases in a shift in retention time. The muteins eluting as peak A contain a methionine sulfoxide at 104. When the muteins were reduced (compounds with subscript red), a shift toward longer retention times was observed, indicating that more hydrophobic regions are exposed to the stationary phase.

The N-terminal heterogeneity and the oxidation states of the methionines were also studied. The species that contains the oxidized methionine group elutes prior to the nonoxidized species (Table II). Oxidation of methionine residues of proteins in general results in more hydrophilic compounds. Frelinger and Zull (1984) described the elucidation of the structure of the three oxidized forms of bovine parathyroid hormone, which could be separated with RPC.

Purified recombinant human granulocyte macrophage–colony stimulating factor (rhGM-CSF) was characterized with RPC. Analysis of the compound

Table II. Retention Times of IL-2 Muteins in RP-HPLC[a]

	Retention time (min)			
Mutein	Peak B_{red}	Peak A_{red}	Peak B_{ox}	Peak A_{ox}
Ala^{1}Cys^{125} (parent)	39.0	36.3	35.0	33.2
Ala^{1}Ser^{125}	34.6	31.9	31.4	29.6
desAla^{1}Cys^{125}	38.8	36.1	34.9	33.0
desAla^{1}Ser^{125}	34.6	32.1	31.4	29.7
desAla^{1}Ala^{125}	40.9	38.1	36.6	34.6
desAla^{1}Ala^{104}Ser^{125}	32.8	—	30.0	—
desAla^{1}Ser^{58}	38.9	36.4	19.9	—
desAla^{1}Ser^{105}	38.0	35.2	20.3	—

[a]From Kunitani *et al.* (1986).

showed the presence of the major peak and five small peaks (see Fig. 2). The six compounds were isolated by RPC and subjected to proteolytic digest with trypsin and protease V8. The digests were analyzed by RPC and the separated peptides were subjected to fast-atom bombardment mass spectrometry (FAB-MS). Fractions 1 to 4 proved to be rhGM-CSF derivatives of which different methionines are oxidized. Fraction 6 proved to be rhGM-CSF of which two amino acids at the N-terminal were deleted (Met-Ala) (Ohgami *et al.*, 1989). Peak 5 is the native compound.

RPC is the method of choice for studying the position, number, and oxidation status of the cysteine residues. The presence of S-S bridges strongly contributes to the conformation and refolding of a protein. Direct analysis of the disulfide content of proteins can be a method for monitoring the stability and refolding process of cysteine-containing proteins. The general approach of the determination of the number of cysteines and cystines involves the carboxymethylation of the protein with iodoacetic acid and radiolabeled iodoacetic acid. The first carboxymethylation is performed prior to reduction of the protein, and the second carboxymethylation is performed following reduction. The samples are analyzed by RPC with UV and radiochemical detection.

A compound whose refolding process was studied extensively by RPC is rhIL-2. Recombinant hIL-2 is a hydrophobic protein with a limited aqueous solubility and can be chromatographed using aqueous acetonitrile and TFA as eluent. Due to the presence of three cysteines in endogenous rhIL-2, three isomers can be formed each with a different disulfide bridge. Weir and Sparks (1987) studied the refolding of rhIL-2. Recombinant hIL-2, expressed in *E. coli*, was isolated as inclusion bodies after cell breakage. Recombinant hIL-2 and contaminants were dissolved and further purified in reduced and denatured form by gel-permeation chromatography. Refolding took place by diluting the solvent. Renaturation/auto-

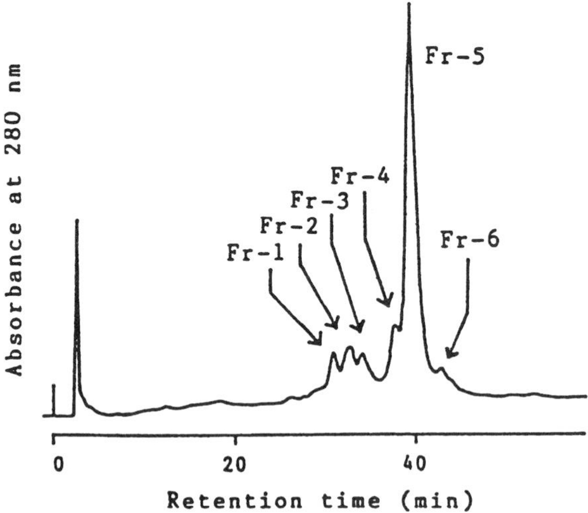

Figure 2. HPLC profile on a reversed-phase column of renatured rhGM-CSF (Ohgami *et al.*, 1989).

oxidation was initiated and two peaks were observed on RPC (C_{18}). The first peak to elute was oxidized (folded) IL-2, and the second peak was the less bioactive reduced IL-2. When the renaturation time was increased, peak 1 grew at the expense of peak 2.

Browning and co-workers (1986) also used RPC (C_4) to clarify the structure and disulfide scramble of nonglycosylated rhIL-2. In the presence of a chaotrope the proper disulfide bond is scrambled, resulting in a mixture of the three disulfide isomers. The identity of each of the three forms was determined by labeling of the free cysteines with [^{3}H]iodoacetic acid and subsequent peptide mapping. The nonnative isomers showed a less hydrophobic behavior when compared with the native form. The fully reduced rhIL-2 is more hydrophobic than the native form. From this it was concluded that the nonnative isomers retain some folded structure to mask a hydrophobic region of the molecule.

Kunitani and co-workers (1986) studied variants of rhIL-2. Endogenous IL-2 has only one disulfide bridge between amino acid 58 and 105 but it contains three cysteines, at positions 58, 105, and 125, respectively. Recombinant desAlaSer125IL-2 is a protein that can only form one cystine disulfide. The cys-125 is replaced by a serine by site-specific mutagenesis, which resulted in an improved stability. The complete reduction of this compound and the complete reduced intact native molecule result in an increase of retention (Table II). The complete reduction probably results in an exposure of the hydrophobic regions to the stationary phase. The compounds with intact cystine bridges are less hydrophobic because the hydrophobic regions are covered in the tertiary structure. The other muteins exhibit the same behavior: Reduction results in an increase of retention.

O'Keefe and co-workers (1992) determined the number of cysteines in transforming growth factor-α-*Pseudomonas aeruginosa* exotoxin A40 (TGFα-PE40), a fusion protein expressed in *E. coli*. The number of cysteines was determined by derivatization of the cysteine moieties in reduced protein with monobromobimane. The resulting derivative can be selectively detected with fluorescence detection. It not possible to separate PE40 (M_r = 40,000) from the intact fusion protein (TGFα-PE40, M_r = 44,960). However, with derivatization of the cysteines in TGFα-PE40 prior to RPC, it was possible to discriminate the cysteine free PE40 from TGFα-PE40 (O'Keefe *et al.*, 1992).

Chromatography of pituitary-derived hGH (P-hGH) and biosynthetic hGH (B-hGH) was described (Christensen *et al.*, 1990). Small peaks were visible in the chromatograms of both compounds in addition to the dominating hGH peak. B-hGH consistently elutes as a slightly sharper peak when compared with P-hGH under various isocratic conditions. This is caused by the absence of a 20-kDa fraction in the B-hGH. The 20-kDa compound is a decomposition product of hGH obtained after deletion of amino acid residues 32-46 from the 22-K hGH. Although both compounds differ significantly in molecular weight, separation could not be obtained with RPC.

Different forms of recombinant human interferons (rhINF) were analyzed on a reversed system (Felix *et al.*, 1985). The method was able to separate rhIFN-αA, rhIFN-αD, and the hybrid rhIFN-αA/D. Recombinant hIFN-αA and rhIFN-αD differ about 15% in their amino acids. In addition, it was possible to separate the rhIFN-αA fast-migrating monomer (Cys^{1}–Cys^{98} and Cys^{29}–Cys^{138}) from the slow-migrating monomer (Cys^{29}–Cys^{138}) and from the oligomers. When the Cys^{1}–Cys^{98} bridge was replaced, only the slow-migrating monomer was found.

2.4.3. PEPTIDE MAPPING

Recombinant hIL-2 expressed in *E. coli* was identified by N-terminal sequence and by peptide mapping (Lahm and Stein, 1985). Disulfides were determined by using a double-label S-carboxymethylation procedure with tritiated iodoacetic acid and cold iodoacetic acid. The labeled compound was digested with trypsin and the peptides were chromatographed. The retention of labeled peptides was detected and indicated the presence of cysteine residue. A disulfide linkage between cysteine residues at positions 58 and 105 was confirmed. About 90% of the isolated rhIL-2 contained a methionine residue at the N-terminus.

Renlund *et al.* (1990) described the routine peptide mapping of recombinant human immunodeficiency virus (HIV) proteins p24 core and p24-gp41, expressed in *E. coli*. These proteins were dissolved and reduced and derivatized using

4-vinylpyridine, generating pyridylated cysteine residues. Following this sample treatment the protein was digested with trypsin. The peptides were analyzed by RPC with UV detection at 214, 254, and 280 nm. Due to the derivatization with 4-vinylpyridine the cysteine residues were easily located at 254 nm. In particular, the ratios of A254/A280 and A280/A215 of different peaks give information on the presence of cysteine residues. Figure 3 presents the tryptic maps of three batches of the p24-gp41 protein. The fragments eluting at 43.4, and 76.1 min contain one cysteine residue, while the peak at 124 min contains two cysteine residues. It was concluded that peptide mapping is a useful technique for assessing the equivalency between different batches of these proteins.

Chloupek *et al.* (1989) determined the primary structure of recombinant tissue plasminogen activator (rtPA), a glycoprotein with a molecular weight of 59,042 Da (527 amino acids). Recombinant tPA contains variants that have either two or three carbohydrate side chains. The peptides obtained after reduction, carboxymethylation, and digestion (RCM rtPA) with trypsin were chromatographed on two different reversed-phase systems (see Figs. 4 and 5). Both systems have their own selectivity resulting in a different elution order of the peptides. The amino acid content of the peptides was determined by amino acid analysis. The accuracy of the tryptic maps was confirmed by the tryptic maps of two domains, 1–275 and 276–527 amino acids. The glycopeptides were isolated by affinity

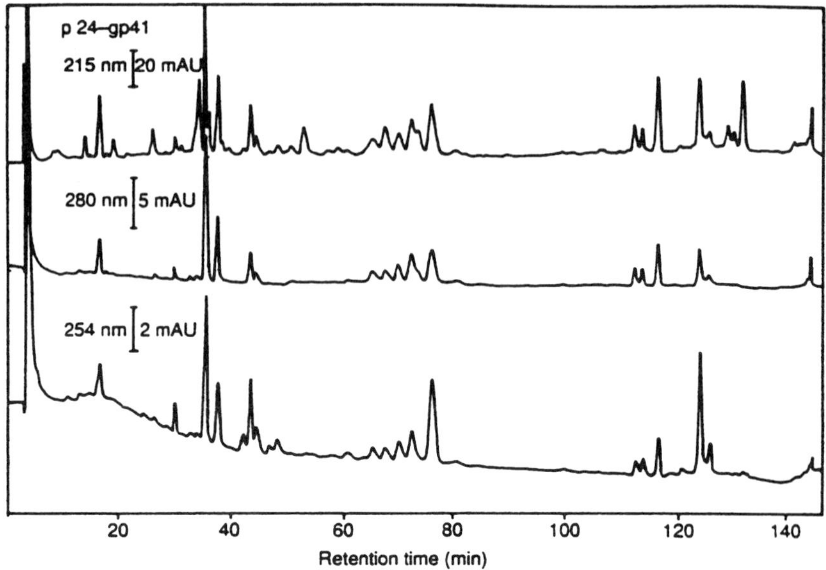

Figure 3. Three tryptic maps of p24-gp41 (2 nmole) monitored at 215, 254, and 280 nm (Renlund *et al.*, 1990).

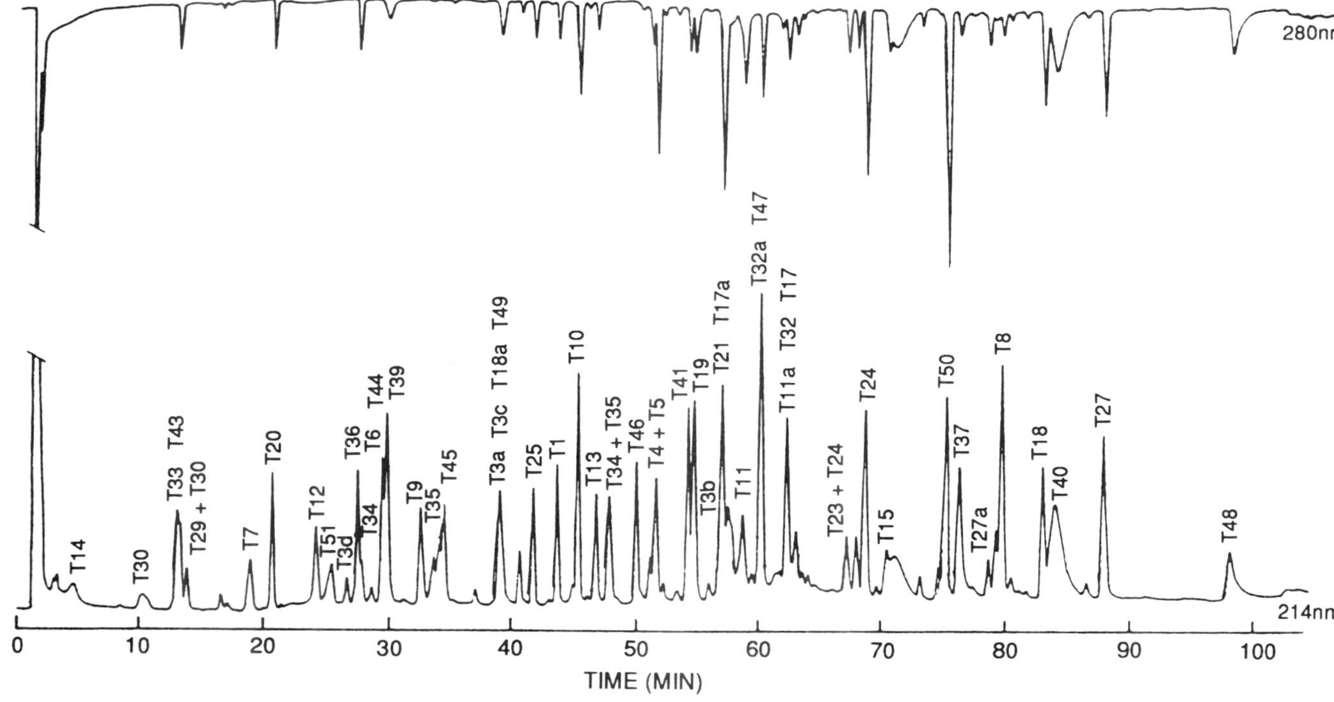

Figure 4. Sodium-phosphate-based tryptic map of RCM rt-PA. This separation was performed on a 5-µm Nova-Pak C_{18} column (15 × 0.46 cm inner diameter). Mobile phase A consisted of 0.05 M sodium phosphate (pH 2.85) and mobile phase B was acetonitrile. A linear gradient of 0–30% mobile phase B was run in 90 min followed by 30–60% mobile phase B in 30 min. The 70-µg sample was loaded in 0.1 M ammonium bicarbonate and monitored at both 214 and 280 nm (Chloupek *et al.*, 1989).

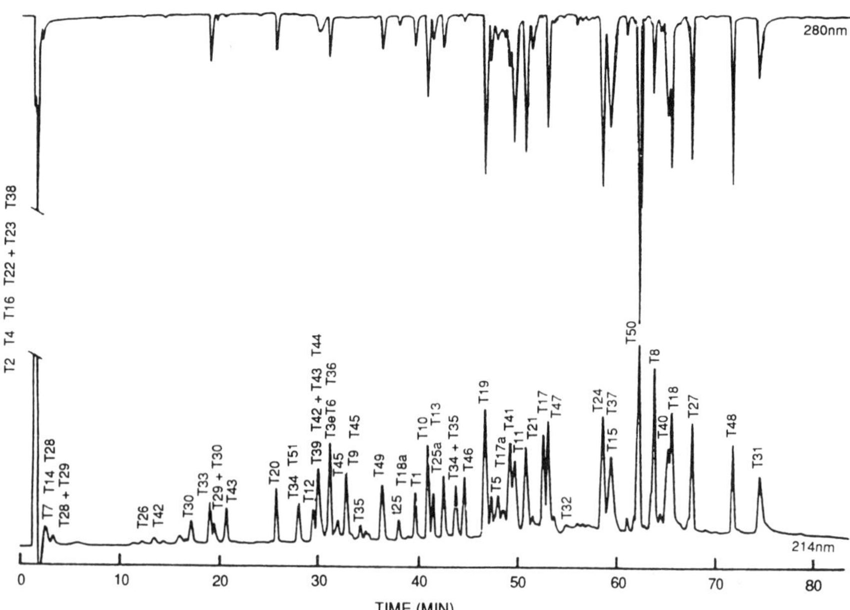

Figure 5. TFA-based tryptic map of RCM rt-PA. This separation was performed on a 5-μm Nova-Pak C_{18} column (15 × 0.46 cm inner diameter). Mobile phase A consisted of aqueous 0.1% TFA and mobile phase B was acetonitrile with 0.08% TFA. A linear gradient of 0–25% mobile phase B was run in 50 min followed by 25–60% mobile phase B in 35 min. The 200-μg sample was loaded in 0.2 ml of 0.1 M ammonium bicarbonate and monitored at both 214 and 280 nm (Chloupek *et al.*, 1989).

chromatography on a Con-A-Sepharose lectin column. Specific enzymes such as Endo H glycosidase and PNGase F were used to cleave the mannose structures and all carbohydrate side chains, respectively. Using a combination of enzymes and two RPC methods it was possible to obtain information on the primary structure of this compound. Ling *et al.* (1991) used electron spray mass spectrometry (ES-MS) as an on-line high-performance liquid chromatography (HPLC) detector for the tryptic mapping of rtPA (see also Chloupek *et al.*, 1989). The glycopeptides and the carbohydrate heterogeneity of rtPA could be identified.

Another approach for structure elucidation was followed for the detection of a variant of human serum albumin (HSA) (Iadarola *et al.*, 1990). The Castel di Sangro variant of HSA and normal HSA were treated with CNBr and the digests were analyzed by RPC. One of the fragments of the variant showed a different elution behavior when compared with the fragment obtained from the normal HSA. This fragment was isolated and subjected to a digestion with V8 protease and trypsin. Abnormal peptides were isolated and analyzed for amino acid content and sequence. This analysis revealed that in the case of the Castel di Sangro variant of HSA Lys 536 is replaced by Glu.

Peptide mapping in combination with RPC is in principle a powerful tool in the elucidation of the primary structure of proteins. However, large amounts of the protein are required. Mass spectrometric analysis following chromatographic separation is a technique that is gaining popularity for the identification of proteins, especially through the analysis of peptides generated during the digestion.

2.5. Detection

UV detection is the most frequently used detection technique. Proteins containing fluorescent moieties, e.g., tryptophan, can be detected by fluorescence detection. Dou *et al.* (1991) used electrochemical oxidation for the detection of proteins (e.g., albumin and insulin) after photolysis of the protein at 254 nm. The photolysis includes the cleavage of disulfide bonds, resulting in the formation of oxidizable sulfhydryl species and cleavage of amino acid side chains. Due to these changes the conformation is altered and an increased exposure of electroactive amino acids to the electroactive surface occurs, resulting in a higher sensitivity.

For a precise determination of the molecular weight of a biopolymer, classical light-scattering detection following RPC can be used on-line. The method is straightforward and accurate for the determination of biopolymers up to 10^3 kDa (Dollinger *et al.*, 1992). The combination of RPC and mass spectrometry will be discussed in Section 7.4.

3. HYDROPHOBIC INTERACTION CHROMATOGRAPHY

3.1. General

The separation of proteins with hydrophobic interaction chromatography (HIC) is based as in RPC on the interaction of the hydrophobic areas of the protein with the hydrophobic ligand. HIC operates under milder conditions due to the general absence of organic solvents and the use of less hydrophobic supports and a lower ligand density on the column. Due to the absence of organic solvents the biological activity is preserved and high recoveries of proteins are obtained (Fausnaugh *et al.*, 1984). Protein separation on the HIC column is based on native hydrophobicity and not on that of the unfolded protein as with RPC. Retention time and selectively are ligand-dependent in HIC because the ligands generally interact with hydrophobic amino acid residues that are accessible on the surface of the folded native molecule. In contrast, with RPC bonding interactions occur throughout the length of the denatured protein, including those normally buried in the hydrophobic protein core. HIC is mostly used in the preparative mode in order to obtain large amounts of native proteins. The analytical scale of HIC is often used for troubleshooting of a preparative separation or for process validation purposes. HIC is of limited use as an analytical technique because of a marginal exposure of hydrophobic groups to the stationary phase, which decreases the possibility of interaction, long retention times, and resolution.

In HIC the solutes are eluted in order of increasing hydrophobicity as in RPC. This is achieved by eluting isocratically or, more generally, by a descending salt concentration. Lowering the salt concentration weakens the hydrophobic interactions and causes the solute to be released from the column.

3.2. Stationary Phase

The column packings available for HIC are based on silica, hydroxylated-polyether, polymer and cross-linked agarose. The silica-based material for HIC has the same disadvantages as the silica-based material used for RPC. The instability of silica-based material at high pH is solved by the introduction of polymer-based columns, e.g., hydroxylated polyether (e.g., Bio-gel TSK Phenyl-5PW). The polymer columns are stable in the pH range of 2 to 12 (Mant and Hodges, 1991; Kennedy *et al.*, 1989).

The properties of the ligand are of major importance for the hydrophobic interaction between protein and ligand. The polymeric-bound alkyl and aryl ligands that are generally used in HIC are less hydrophobic and less dense when compared with typical RPC column material (such as C_8 and C_{18}). The retention of proteins increases in the order of: hydroxypropyl $<$ methyl $<$ benzyl = propyl $<$ isopropyl $<$ phenyl $<$ pentyl ligand (Gooding *et al.*, 1984). The retention behavior in HIC can be manipulated by varying the alkyl chain length as well as the ligand density. In general a lower density results in a decrease of retention. Longer alkyl chains result in an increase in retention, the extent of which is dependent on the surface hydrophobicity and size of the protein.

It is also possible to perform HIC on reversed-phase columns. Sing *et al.* (1992) prepared a stationary phase suitable for HIC by coating a commercial reversed-phase column with a nonionic surfactant of the polyoxyethylene type (Brij 76). The surfactant was adsorbed on the reversed-phase material by interaction of the C_{18} chains of the stationary phase with the C_{18} chains of the surfactant, thus exposing the polyoxyethylene part of the surfactant to the aqueous mobile phase. The coated column proved to be stable and behaves like a HIC column without denaturating properties. An important guideline for choosing a column is to use more hydrophobic sorbents for proteins having little surface hydrophobicity and sorbents with either shorter alkyl chains or lower ligand density for more hydrophobic or labile samples.

3.3. Mobile Phase

Initial salt concentrations generally range from 1 to 3 M. Initial ionic strength can significantly affect selectivity, suggesting that by altering initial loading conditions, protein resolution may be enhanced. This effect is the greatest for the more hydrophilic, early-eluted components of a mixture. Because the effect on resolution is less for more hydrophobic proteins, a potential benefit of using a lower initial ionic strength is that such proteins will be eluted earlier, resulting in shorter analysis times.

Mobile phases generally have a pH between 5 and 7, buffered with sodium or potassium phosphate. The effect of the pH on the retention time is protein dependent and is due to the ionization state of the amino acid residues.

In HIC, volatile buffers can be used in order to simplify the subsequent purification procedures (Konishi and Kamada, 1990; Nakamura *et al.*, 1990). Retention may be influenced by adding modifiers to the mobile phase. Addition of 5–10% methanol or isopropanol, 20% ethylene glycol, or subdenaturing (1–2 M) concentrations of urea or guanidine–HCl promotes desorption of the protein without necessarily causing denaturation and loss of activity. Solvent modulation in HIC has been reviewed recently (Arakawa and Owers Narhi, 1991).

The selectivity of the separation can be changed by the addition of the nonionic detergent CHAPS. Buckley and Wetlaufer (1990) studied the influence of CHAPS on the elution of several globular proteins. The presence of CHAPS had a positive effect on the selectivity of the separation but did not have a significant effect on the peak shape. Only bovine serum albumin yielded a less distorted peak in the presence of CHAPS, probably because of a conformational change.

Since the strength of hydrophobic interactions increases with temperature, protein retention times generally increase with temperature. If the temperature is too high, the protein unfolds, which depends on the protein. The degree to which temperature affects retention time may vary significantly between proteins. Hence, utilizing temperature variation during HIC may be used to obtain better resolution. On the other hand, lowering the column temperature often results in peak sharpening and better resolution. Coupling this factor with the decreased liability of proteins to degrade at lower temperature suggests that operation at subambient temperatures may often be beneficial. A typical gradient time used in

HIC is 30 min. The effect of gradient rate on the resolution is very diverse and is dependent on protein and eluent. A lower gradient rate can result in a decreased resolution and/or higher retention time of the protein. Comparing the selectivity of HIC and RPC by elution order showed similar or identical patterns. RPC exhibited a better resolution. HIC is a better choice for purification of extremely hydrophobic peptides. HIC may also give a better resolution for small proteins or peptides large enough to possess secondary or tertiary structure.

Another area where HIC can supplement RPC is purification of hydrophilic peptides, which cannot be resolved by RPC, due to too weak hydrophobic properties. Protein retention on a HIC column can be manipulated by variation of the mobile phase. These variables are salt type, salt concentration, concentration organic modifier, gradient steepness, and pH.

3.4. Examples

Wu *et al.* (1986a) studied the conformational behavior of α-lactalbumin (α-LACT) under HIC. The influence of Ca^{2+} and Mg^{2+} added to the mobile phase was investigated. Since α-LACT is a calcium-binding protein, addition of this metal leads to stabilization, i.e., higher column temperatures are required for conformational changes of the protein. On the other hand, addition of Mg^{2+} appears to destabilize the protein.

Using a hydrophobic support, C_2-(ethyl) ether phase, it was possible to separate the native and the unfolded species. The experiments showed that when the column temperature is raised or the contact time with the column increases, the peak of the unfolded protein grows at the expense of the native peak, indicating a conformational change of α-LACT on the column. Reinjection of the late eluted fraction, the unfolded protein, reveals that rapid reformation of the native species takes place in solution.

Wu *et al.* (1986b) studied the conformational changes of several proteins on a weakly hydrophobic ether-bonded phase column. Changes in the *Z*-value (which represents the slope of the graph of log capacity factor vs. the log of the percentage of ammonium sulfate in the mobile phase) with temperature are a measure for recognizing conformational changes. The *Z*-value is a factor that characterizes the protein retention as a function of salt concentration and is sensitive to conformational changes. The conformational stability proved to be dependent on protein and temperature.

The condition under which the protein first contacts the chromatographic column may play an important role in determining the behavior in protein chromatography (Karger and Blanco, 1989). Experiments showed that β-lactoglobulin A, normally present as a dimer at pH 6, aggregates when it comes into contact with the mobile phase at pH 4.5. The aggregation is the result of a hydrophobic interaction. However, it was hypothesized that when the species comes into contact with the column, the aggregate distribution is frozen because the hydrophobic site of the molecule is bound to the stationary phase. Large aggregates were formed when the compound was dissolved in 6 M ammonium sulfate. The interaction of the hydrophobic column with the aggregates prevented the formation of larger aggregates on the column.

Kunitani and co-workers (1988) studied the behavior of recombinant tumor necrosis factor (rTNF) during HIC. Chromatography of rTNF results in reversible dissociation of the quaternary protein structure, yielding trimer and monomer peaks. The relative amounts of these peaks are influenced by contact time with the column as well as the temperature (Fig. 6). In Fig. 6, peak 3 is the monomer that is

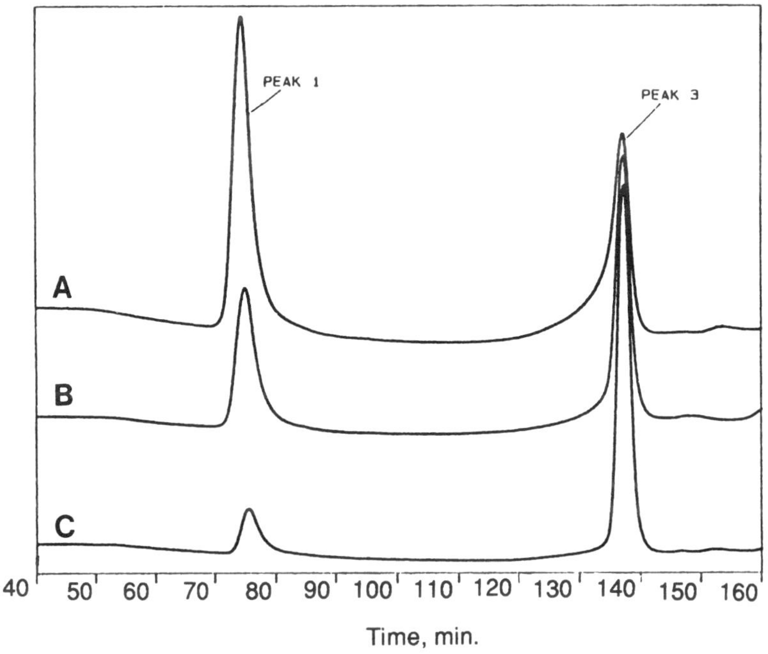

Figure 6. HIC profile of TNF after various on-column incubation periods. (A) No on-column incubation; (B) 2-hr on-column incubation; and (C) 7-hr on-column incubation. After 5 min of the starting eluent the flow was stopped and restarted after the delay indicated. For clarity, chromatograms have been adjusted to align the peaks. Peak 1 represents the trimer and peak 3 the monomer (Kunitani *et al.*, 1988).

strongly retained on the column and as such not available to form the trimer (peak 1). The dissociation of the trimer makes the hydrophobic region accessible for the sorbent surface. From the experiments it was concluded that rTNF is most likely a trimer. After elution the monomers reassociate to the native trimer (Kunitani *et al.*, 1988).

4. ION-EXCHANGE CHROMATOGRAPHY

4.1. General

Ion-exchange chromatography (IEC) is of major importance in the isolation, purification, and analysis of peptides and proteins. As relatively mild separation conditions are used, the biological activity of these compounds is usually maintained. Ion-exchange HPLC employs a charged stationary phase that interacts with charged amino acid residues on the surface of a protein. In addition to the overall charge of the protein, local concentrations of charged residues on the protein surface are believed to influence the binding of the protein to the stationary phase. Since proteins assume tertiary structures, not all charged residues are necessarily exposed to the protein surface. Hence, the overall net charge of the native protein depends on the sum of the exposed surface positively and negatively charged residues. This implies that the net charge and hence the chromatographic behavior will change upon denaturation. Modifications in the protein that cause alterations of the surface charge (such as oxidation, deamidation) will also lead to changes in the interaction of the protein with the stationary phase. Elution usually occurs by gradually increasing the ionic strength of the eluting mobile phase. The salt ions in the mobile phase compete with the ionized ligands on the stationary phase for the ionic interactions with the protein. The net charge on a protein and its surface distribution (expressed as local pK values) determines the pH and salt concentration at which it will be eluted. Selectivity can be tuned by subtle pH changes as well as by varying the type and concentration of the eluting salt. As outlined earlier, proteins may be structurally altered by various mechanisms (see Table I). These processes lead to proteins that are structurally closely related to the parent protein, and their appearance as impurities in the final purified product is possible because they copurify with the native protein. Alterations that cause a change in the net charge may be detected with IEC. Protein charge isoforms differ structurally only in their net charge.

The charged stationary phase forms the basis for the two forms of IEC: anion-exchange chromatography and cation-exchange chromatography. A special form of IEC is chromatofocusing. In this technique a pH gradient can be generated in the column, e.g., Mono P, by using mixtures of electrolytes. When a buffer of different pH is applied, the gradient moves through the column. Proteins will desorb at pH $<$ pI (isoelectric point) and readsorb at pH $>$ pI; the proteins will move through together with the moving gradient at the pI values. Proteins of identical pI will elute together.

4.2. Stationary Phase

4.2.1. GENERAL

The surface of the stationary phase bears a fixed and well-defined charge. The stationary phase interacts electrostatically with the oppositely charged analyte of interest and releases the analyte if a competing salt present in the mobile phase becomes more favorable. In addition to the electrostatic interaction, other effects may influence protein elution. Ståhlberg *et al.* (1991, 1992) describe models that include either only electrostatic interactions or both electrostatic and non-electrostatic interactions (van der Waals, between nonpolar moieties of protein and stationary phase). Both interactions affect protein retention, with the non-electrostatic interactions being unwanted.

The base material of the column packing used is silica or a polymer. Both supports differ in mechanical and pH stability. Silica has excellent rigidity, but it is only applicable in the pH range 2 to 8 since it dissolves in alkaline media. This does not have to be a problem since most proteins can be chromatographed in this pH range. Polymeric supports have a wider pH stability range, especially on the basic side, but their mechanical stability is less than that of silica. Another difference is that silica is available in more pore diameters and is less expensive than comparable polymeric matrices.

Complete surface coverage is necessary to avoid silanophilic or hydrophobic interactions of the protein with the support matrix. This coverage is usually provided by a polymeric layer that is more stable than simple bonded phases. The ion-exchange properties are determined by the specific ionic moieties incorporated into the bonded phase and the arrangement of these moieties.

Both types of ion-exchange materials are subdivided into strong and weak types; weak anion-exchange sorbents contain diethylaminoethanol, ethylamine, polyethyleneimine, and strong anion-exchange sorbents contain quaternized amines. Weak cation-exchange sorbents usually contain carboxymethyl groups, and strong cation-exchange sorbents contain sulfonyl, sulfonpropyl, or phosphonyl ionic groups. Due to charge permanence, the ion-exchange capacity of strong ion-exchange sorbents remains high at extreme pH, whereas for weak anion-exchange and cation-exchange sorbents it declines at high and low pH, respectively.

4.2.2. SILICA-BASED STATIONARY PHASE

To reduce deterioration of the column material, as is observed for silica columns (Stout *et al.*, 1986), a deactivation process of the silica phase was introduced and synthetic organic polymer-based columns were designed. Deactivation of the silica packing was done by hydrophilic polymer coating or by completely covering the silica surface with a metal oxide such as zirconium oxide, which improved the water resistance of the stationary phase (Stout *et al.*, 1986, and references therein). Zirconium-oxide-modified silicas were found to be stable at pH > 9 (Stout and DeStefano, 1985).

Based on this material, new ion-exchange columns were developed. The diol functionality of this material was derivatized with strong and weak ionogenic functional groups at two pore sizes (150 and 300 Å), such as $-SO_3H$, $-COOH$, $-CH_62NR_3$, and $-CH_2NH_2$. These columns were compared with two columns with similarly charged surfaces but linked to an organic substrate, namely Mono S ($-SO_3H$ functional group) and Mono Q ($-NR_3$) columns (both with pore size 800 Å). The organic substrate was expected to show some hydrophobic interaction phenomena, whereas the silica-based columns were expected to show silanophilic interactions. Similarities as well as significant differences in selectivity and resolution were observed between the two column types for a variety of test proteins.

Schafer and Carr (1991) adressed the problematic separation of phosphate-containing proteins on zirconium-modified stationary phases. They used the strong affinity of zirconium for phosphate to chemically modify the zirconium to its phosphate while maintaining the mechanical stability necessary for HPLC conditions. Phosphate present in the mobile phase (20 mM) improved the stability of the column to avoid desorption of phosphate from the stationary phase. The latter occurred at pH > 10, in spite of the presence of phosphate in the mobile phase. The authors concluded that phosphate-modified zirconium is a useful support for the separaticn of proteins.

4.2.3. POLYMER-BASED STATIONARY PHASE

A significant effort is put into the development of rigid stationary phases that show an ideal ion-exchange behavior with minimal unwanted interactions, are usable in a broad pH range, and do not have denaturating properties. Kato *et al.* (1983) introduced the TSK-GEL IEX-645 DEAE column, which was manufactured by the coupling of diethylaminoethanol groups onto the stationary phase of the G5000PW gel filtration column. Nelson and Kitagawa (1990) introduced two new HPLC ion-exchange columns, based on a macroporous hydrophilic polymer core, to which quaternary nitrogen groups were covalently attached for strong anion-exchange applications or to which sulfonyl groups were bound to the polymer. Both columns are useful in the pH-range from 2 to 12. Müller (1990) emphasized the importance of the arrangement of the ionic groups on or in the matrix. In traditional matrices, the ionic groups are fixed via short arms on the support surface, thus forming a rigid array of charged binding sites for the analyte. To maximize the number of ion pairs formed at low ionic strength when the protein is bound to the stationary phase, the protein may be distorted which may result in denaturation of the protein by exposing hydrophobic areas of the protein to the outside. According to Müller (1990), this effect can be avoided by arranging the ionic groups in a more "tentacle"-like structure as linear polyelectrolytes bound to the matrix, which are cross-linked to each other. A set of new ion-exchangers was used having the same stationary phase material (Fractogel, vinyl polymer) that was prepared in the conventional way as well as in the tentacle way. Functional groups attached to the polymer included $-N(CH_3)_2H^+$ and $-N(C_2H_5)_2H^+$ for weak anion exchangers, $-N^+(CH_3)_3$ for strong anion exchangers, $-COO^-$ for weak cation exchanger, and $-SO_3^{2-}$ for a strong cation exchange activity. A substantial increase in binding capacity for the tentacle stationary phase was observed using bovine serum albumin, lysozyme, and hemoglobin as test proteins. The nonspecific interaction of the protein with the tentacle stationary phase was reduced and resulted in sharper peaks and enhanced selectivity when compared with the conventional stationary phase. The tentacle structure makes the overall distribution of the charges on the protein more important, because due to the increased flexibility of the polyelectrolyte chains the electrostatic interaction with the analyte is easier. The charges on the analyte can be approached in a more efficient way.

4.3. Mobile Phase

Salts influence the elution by displacing solute ions from the charged sites on the stationary phase and the proteins and by changing the tertiary structure of proteins. In general, divalent ions tend to be a stronger displacer than monovalent ions, and smaller ions tend to have a higher elution strength when compared with larger ions of the same group. The ions can be arranged in order of decreasing solute retention: $K^+ > Na^+ > NH_4^+ > Ca^{2+} > Mg^{2+}$ for cations and $CH_3COO^- > Cl^- > HPO_4^{2-} > SO_4^{2-}$ for anions. To achieve adequate binding, the pH of the mobile phase should generally be at least 0.5 units above the pI of the solute for AEC and below the pI for cation-exchange chromatography. Buffer concentration should range from 10 to 100 mM to supply enough buffering capacity without contributing excessive ionic strength. Commonly used buffers are, e.g., phosphate and citrate for cation-exchange chromatography and Tris buffer for anion-exchange chromatography.

Because of the numerous charged sites on a protein's surface, a pH or salt gradient must be used to release the protein in narrow bands. Ion-exchange chromatography often utilizes salt gradients up to 1 M for monovalent salts or 0.5 M for divalent salts. Sodium acetate and sodium chloride have been used frequently. Salts for IEC should be free of UV-absorbing impurities that may interfere with detection or destabilize baselines.

4.4. Examples

4.4.1. ANION-EXCHANGE HPLC

The number of publications on the application of IEC for the analysis of pharmaceutical proteins that can be found in recent literature is limited, most likely due to the fact that most pharmaceutical and biotechnical companies do not publish their processes of manufacture and quality control in scientific literature.

IEC is in principle the method of choice for characterizing proteins that differ in their charge. Alterations that cause a change in net charge frequently occur during storage and can become an important factor for the evaluation of the stability of proteins (Clogston *et al.*, 1992). Frenz and Hancock (1989) used anion-exchange HPLC (AE-HPLC) and capillary electrophoresis to study deamidation of human growth hormone.

Because IEC does not affect the protein conformation in general, it can be used for studying protein conformations. Withka *et al.* (1987) used size-exclusion HPLC (SE-HPLC), AE-HPLC, cation-exchange-HPLC (CE-HPLC), and HIC to monitor protein conformational changes and stability. They studied three globular proteins: bovine serum albumin, lysozyme, and trypsin. Weak AE-HPLC was used for bovine serum albumin and CE-HPLC for lysozyme and trypsin. Alterations in the configuration became visible by changes in peak height and retention and by the appearance of multiple peaks. The changes in the structure were confirmed with classical physical and biochemical methods. These alterations were sometimes accompanied by loss of biological activity.

Hearn (1991) and Malmquist and Lundell (1992) studied the influence of displacing salts on retention in strong AE-HPLC analysis of proteins, using gradient elution. A series of anions and cations were used and all data collected were analyzed using principal component analysis. Most of the data could be explained by elution strength characteristics of the starting buffer and the elution buffer. The results can be used to optimize a particular separation.

Cacia *et al.* (1993) used AE-HPLC, CE-HPLC, affinity chromatography, and tentacle CE-HPLC to sort proteins. The AE-HPLC column used was a Polymer Labs SAX column. Using AE-HPLC, the charge difference between deamidated recombinant human deoxyribonuclease (rhDNase) and the parent molecule could not be detected, probably because of steric hindrance. Using a tentacle-type of stationary phase, Fang *et al.* (1992) studied the bandwidth behavior of proteins analyzed on tentacle-type anion exchangers. It was concluded that the tentacle structure results in a dynamic interaction between protein and ligand (thus affecting conformation and retention) rather than a well-predicted migration of a rigid protein. This is an advantage because the flexible tentacle structure can adopt the most optimal configuration for the interaction with the analyte.

In principle, ion exchange is not applied in combination with the use of surfactants, because they can denature the protein and influence its retention behavior. However, sometimes it is necessary to use surfactants in order to solubilize very hydrophobic proteins, e.g., membrane proteins. Welling *et al.* (1992) compared three detergents for the extraction of Sendai virus membrane proteins, followed by AE-HPLC analysis using 0.1% of a detergent in the mobile phase. It was shown that the addition of a nonionic detergent to the mobile phase was essential to preserve the native conformation of Sendai virus membrane proteins during AE-HPLC analysis.

4.4.3. CATION-EXCHANGE HPLC

Snider *et al.* (1992) used cation-exchange HPLC (CE-HPLC) for the characterization of PEG-modified superoxide dismutase. Due to the inherent polydispersity of the PEG polymer and the highly variable stoichiometry of the pegylation reaction, complex peak patterns were obtained and complete resolution or assignment of the individual peaks was not possible. Nevertheless, the technique was demonstrated to be useful for the assessment of product consistency.

Kumagaye *et al.* (1985) compared the ability of CE-HPLC and reversed-phase chromatography to separate closely related peptides. In addition, two closely related proteins of human parathyroid hormone (hPTH), i.e., Asp^{76}-hPTH and Asn^{76}-hPTH, were studied. These two residues could not be separated by RPC. However, a good separation was obtained with CE-HPLC, using a Toyo Soda TSK gel CM-2SW cation-exchange column and a phosphate buffer gradient from pH 6.0 to 7.0.

Clogston *et al.* (1992) compared different chromatographic methods for the analysis of rhG-CSF charge isoforms in the purified protein product. Isoelectric focusing gel electrophoresis and peptide mapping were both not capable of separating the charged isoforms. The cation-exchange method proved to have the best resolution and highest reproducibility. The method was evaluated and validated using the formylmethionyl isoform and several deamidated (Gln→Glu) analogues generated through site-directed mutagenesis. A strong CE-HPLC column was used (Poly LC, polysulfoethyl aspartamide silica-based packing). The chromatographic system used was able to separate *N*-formyl methioninyl rhG-CSF and three deamidated analogues (Gln for Glu) of rhG-CSF [rhG-CSF(Gln67→Glu), rhG-CSF(Gln11,20→Glu), rhG-CSF(Gln11,20,67→Glu)]. A sodium chloride gradient in sodium acetate (pH 5.4, about 0.4–0.7 pH units below the pI) was used to separate the different forms. Baseline separation of the five species was obtained, as shown in Fig. 7.

Cacia *et al.* (1993) compared CE-HPLC and tentacle CE-HPLC for the sorting of two variants of rhDNase that differ in the occurrence of deamidation at a single residue. On a TSK SP 5PW column for strong CE-HPLC, a poor resolution of the two variants was obtained (see Fig. 8). In contrast, when using a LiChrosphere SO_3^{2-} column for tentacle CE-HPLC, under otherwise identical conditions, the two species were fully separated (see Fig. 9).

Patrick and Lagu (1992) combined Zorbax GF-250 SE-HPLC column with CE-HPLC. The eluate of the SE-HPLC was loaded on the CE-HPLC. The usefulness of the column-switching method was demonstrated for trypsinogen (24 kDa), using a mobile phase of monobasic potassium phosphate in 7 M urea adjusted to pH 3.5 and a Supelco TSK SP-5PW column. The column-switching method can be used for other proteins by changing the SE-HPLC fraction that is loaded onto the column as well as by changing the ion-exchange conditions (pH and salt).

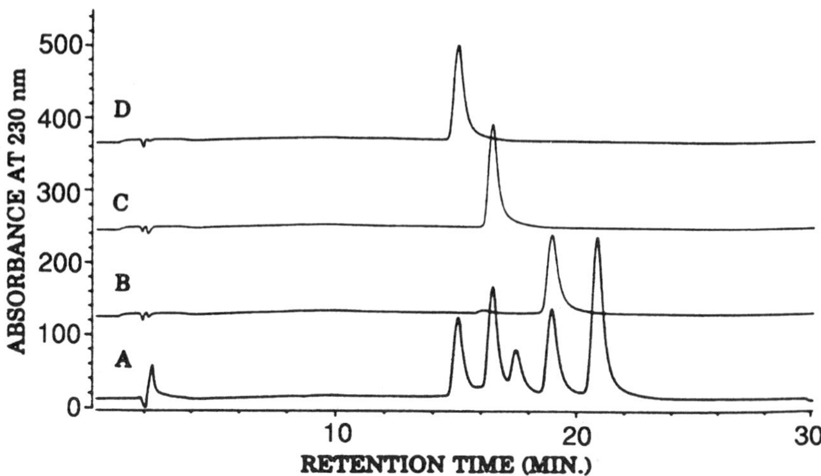

Figure 7. Ion-exchange HPLC of rhG-CSF and the deamidated analogues. (A) $Gln^{11,20,67}$-Glu analogue, Gln^{67}-Glu analogue, f-met rhG-CSF, $Gln^{11,20}$-Glu analogue, and met rhG-CSF (from left); (B) $Gln^{12,21}$-Glu analogue; (C) Gln^{67}-Glu analogue; and (D) $Gln^{11,20,67}$-Glu analogue (Clogston *et al.*, 1992).

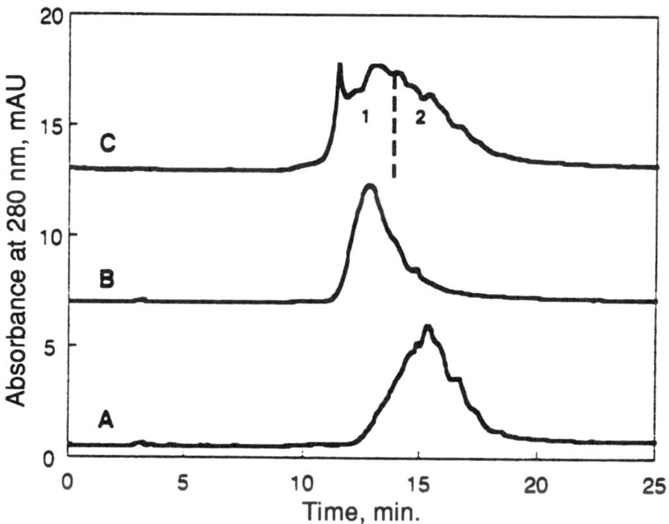

Figure 8. Strong cation-exchange chromatography on a sulfopropyl column (TSK SP 5PW), flow rate 1.0 ml/min. Eluent A: 10 mM sodium acetate, 1 mM $CaCl_2$ pH 4.5; eluent B: 1 M NaCl, 10 mM sodium acetate, 1 mM $CaCl_2$ pH 4.5. Gradient: 4-min hold at 0% B, followed by a linear gradient to 58% B over 26 min. (A) rhDNase, (B) its deamidated variant, and (C) an admixture of the two variants (Cacia *et al.*, 1993).

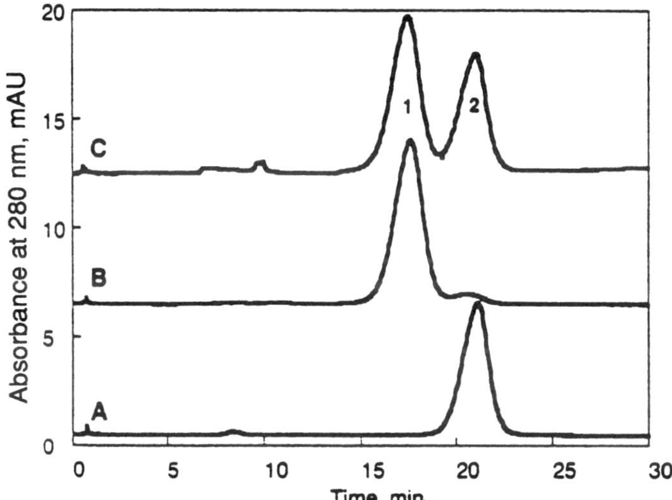

Figure 9. Tentacle cation-exchange chromatography on a LiChrosphere SO_3^{2-}, 5 μm, 1000 Å packed into a 50 × 4.6 mm inner diameter stainless steel column. Chromatographic conditions were identical as described in Fig. 8. (A) rhDNase, (B) its deamidated variant, and (C) an admixture of the two variants (Cacia *et al.*, 1993).

Sample recovery was maximized by using 7 M urea in the buffer to denature the protein.

Mhatre and Krull (1992) demonstrated the possibility of using low-angle laser light-scattering photometry in CE-HPLC determination of protein molecular weight. A HP-SCX strong cation-exchange column was used to determine the molecular weights of ribonuclease A, α-chymotrypsinogen A, and trypsinogen; various gradients of 5mM Bis-Tris (pH 5.8) were used.

For β-lactoglobulin A, gradients using 10 mM monobasic sodium phosphate (pH 4.0) and 10 mM monobasic sodium phosphate (pH 4.0) plus 0.5M NaCl were used. Accurate molecular weights were determined, except for β-lactoglobulin A, which aggregated on-column as a function of the gradient.

4.4.3. CHROMATOFOCUSING

Yamada *et al.* (1986) used chromatofocusing to obtain a separation between IL-2 and methionyl-IL-2 (Met-IL-2). Met-IL-2 was produced by *E. coli* harboring the gene code for IL-2. The *E. coli* produces a mixture of met-IL-2 as well as the IL-2 with a terminal alanine. It was not possible to separate these compounds using IEC, size-exclusion chromatography, affinity chromatography, or RPC. Complete separation was obtained on a Mono-P column based on the difference in isoelectric point, namely 7.7 for IL-2 and 7.5 for Met-IL-2. Both compounds were further purified on RPC and identified by amino acid sequencing.

5. SIZE-EXCLUSION CHROMATOGRAPHY

5.1. General

In size-exclusion HPLC (SE-HPLC) or gel permeation chromatography, noninteractive porous solids are used as stationary phase and the mobile phase consists in general of an aqueous buffer with some additional salt. The method separates proteins based on their hydrodynamic properties and molecular shape and size. Small molecules are eluted later than larger molecules. The smaller the molecules, the easier their penetration into the porous solid phase, and hence the more they are retained. In addition to molecular size and shape, other interactions may influence retention and resolution on the column. These unwanted interactions can be prevented as well as utilized by choosing optimal column material and mobile phase. Residual charged ionic surface sites result in ion-exchange activity of the column. Increasing the ionic strength of the isocratic mobile phase, combined with lowering the pH, results in the protonation of the SiO^- and prevents or decreases this electrostatic interaction. Hydrophobic interactions may also influence elution time and can be decreased or eliminated by decreasing the ionic strength of the mobile phase and by the addition of organic solvents to the mobile phase. Adaptation of the mobile phase to prevent these interactions, however, may hamper protein stability.

Under properly chosen experimental conditions, SE-HPLC generally does not denature the protein of interest, but rather evaluates the protein in its conformation as it exists in the bulk buffer. It is a low-resolution method but is still capable of separating monomer from aggregates (e.g., dimers). For globular proteins there is a correlation between molecular weight and retention time. A particular column can be calibrated using a mixture of well-characterized proteins with known molecular weight. The molecular weight of a protein sample can thus be assessed.

5.2. Stationary Phase

Since its introduction, size-exclusion chromatography mostly has been performed on carbohydrate gel columns (Unger, 1983, and references therein; Unger *et al.*, 1984). Columns of cross-linked dextran and agarose enabled partitioning of the analyte between the mobile phase and the stationary liquid in the gel phase without significant interaction with the gel-forming polymer matrix. However, the lack of physical and chemical stability made them incompatible with the high pressure in HPLC analysis. The evolution of the technique during the last decades resulted in more stable column materials, e.g., surface-modified silica and hydrophilic cross-linked organic polymers (Unger, 1983) and covalently cross-linking an aggregated agarose polymer. Now the materials are available in a range of pore sizes, thus enabling a variety of fractionation ranges. Typical fractionation ranges are 1 to 500 kDa and 5 to 5000 kDa (Andersson *et al.*, 1985; Yang and Verzele, 1987).

Narrower ranges are also available, yielding higher-resolution characteristics. Solutions were found in the synthesis of cross-linked agarose (Andersson *et al.*, 1985) and cross-linked polystyrene-divinylbenzene resins containing neutral hydrophilic functionalities chemically bound to the polystyrene-divinylbenzene by ether linkages (Yang and Verzele, 1987). The latter material of Yang and Verzele proved to be rigid and stable but also possessed some hydrophobic properties which made it necessary to add acetonitril and TFA to the mobile phase.

Stout and DeStefano (1985) introduced a new generation of zirconium-oxide-modified silicas that were found to be stable at pH > 9. This column material was used by Kennedy and Jorgenson (1990) to pack columns with inner diameters of 50 and 28 μm (diameter, 4–5 μm; pore size, 150 Å). These columns were compared with commercially available 9.4 mm inner diameter columns packed with the same material. These microcolumns can be made in large lengths (> 1 m).

They have high plate numbers and efficiency, they allow the use of small sample volumes, and they can use low flow-rates, thus enabling the coupling to other methods such as mass spectroscopy. Better resolution, less tailing, and narrower and more symmetrical peaks were observed for bovine serum albumin and chicken ovalbumin using the microcolumns as compared with the large-bore packed column.

Ahmed and Modrek (1992) reported the development of a modified silica-based stationary phase that demonstrated ideal size-exclusion behavior for a series of test peptide standards with increasing hydrophobicity, cationic charge, and molecular weight. The Biosep-SEC-S columns were prepared by bonding silica of pore sizes 145, 290, and 500 Å with a hydrophilic coating. The elution patterns obtained for peptide mixtures were independent of the composition of the mobile phase, thus demonstrating the absence of residual charged silica functions as well as the absence of hydrophobic interaction of the peptides with the stationary phase.

5.3. Mobile Phase

In general, the mobile phase is a buffer with pH 2–8 (0.05–0.1 M phosphate, Tris, citrate, acetate) with salt added to increase the ionic strength (sodium chloride or sulfate, ammonium acetate or formate). Sometimes a stabilizer such as TFA (to suppress the ionic interaction with the stationary phase) or an organic modifier such as methanol or acetonitrile is added, if the column type allows it. Sometimes the protein is analyzed after denaturation by, e.g., 6 M urea, 6 M guanidine hydrochloride, or 0.1% SDS. Chang (1984) studied the effect of adding nonionic surfactants such as Tween and Triton X-100 to the aqueous buffer on the retention behavior of proteins on diphenyl-bonded silica columns. A hydrophilic surface is formed by the strong interaction of the alkyl chains of the surfactant with the diphenyl groups of the stationary phase, which enabled the size-exclusion separation of proteins.

The use of salts, TFA, and organic modifiers depends on whether analyte–stationary phase interactions are observed or not. Dubin and Principi (1989) described an empirical and efficient procedure to identify ideal SE-HPLC conditions.

5.4. Examples

The number of publications of SE-HPLC on pharmaceutical proteins is limited. Therefore, several applications on other proteins will be described to illustrate the development and state of the art of the SE-HPLC for proteins in general. Determination of the purity of proteins is not often performed by SE-HPLC because of its low resolution.

Schröder *et al.* (1990) quantitated recombinant factor VIII (220 kDa) on an Alltech LiChrosorb Si-100 column. Substances that are known to interfere with common protein assays (such as trypthophan, SDS, Tween 80) were evaluated and were found not to interfere with the present method. The use of 90% formic acid was highly recommended because it dissociates noncovalent bonding (protein–protein as well as protein–stabilizer/detergent) and causes the protein to elute in one single peak.

Kunitani and co-workers described on-line SE-HPLC characterization methods for PEG-modified proteins (Kunitani *et al.*, 1991) and for glycoproteins (Kunitani and Kresin, 1993). Using UV and radioimmunoassay detection in series, molecular size, polymer distribution, and weight composition of PEG-modified proteins were determined (Kunitani *et al.*, 1991). A series of PEG standards, protein standards, and pegylated IL-2 (PEG IL-2) were used. The method was compared with SDS-polyacrylamide gel electrophoresis analysis, the latter technique showing misleading results in identification and quantification of PEG–protein bands. The same method was applied to the analysis of carbohydrate mass composition in glycoproteins (mass carbohydrate/mass protein) (Kunitani and Kresin, 1993). The data indicate the degree of glycosylation in glycoprotein pharmaceuticals, without providing detailed information on the characterization of the glycoprotein structure.

SE-HPLC is in most cases combined with UV detection at 280 nm. Other detection techniques used are refractive index detection, low-angle laser light scattering detection, 90° laser light scattering detection, circular dichroism spectrophotometric detection, scanning diode array detection, and (second) derivative UV absorption detection. Hearn *et al.* (1988) applied derivative spectroscopy to study column residency effects in RP- and SE-HPLC analysis of proteins. Stationary phase-induced effects on protein conformation of hGH and bGH were studied. Second-derivative UV detection as well as circular dichroism spectrophotometric detection were applied by Kurosu *et al.* (1990) to monitor the influence of stationary and mobile phase on protein conformation, in particular the α-helix. Ackland *et al.* (1991) used scanning diode array second-derivative UV absorption spectroscopy to study the degree of aggregation as well as conformational perturbation of a series of IL-2 structural mutants with the same aromatic amino acid composition. The conformation of IL-2 was monitored during the isolation of the protein from inclusion bodies. To obtain the native protein it is necessary to solubilize and refold the inclusion body protein. This process can lead to the formation of aggregates. The presence of these aggregates was confirmed using size-exclusion chromatography. The second-derivative UV absorption characteristics of the aggregate peaks were different from the spectra of the monomers. Circular dichroism spectrophotometric detection was also used by Kato *et al.* (1992) when they studied the composition of dextran–ovalbumin and dextran–

lysozyme conjugate samples by SE-HPLC. Detection was by low-angle laser light scattering (LALLS), refractive index, and UV detection. The molecular weight of the conjugates was determined as well as the number of dextran chains per protein. LALLS detection is reviewed in a publication of Tagaki (1990). Dollinger *et al.* (1992) used 90° light-scattering instead of LALLS detection, which is possible for molecules that are small relative to the wavelength of light used. A simple 90° HPLC fluorimeter can be used, which is not as sensitive to dust and other particles as the LALLS detector. If a multiwavelength fluorimeter is used, the detection wavelength can be optimized to minimize undesirable effects of scatter or solvent absorption or to match the wavelength to that used in the refractive index detector. The suitability of the method was proved by the determination of the molecular weight of several pharmaceutical proteins, e.g., recombinant tumor necrosis factor (rTNF, non-covalent trimer of 51 kDa) and macrophage colony stimulating factor (covalent dimer of 49.7 kDa) and human Glu-plasminogen (94 kDa).

Watson and Kenney (1988) studied IL-2 and interferon analogues with SE-HPLC. At neutral pH, aggregates did not elute and monomers showed nonideal behavior. At pH 2.5, both monomer and aggregates eluted properly and the rate of formation of aggregates was found to correspond to the rate of degradation of the monomeric protein.

Brems *et al.* (1988) used SE-HPLC to examine the folding kinetics of a bGH mutant made by site-directed mutagenesis. The results obtained with various unfolding conditions demonstrated the presence of two species that are in slow equilibrium, each species appearing as a peak in the chromatogram with both species visible. It was concluded that the mutation indeed slows the kinetics of the conversions.

Utsumi *et al.* (1989a) described the stability of INF-β1 (25 kDa), which was studied using SE-HPLC with and without 0.1% SDS. The monomeric form was separated from the inactive oligomeric (tetramer) form. This study demonstrated that INF-β1 is unstable in the presence of saline but that the bioactivity can be recovered by treatment with a detergent. The loss in bioactivity is caused by oligomerization to mainly tetramers. The oligomers can be reactivated by the addition of SDS.

Shalongo *et al.* (1993a) described a model that can simulate the elution profiles obtained for guanidine-HCl-induced unfolding of model proteins. The model requires three experimentally determinable parameters. The model was applied to the conformational analysis of ribonuclease A in guanidine hydrochloride and resulted in reliable kinetic information (Shalongo *et al.*, 1993b).

Snider *et al.* (1992) studied the characterization of pegylated superoxide dismutase (SOD). Each attached PEG chain will contribute on average 5000 Da to the total mass of SOD, which is 16% of the native protein weight. SOD has 20 possible PEG attachment sites per dimer. The theoretical maximum number of species produced on reaction with PEG is approximately 2020, which does not include the polydispersity of the PEG reagent. This number is a theoretical number, however, because not all the lysine moieties can be derivatized (due to steric hindrance) and the reaction conditions can be manipulated in order to control the product composition. By using SE-HPLC it was possible to separate the underivatized compound, mono-, dimer-, trimer-, and tetra-PEG-SOD (Fig. 10). When more than four PEGs per SOD dimer are introduced, the resolution drops significantly because of the decreasing relative difference in molecular size between the individual species (Snider *et al.*, 1992).

6. AFFINITY AND IMMUNOAFFINITY CHROMATOGRAPHY

6.1. General

Affinity chromatographic (AC) techniques are frequently used mostly in the purification of proteins. In addition, AC is being used more often in the analysis and characterization of proteins, either in the sample preparation steps or in the analysis itself. Several overviews on these techniques have been published (Ohlson *et al.*, 1989; Scouten, 1991; Formosa *et al.*, 1991; Nau, 1989; Chaiken, 1990). High-performance affinity chromatography (HPAC) combines the speed of HPLC with the selectivity of AC.

AC is based on the specific interaction between the analyte (e.g., protein) and an immobilized ligand, which is frequently another biomolecule such as an antigen, substrate, or receptor. The choice of the ligand is critical for the initial highly specific interaction. Elution is achieved by choosing an elution buffer that markedly reduces the affinity of the molecule of interest for the ligand. Proteins having a similar affinity toward the ligand may be separated during the elution step after careful selection of the elution conditions, during which the affinity of the various proteins toward the immobilized ligand are modulated. Nondenaturing conditions during the separation are required to ensure that the ligand and/or solute molecule do not lose their biospecificity for each other. Ligands can be specific for a particular molecule or may bind to a group of related molecules. A ligand must bind reversibly to the solute molecule, it must be stable and it must contain a functional group that is coupled to the support in such a way that it can not interfere with the solute-binding site.

The advantage of AC is the inherent biospecifity. The critical factors for the separation are the choice of the ligand, its immobilization to the inert column support, the type of desorption agent (elution buffer) and whether it is applied as a step gradient or as linear gradient, and the choice of flow rate. These factors will dictate the type of column that can be used, the efficiency of the separation, peak width, peak height, and peak elution volume.

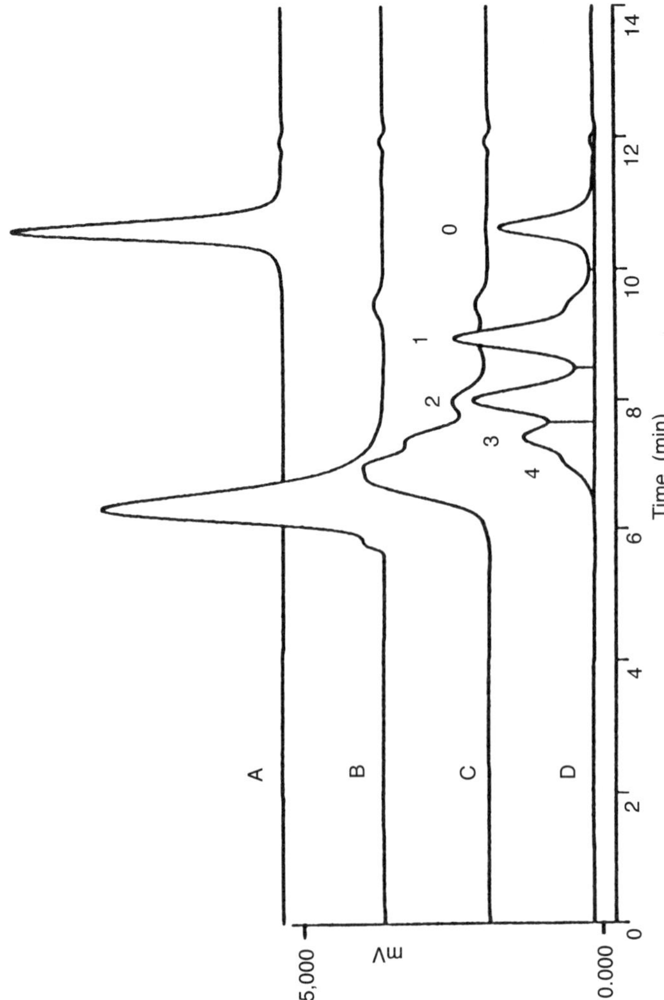

Figure 10. HPSEC chromatograms of (A) unmodified SOD, (B) 50-min PEG-SOD sample, (C) 4-min PEG-SOD sample, and (D) 1-min PEG-SOD sample. Numbers over the peaks refer to the predicted number of PEG chains per SOD dimer (Snider *et al.*, 1992). SE-HPLC was performed on a Progel TSK G3000SWXL column using a mobile phase of methanol and phosphate buffer/NaCl pH 6.8. Detection was by UV at 214 nm.

Several examples of AC with a variety of affinity columns can be found in the literature, such as methods based on immunoaffinity chromatography (IAC), immobilized metal AC (IMAC), immobilized dye chromatography, and methods using a variety of immobilized ligands. IAC depends on the interaction between an antibody and an antigen. Methods using immobilized heparin, DNA, metal ions, and dye are based on the same principle: binding of the analyte of interest to the ligand immobilized on the column matrix followed by subsequent elution of the analyte. The efficiency of the AC separation depends on the selectivity of the stationary phase and the choice of elution conditions determines its resolution. It is the combination of the two that determines the successful use of this technique.

6.2. Stationary Phase

6.2.1. GENERAL

Ohlson *et al.* (1989) present an overview of the various aspects that are of importance in selecting a stationary phase, such as porosity, surface area, mechanical and chemical stability, pore size, immobilization procedures available, stability of bond between ligand and surface, reuseability, capacity, nonspecific binding, cost, and so on. The pores need to be sufficiently large to enable the protein of interest to penetrate, yet further increase of the pore size decreased the surface area required for optimum binding.

6.2.2. IMMUNOAFFINITY CHROMATOGRAPHY

According to Ohlson *et al.* (1989), the material of choice is silica, either in the form of irregular particles or as glass beads, because of its high mechanical stability. Reactive side chains for protein immobilization are *N*-hydroxysuccinimide, carbonylimidazole, epoxy or epoxide, and thiol or carboxyl groups. Synthetic polymers are becoming more popular because of their high chemical stability.

Formosa *et al.* (1991) used protein AC to characterize and isolate proteins by using activated agarose. The techniques to construct agarose-based affinity matrices are summarized as well as the methods to analyze binding fractions. Formosa and co-workers emphasize that an optimal sensitivity is obtained when a maximal amount of protein is coupled in a minimal volume with no denaturing side reactions. The level of nonspecific binding is of importance in selecting the column matrix: The activated matrix (e.g., agarose treated at high pH with CNBr) may react nonspecifically with proteins due to interactions with charged residues on the matrix that obscure the interaction with the covalently bound protein. The coupling process, techniques to measure coupling efficiency as well as methods to prepare micro-, standard, and preparative scale columns are described. The successful use of N-hydroxysuccinimide-activated agarose in coupling proteins (Bio-Rad Affi-gel 10) is reported to result in low background binding with high enough sensitivity. The limited alkaline stability and nonspecific adsorption have been eliminated by coating the silica surface with hydrophilic layers, which also facilitates immobilization.

New trends in stationary phase development include (1) the use of preactivated columns, where the ligand is directly coupled to the matrix during its passage through the column; this replaces activation and packing of the column by the user; and (2) the use of recombinant DNA techniques in the immobilization steps, such as the fusion of specific molecules into proteins to enable a certain AC application (e.g., histidine-containing peptide for IMAC) or to improve the orientation of antibodies upon immobilization for an IMAC application (see the work of Evans and Loetscher discussed in the following sections).

Philips (1991) gives an overview of the practical aspects of immunoaffinity chromatography including the different techniques available for the preparation of columns and guidelines for elution and regeneration procedures. Silica particles and glass beads are used for the immobilization of ligands. The matrix is often treated in order to prevent later nonspecific interactions between glass and biological material. These column materials can be coated (covalently or by adsorption) with, e.g., protein A or protein G, avidin, and streptavidin to enable efficient binding of antibodies to their surfaces. The antibodies attached to the column material via the abovementioned coating can be intact antibodies from the IgG, IgE, or IgA class or can be the Fab fragment of an antibody.

6.2.3. IMMOBILIZED METAL AFFINITY CHROMATOGRAPHY

IMAC, the silica- or polymer-based stationary phase is linked to a spacer that is coupled to the actual chelator, traditionally iminodiacetate (IDA), with a metal ion that is bound to the IDA carboxyl groups and that binds to the peptide or protein. Active sites in the peptide or protein are histidine, cysteine, and tryptophan, with histidine being the primary target function for metal ions such as Cu^{2+}, Ni^{2+}, and Zn^{2+}. Applications for several peptides using Cu^{2+}, Ni^{2+}, Zn^{2+}, Fe^{2+}, and Fe^{3+} have been described (e.g., Porath, 1988, and references therein). The length of the spacer may influence the efficacy of the affinity process, as demonstrated by Utsumi *et al.* (1989b). The influence of the ligand density on the binding properties of the IDA-Cu^{2+} chelator was studied by Wirth *et al.* (1993). Several other chelators have been reported, e.g., nitrilotriacetic acid (Hochuli *et al.*, 1987; Loetscher *et al.*, 1992), ethylenediamine-*N,N'*-diacetic acid (Bacolod and El Rassi, 1990), and 8-hydroxyquinoline-metal^{3+} chelate (Zachariou and Hearn, 1992).

6.2.4. IMMOBILIZED DYE AFFINITY CHROMATOGRAPHY

The use of dye AC in the purification of proteins has been recently discussed by Scawen (1991). Textile dyes are inexpensive bulk chemicals and are easy to couple to agarose stationary phase. Examples of dyes include Procion Blue/Brown/Red/Yellow, and Cibacron Blue. The dye is usually not selective for a particular protein. It is difficult to predict beforehand which dyes are most valuable in the purification of a particular protein. Therefore, an empirical approach is usually taken, screening the protein with a variety of immobilized dye columns. Giuliano (1992) describes the polyvinylpyrrolidone column to which the dye adsorbs (results for Procion Yellow and Brown are presented) without losing its protein binding properties. An easy and rapid screening system is described, including recommended elution conditions.

6.2.5. MISCELLANEOUS IMMOBILIZED LIGANDS

A wide variety of immobilized ligands can be found in the literature, such as dextran-coated silica columns grafted with heparin (Jacquot-Dourges *et al.*, 1991), immobilized heparin and DNA columns (Cacia *et al.*, 1993), immobilized synthetic peptide ligand column (Welling *et al.*, 1990), immobilized mellitin columns (a bee venom peptide) (Fleminger *et al.*, 1992), and immobilized DNA polymerase α (Miles and Formosa, 1992), Ohlson *et al.* (1989) present an overview of commercially available columns including a variety of ligands.

6.3. Mobile Phase

The choice of mobile phase is critical for the successful use of AC (Scouten, 1991). Elution is based on the type of interaction between the protein of interest and the immobilized ligand on the column. If the interaction has an ionic component, then the elution of the protein can be accomplished by salt. Increasing the ionic strength in a gradient usually gives best results. Other methods applied include pH changes, temperature changes, centrifugal force, chelating agents, and so on (Scouten, 1991, and references cited therein). Washing of nonspecifically bound matrix components is highly recommended.

6.3.1. IMMUNOAFFINITY CHROMATOGRAPHY

The elution of the isolated material is usually done by acid or (chaotropic) ion elution. Examples of acidic solutions are glycine, Tris/HCl, citric acid, and acetic acid. Low pH lowers the antibody–antigen bond interactions and the isolated solute can be eluted, preferably by using a gradient. Solutions of sodium and potassium hydroxide are also used. Chaotropic ions are sodium thiocyanate, sodium chloride, and polyvinylpyrrolidone–iodide complex (Philips, 1991). The chaotropic ions interfere with the organization of the ionic interactive forces and cause dissociation of the antibody–antigen complex. The effective dissociation parameter of these ions is as follows: Cl^-, I^-, ClO_4^- < CF_3COO^- < SCN^- < CCl_3COO^-. Denaturing agents, e.g., urea, guanidine HCl, and the polarity-reducing agents dioxane and ethylene glycol, are not very popular for the elution.

6.3.2. MISCELLANEOUS IMMOBILIZED LIGANDS

Elution from IMAC matrices is achieved by biospecific elution, by a change in ionic strength/pH/polarity, or by a chaotropic or denaturing agent. Fe^{3+} AC is typically used for the purification and analysis of phosphoproteins/peptides (Hjerten *et al.*, 1989; Muszynska *et al.*, 1992). The presence of magnesium ions in the elution buffer influences the elution of phosphocompounds bound to the Fe^{3+} matrix by an interaction with the phosphate groups. It has been suggested that magnesium ions can be useful for separating proteins that differ in the number of phosphate groups (Muszynska *et al.*, 1992).

The methods of elution from dye affinity matrices are biospecific elution (cofactor, substrate, inhibitor, free ligand), a change in ionic strength/pH/polarity, and a chaotropic or denaturing agent (Scawen, 1991).

6.4. Examples

6.4.1. IMMUNOAFFINITY CHROMATOGRAPHY

Monoclonal antibodies are widely used in the purification and analysis of proteins. The use in analysis of pharmaceutical proteins, however, is limited. In the study reported by Hayashi *et al.* (1988), rhEGF was determined using an rhEGF antibody precolumn. Loetscher *et al.* (1992) described a novel procedure for oriented immobilization of monoclonal antibodies using Ni^{2+} chelate chemistry and a nitrilotriacetic acid resin. A hexahistidine chelating peptide [Lys-Gly-$(His)_6$] is chemically conjugated to the aldehyde groups generated on the carbohydrate side chains of the monoclonal antibody (anti-human leukocyte interferon LI-8). The interaction of the histidine-containing peptide with the Ni^{2+} ions bound to the nitrilotriacetic acid resin results in an oriented immobilization of the antibodies. An interesting approach for selective purification and analysis of proteins is the use of a synthetic antibody fragment that mimics the antigen-binding site as ligand (Welling *et al.*, 1990).

Fusion proteins aiming for a selective IAC purification process for β-galactosidase were used by Downham *et al.* (1992). An example of fusion proteins to improve IMAC purification is described by Scouten (1991) and to improve their detection by Evans *et al.* (1992). Evans and co-workers used antibodies directed against the metal-binding peptide portion of a fusion protein for the off-line detection. A dual-column IAC method for the simultaneous determination of albumin (using immobilized anti-albumin antibodies on first column) and immunoglobulin G (using protein A column as second column) is described by Hage and Walters (1987).

6.4.2. IMMOBILIZED METAL AFFINITY CHROMATOGRAPHY

Evans *et al.* (1992) developed a genetically engineered metal-binding peptide for the IMAC purification of recombinant proteins. The peptide (His-Asp-His-Asp-His) is highly reactive to immobilized metal ions such as Ni^{2+}, Zn^{2+}, and Cu^{2+}. The IMAC purification is described by Vosters *et al.* (1992). The metal-binding peptide sequence was fused into the DNA of two model proteins. High selectivity and strong interaction for Ni^{2+}-chelating Sepharose was observed for the modified proteins as compared to the wild types. This approach, namely, tailoring recombinant proteins with an alternating histidine moiety, could result in highly specific IMAC retention of the protein, and thus could be a valuable tool in the purification and analysis of recombinant proteins.

Hochuli *et al.* (1987) used nitrilotriacetic acid as a chelator, which was found to be more stable than Ni^{2+}–iminodiacetate adsorbent and showed a remarkable selectivity for peptides and proteins containing neighboring histidine residues. Bacolod and El Rassi (1990) used silica-bound ethylenediamine-*N,N'*-diacetic acid as chelating agent. A different selectivity and retentivity was observed as compared to an iminodiacetic acid stationary phase.

Another new chelate, namely, 8-hydroxyquinoline-metal^{3+} chelate, was used by Zachariou and Hearn (1992). It was observed that metal ions such as Fe^{3+}, Al^{3+}, and Ca^{2+} could result in selective binding of a protein, even in absence of histidine, trypthophan, and cysteine.

Hjerten *et al.* (1989) used immobilized ferric oxide and ferric oxyhydroxide by direct precipitation on beads of nonporous cross-linked agarose. Figure 11 presents the separation of a standard mixture of proteins using the ferric column. Muszynska *et al.* (1992) studied the chromatographic properties of phosphorylated and nonphosphorylated amino acids, peptides, and proteins using Fe^{3+}-chelating Sepharose and Fe^{3+}-chelating Superose. Their data indicate that in addition to the phosphate function, carboxylic and phenolic groups of the amino acid residues are also involved in the metal ion interaction.

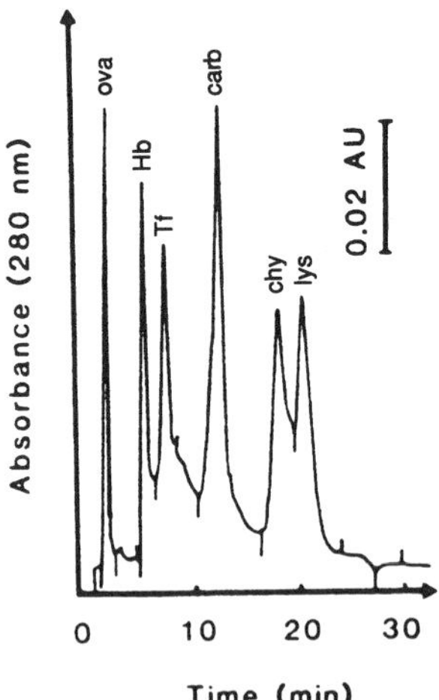

Figure 11. Chromatography of a model protein mixture on a $FeO(OH)(PO_4)$-agarose column. Sample: 75 μg of each of the following proteins in a total volume of 60 μl: ovalbumin (ova), hemoglobin A_{1c} (Hb), human transferrin (Tf), carbonic anhydrase (carb), chymotrypsinogen A (chy), and lysozyme (lys). Elution: a linear gradient in potassium phosphate from 0 to 0.2 M over 30 min in 0.01 M sodium cacodylate, pH 6.0 (Hjerten *et al.*, 1989).

6.4.3. IMMOBILIZED DYE AFFINITY CHROMATOGRAPHY

Giuliano (1992) described an easy and rapid screening system using a polyvinylpolypyrrolidone column. Yang *et al.* (1992) used Blue-Sepharose affinity columns to purify 17β-hydroxysteroid dehydrogenase in a fast and efficient process.

6.4.4. MISCELLANEOUS IMMOBILIZED LIGANDS

Jacquot-Dourges *et al.* (1991) described the use of dextran-coated silica columns grafted with heparin in the analysis of fibroblast growth factors. Immobilized heparin and homemade DNA columns were used by Cacia *et al.* (1993) in the analysis of rhDNase I. Concanavalin A AC can be used for the analysis of glycoproteins and peptides. An example can be found in Josic *et al.* (1988). Ohlson *et al.* (1989) present an overview of commercially available columns including a variety of ligands, such as boronic acid for the analysis of carbohydrates, cephalosporin for the analysis of β-lactamase, and glucosamine for the analysis of concanavalin A. An application of boronate affinity chromatography can be found in Yasukawa *et al.* (1992), who describe a rapid assay of glycosylated albumin.

Immobilized peptides have also been used as stationary phase (Welling *et al.*, 1990; Fleminger *et al.*, 1992). DNA polymerase-α was immobilized on Affi-Gel 10-activated agarose by Miles and Formosa (1992).

The continuous improvement of immobilization chemistry and the more readily availability of ligands will further increase the wide application of AC and HPAC in purification and analysis of proteins.

7. RECENT DEVELOPMENTS

7.1. Perfusion Chromatography

Perfusion chromatography was described for the first time in 1990 (Afeyan *et al.*, 1990). In this technique a special type of column material is used. The liquid flows through the particles. The particles contain 6000- to 8000-Å pores that transect the particle. The surface area of the large-pore diameter material is enhanced by using a network of smaller, 500- to 1500-Å pores. Data from electron microscopy, column efficiency, and frontal analysis suggest that the mobile phase will flow through these large pores. The solutes enter the interior of the particle through a combination of convective and diffusional transport to the active surfaces in the interior of the sorbent. The convective transport is directly related to the bed mobile phase velocity. An increase of fluid velocity will increase the solute flux into the particle. The flow through particles affects resolution and dynamic loading capacity. Scanning electron micrographs show that the pore network is continuous and that no point in the matrix is more than 5000 to 10000 Å from a through-pore.

As a consequence, diffusional path lengths are minimized, and the large porous particles have transport characteristics of much smaller particles but with a fraction of the pressure drop. Capacity and resolution studies show that these materials bind and separate an amount of protein equivalent to that of conventional HPLC as well as low-performance agarose-based media at greater than 10–100 times higher mobile phase velocity without loss of resolution. A mathematical model of this type of chromatography is described by Liapis *et al.* (1992).

The material produced for this technique, poly(styrene-divinylbenzene), allows perfusion chromatography to be performed in the reversed-phase, ion-exchange, hydrophobic interaction and affinity modes (Afeyan *et al.*, 1990, 1991). Figure 12 shows two chromatograms presenting the separation of a mixture of model proteins eluting in the reversed-phase mode at two different elution velocities. The complete separation is obtained within 5 min at 5.0 ml/min. Although this technique is not yet widespread, its potential for the analysis of (bio)polymers is obvious.

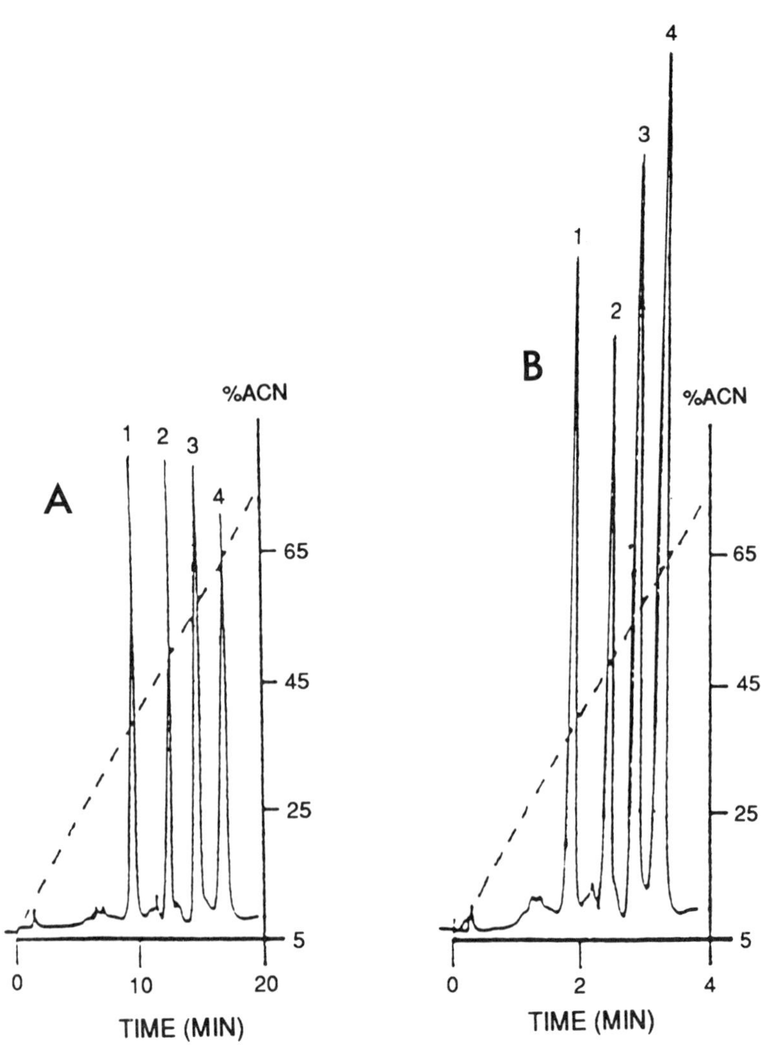

Figure 12. Separation of model proteins on POROS R/M (reversed-phase mode); 100-μg load of a protein mixture containing (1) ribonuclease A, (2) lysozyme, (3) β-lactoglobulin (A and B), and (4) ovalbumin; detection at 280 nm; column, 100 × 4.6 mm inner diameter; 5% acetonitrile plus 0.1% trifluoroacetic acid in water, (A) 1 ml/min, 20-min gradient to 70% acetonitrile; (B) 5 ml/min; 4-min gradient to 70% acetonitrile (Afeyan *et al.*, 1990).

7.2. Hydrophylic Interaction Chromatography

An interesting new development is the use of hydrophylic interaction chromatography (HILIC) in the separation of proteins (Alpert, 1990; Boutin *et al.*, 1992). The technique is complementary to or an alternative for other techniques such as RPC. A hydrophilic column is used, which is eluted with a hydrophobic mobile phase. Retention increases with increasing protein hydrophilicity, and as a result the order of elution is opposite to that in RP-HPLC. The technique enables the separation of basic or highly polar molecules that are difficult to separate by RP-HLPC.

Phosphorylated peptides were analyzed on a polyhydroxyethyl aspartamide column. It was possible to separate plain peptides from phosphorylated peptides and inorganic phosphate species. The elution was performed with a gradient of 4 mM TEAP, pH 2.8, to 800 mM TEAP in 90% acetonitrile. HILIC proved to be very useful in separating a range of polar peptides. Boutin *et al.* (1992) used HILIC for the study of tyrosine protein kinase specificity. Using a polyhydroxyethyl A column, phosphorylated peptide, nonphosphorylated peptide, ATP, and inorganic phosphate were separated.

7.3. High-Performance Affinity Chromatography

A recent development is chromatographic analysis of mixtures of antigen, antibodies, and labeled antibodies described by Cassidy *et al.* (1992). Cassidy actually describes an immunoassay on-column. On a protein A column, the antibody, the sample containing the antigen, and a well-known amount of antigen (the label) are sequentially injected. The antigen–antibody complex is adsorbed on the column. The amount of unbound antigen in the label is proportional to the amount of antigen of the sample bound and is quantitated using UV. The sample antigen and label antigen contact the antibody at different times, and as a result the label antigen needs no tag to distinguish it from the sample label. Figure 13 schematically shows the principles of this technique (kinetic immunochromatographic sequential addition).

Flurer and Novotny (1993) described a dual microcolumn IAC assay to analyse human plasma proteins. The first column was the IAC antibody column, which captures the antigen; the resulting antigen-free sample was then directed to a second (RP-HPLC) column for analysis, followed by desorption of the antigen from the IAC column and RP-HPLC analysis. Sequential immunoaffinity isolation of nine human plasma proteinase inhibitors was described by Dubin *et al.* (1990), using a total of seven different affinity columns of various affinity characteristics placed in series. The method is fast, simple, and yields the proteins with high recovery and purity.

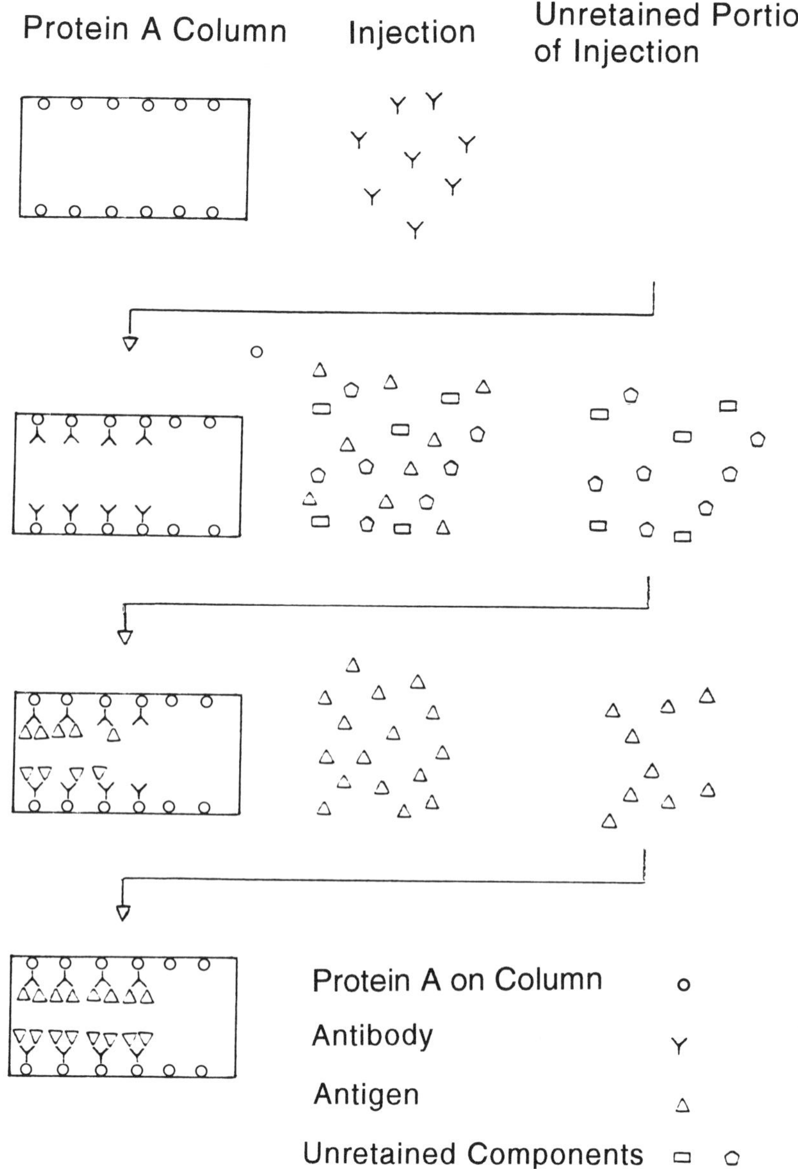

Figure 13. Schematic presentation of KICQA. Row 1 represents injection of the antibody. Row 2 represents injection of the dose (the sample to be analyzed). Row 3 represents injection of the label. The unbound portion of the label (row 3) is the signal for the experiment (Cassidy *et al.*, 1992).

7.4. Liquid Chromatography in Combination with Mass Spectrometry

Peptides obtained after an enzymatic digest of a protein are often isolated by RPC prior to analysis by fast atom bombardment mass spectroscopy (FAB-MS). With this approach, antithrombine III (Zhou and Smith, 1990), rhGH, human pituitary growth hormone, and recombinant methionine human growth hormone (Met-rhGH) (Nakazawa, 1988) were identified.

The combination of RPC and peptide maps was used to determine the primary structure of phosphorylated α-crystallins (Smith *et al.*, 1991), β-casein (Kassel *et al.*, 1991), and Lys(78)-plasminogen (84 kDa) (Bell *et al.*, 1991). The combination of RPC and thermospray mass spectrometry (TS-MS) was used to analyze all the tryptic peptides of the nonglycosylated IL-2 in one run (Blackstock *et al.*, 1988). IL-2 was reduced and carboxymethylated prior to tryptic digestion. It was possible to identify all the fragments by their molecular ions, including those of the large peptides. Due to the presence of multiply charged species, the larger peptides were brought into the range of the equipment. Specific peptides were detected, indicating the presence of a deaminated product and the incomplete removal of theN-initiator methionine.

Electrospray mass spectrometric (ES-MS) and thermospray mass spectrometric (TS-MS) analysis are powerful detection methods for RPC. The extra dimension of mass information allows the easy detection of glycopeptides and peptides that are multiply charged. The glycopeptides are identified without enzymatic cleavage of the carbohydrate moieties from the peptide, which simplifies the identification process. In addition, mass spectrometry gives more detail and enhances the possibility to distinguish coeluting components. The most powerful combination for the elucidation of the primary structure of a protein is the combination of (capillary) HPLC with electron spray tandem mass spectrometry. However, only volatile buffers for the chromatography can be used, since involatile buffers contaminate the mass spectrometer. Microbore HPLC on-line with FAB-MS and ES-MS was used to study the structure of recombinant soluble CD4 glycoprotein and the recombinant hepatitis B surface antigen (Hemling *et al.*, 1990).

Establishing the nature of prosthetic (nonprotein) chemical structures attached to proteins remains one of the most difficult areas of characterization, especially when unexpected or even unknown modification occurs. The role of the mass spectrometric detection is significant in the elucidation of these prosthetic groups. Strömqvist and co-workers (1991) characterized the glycosylation of extracellular superoxide dismutase. This glycoprotein has an apparent molecular weight of 135 kDa. It consists of four identical subunits each of 222 amino acids (24.2 kDa). The protein was isolated from Chinese hamster ovary cells and purified. The protein was reduced, carboxymethylated, and digested with trypsin. The glycopeptide was isolated by RPC. The peptide binds to concanavalin A and lentil lectin. The carbohydrate was cleaved from the peptide by enzymatic hydrolysis and the peptide was identified off-line by FAB-MS.

The potential of the combination of liquid chromatography and mass spectrometry was also illustrated by Covey *et al.* (1991), who used electron spray ionization in combination with HPLC to analyze the tryptic digest of Met-rhGH. Conventional reversed-phase columns as well as microbore columns were used. Molecular weights of the peptides separated on the HPLC columns could be determined. Application of collision-induced dissociation provided structurally useful sequence information. Collision-induced dissociation can also be used to distinguish singly charged ions from the doubly charged ions. The combination of RPC with ES-MS (and MS/MS) supplies additional information on the structure of proteins including posttranslational modifications. The technique enables the detection of minor compounds next to the major compound. The sensitivity of the technique is in the picomole to femtomole range. RPC is necessary to obtain separation of the proteins and peptides, while the choice of enzyme will introduce an additional degree of freedom. The simplicity of the technique will finally result in the routine usage in the identification of quality control of pharmaceutical proteins.

REFERENCES

Ackland, C. E., Berndt, W. G., Frezza, J. E., Landgraf, B. E., Pritchard, K. W., and Ciardelli, T. L., 1991, Monitoring of protein conformation by high-performance size-exclusion liquid chromatography and scanning diode array second-derivative UV absorption spectroscopy, *J. Chromatogr.* **540:**187–198.

Afeyan, N. B., Gordon, N. F., Mazsaroff, I., Varady, L., Fulton, S. P., Yang, Y. B., and Regnier, F. E., 1990, Flow-through particles for the high-performance liquid chromatographic separation of biomolecules: Perfusion chromatography, *J. Chromatogr.* **519:** 1–29.

Afeyan, N. B., Fulton, S. P., and Regnier, F. E., 1991, Perfusion chromatography packing materials for proteins and peptides, *J. Chromatogr.* **544:**267–279.

Ahmed, F., and Modrek, B., 1992, Biosep-SEC-S high performance size-exclusion chromatographic columns for proteins and peptides, *J. Chromatogr.* **599:**25–33.

Alpert, A. J., 1990, Hydrophylic-interaction chromatography for the separation of peptides, nucleic acids and other polar compounds, *J. Chromatogr.* **499:**177–196.

Andersson, T., Carlsson, M., Hagel, L., Pernemalm, P., and Janson, J., 1985, Agarose-based media for high-resolution gel filtration of biopolymers, *J. Chromatogr.* **326:**33–44.

Arakawa, T., and Owers Narhi, L., 1991, Review: Solvent modulation in hydrophobic interaction chromatography, *Biotechnol. Appl. Biochem.* **13:**151–172.

Bacolod, M. D., and El Rassi, Z., 1990, High-performance metal chelate interaction chromatography of proteins with silica-bound ethylenediamine-*N,N'*-diacetic acid, *J. Chromatogr.* **512:**237–247.

Bell, D. J., Brightwell, M. D., Haran, M., Neville, W. A., and West, A., 1991, Routine liquid chromatography/fast atom bombardment mass spectrometry of peptides and enzymatically digested proteins, *Org. Mass Spectrom.* **26:**454–457.

Blackstock, W. P., Dennis, R. J., Lane, S. J., Sparks, J. I., and Weir, M. P., 1988, The analysis of recombinant interleukin-2 by thermospray liquid chromatography-mass spectrometry, *Anal. Biochem.* **175:**319–326.

Boutin, J. A., Ernould, A.-P., Ferry, G., Genton, A., Alpert, A. J., 1992, Use of hydrophylic interaction chromatography for the study of tyrosine protein kinase specificity, *J. Chromatogr.* **583:**137–143.

Brems, D. N., Plaisted, S. M., Havel, H. A., and Tomich, C.-S., 1988, Stabilization of associated folding intermediate of bovine growth hormone by site-directed mutagenesis, *Proc. Natl. Acad. Sci. USA* **85:**3367–3371.

Browning, J. L., Mattaliano, R. J., Pingchang Chow, E., Shu-Mei Liang, Allet, B., Rosa, J., and Smart, J. E., 1986, *Anal. Biochem.* **155:**123–128.

Buckley, J. J., and Wetlaufer, D. B., 1990, Surfactant-mediated hydrophobic interaction chromatography of proteins: Gradient elution, *J. Chromatogr.* **518:**99–110.

Burton, W. G., Nugent, K. D., Slattery, T. K., Summers, B. R., and Snyder, L. R., 1988, Separation of proteins by reversed phase high-performance liquid chromatography. I. Optimizing the column, *J. Chromatogr.* **443:**363–379.

Cacia, J., Quan, C. P., Vasser, M., Sliwkowski, M. B., and Frenz, J., 1993, Protein sorting by high-performance liquid chromatography I. Biomimetic interaction chromatography of recombinant human deoxyribonuclease I on polyionic stationary phases, *J. Chromatogr.* **634:**229–239.

Cassidy, S. A., Janis, L. J., and Regnier, F. E., 1992, Kinetic chromatographic sequential addition immunoassay using protein A affinity chromatography, *Anal. Chem.* **64:** 1973–1977.

Chaiken, I. M., 1990, High performance affinity chromatography: Isolation and analysis of biological macromolecules, in: *High Performance Liquid Chromatography in Biotechnology* (W. S. Hancock, ed.), John Wiley, New York, pp. 289–300.

Chang, J. F., 1984, Effect of surfactants on the separation of proteins by reversed-phase high-performance liquid chromatography. 1. Non-ionic surfactants (Tween), *J. Chromatogr.* **517:**157–163.

Chloupek, R. C., Harris, R. J., Leonard, C. K., Keck, R. G., Keyt, B. A., Spellman, M. W., Jones, A. J. S., and Hancock, W. S., 1989, Study of the primary structure of recombinant tissue plasminogen activator by reversed-phase high-performance liquid chromatographic tryptic mapping, *J. Chromatogr.* **463:**375–396.

Christensen, T., Hansen, J. J., Sørensen, H. H., and Thomsen, J., 1990, RP-HPLC of biosynthetic and hypophyseal human growth hormone, in: *High Performance Liquid Chromatography in Biotechnology* (W. S. Hancock, ed.), John Wiley, New York, pp. 191–204.

Clogston, C. L., Hsu, Y.-R., Boone, T. C., and Lu, H. S., 1992, Detection and quantitation of recombinant granulocyte colony-stimulating factor charge isoforms: Comparitive analysis by cationic exchange chromatography, isoelectric focusing gel electrophoresis, and peptide mapping, *Anal. Biochem.* **202:**375–383.

Cohen, S. A., Schellenberg, K., Benedek, K., Karger, B. L., Greco, B., and Hearn, M. T. W.,

1984, Mobile-phase and temperature effects in the reversed phase chromatographic separation of proteins, *Anal. Biochem.* **149:**223–235.

Cohen, S. A., Benedek, K., Tapuhi, Y., Ford, J. C., and Karger, B. L., 1985, Conformational effects in the reversed-phase liquid chromatography of ribonuclease A, *Anal. Biochem.* **144:**275–284.

Covey, T. R., Huang, E. C., and Henion, J. D., 1991, Structural characterization of protein tryptic peptides via liquid chromatography/mass spectrometry and collision-induced dissociation of their doubly charged molecular ions, *Anal. Chem.* **63:**1193:–1200.

Dollinger, G., Cunico, B., Kunitani, M., Johnson, D., and Jones, R., 1992, Practical on-line determination of biopolymer molecular weights by high-performance liquid chromatography with classical light-scattering detection, *J. Chromatogr.* **592:**215–228.

Dou, L., Holmberg, A., and Krull, I. S., 1991, Electrochemical detection of proteins in high-performance liquid chromatography using on-line, postcolumn photolysis, *Anal. Biochem.* **197:**377–383.

Downham, M., Busby, S., Jefferis, R., and Lyddiatt, A., 1992, Immunoaffinity chromatography in biorecovery: An application of recombinant DNA technology to generic adsorption processes, *J. Chromatogr.* **584:**59–67.

Drake, A. F., Fung, M. A., and Simpson, C. F., 1989, Protein conformation changes as the result of binding to reversed-phase chromatography column material, *J. Chromatogr.* **476:**159–163.

Dubin, A., Potempa, J., and Travis, J., 1990, Isolation of nine human plasma proteinase inhibitors by sequential affinity chromatography, *Prep. Biochem.* **20:**63–74.

Dubin, P. L., and Principi, J. M., 1989, Optimization of size-exclusion separation of proteins on a Superose column, *J. Chromatogr.* **479:**159–164.

Evans, D. B., Vosters, A. F., Carter, J. B., and Sharma, S. K., 1992, Immunodetection of recombinant proteins based on antibodies directed against a metal binding peptide engineered for purification by immobilized metal affinity chromatography, *J. Immunol. Methods* **156:**231–238.

Fang, F. W., Aguilar, M. I., and Hearn, M. T. W., 1992, High-performance liquid chromatography of amino acids, peptides and proteins. CXX. Evaluation of bandwidth behaviour of proteins chromatographed on tentacle-type anion exchangers, *J. Chromatogr.* **599:**163–170.

Fausnaugh, J. L., Kennedy, L. A., and Regnier, F. E., 1984, Comparison of hydrophobic-interaction and reversed-phase chromatography of proteins, *J. Chromatogr.* **317:** 464–472.

Felix, A. M., Heimer, E. P., Lambros, T. J., Swistok, J., Tarnowski, S. J., and Wang, C.-T., 1985, Analysis of different forms of recombinant human leukocyte interferons and synthetic fragments by high-performance liquid chromatography, *J. Chromatogr.* **327:** 359–368.

Fleminger, G., Neufeld, T., Star-Weinstock, M., Litvak, M., and Solomon, B., 1992, Calcium-modulated conformational affinity chromatography. Application to the purification of calmodulin and S100 proteins, *J. Chromatogr.* **597:**263–270.

Flurer, C. L., and Novotny, M., 1993, Dual microcolumn immunoaffinity liquid chromatography: An analytical application to human plasma proteins, *Anal. Chem.* **65:** 817–821.

Formosa, T., Barry, J., Alberts, B. M., and Greenblatt, J., 1991, Using protein affinity chromatography to probe structure of protein machines, *Methods Enzymol.* **208:** 24–45.

Frelinger, A. L., and Zull, J. E., 1984, Oxidized forms of parathyroid hormone with biological activity, *J. Biol. Chem.* **259:**5507–5513.

Frenz, J., Wu, S.-L., and Hancock, W. S., 1989, Characterization of human growth hormone by capillary electrophoresis, *J. Chromatogr.* **480:**379–391.

Garnick, R. L., Solli, N. J., and Papa, P. A., 1988, The role of quality control in biotechnology: An analytical perspective, *Anal. Chem.* **60:**2546–2557.

Geigert, J., 1989, Overview of the stability and handling of recombinant protein drugs, *J. Parenter. Sci. Technol.* **43:**220–224.

Giuliano, K. A., 1992, Chromatography of proteins on columns of polyvinylpolypyrrolidone using adsorbed textile dyes as affinity ligands, *Anal. Biochem.* **200:**370–375.

Gooding, D. L., Schmuck, M. N., and Gooding, K. M., 1984, Analysis of proteins with new, mildly hydrophobic high-performance liquid chromatography packing materials, *J. Chromatogr.* **296:**107–114.

Hage, D. S., and Walters, R. R., 1987, Dual-column determination of albumin and immunoglobulin G in serum by high-performance affinity chromatography, *J. Chromatogr.* **386:**31–49.

Hanson, M., Unger, K. K., Mant, C. T., and Hodges, R. S., 1992, Polymer-coated reversed-phase packings with controlled hydrophobic properties. I. Effect on the selectivity of protein separations, *J. Chromatogr.* **599:**65–75.

Hayashi, T., Sakamoto, S., Fuwa, T., Morita, I., and Yoshida, H., 1987, Determination of human epidermal growth factors in cultured media of *E. coli* by high performance liquid chromatography, *Anal. Sci.* **3:**445–448.

Hayashi, Y., Sakamoto, S., Fuwa, T., Wada, I., and Yoshida, H., 1988, Determination of epidermal growth factors in human urine by high performance liquid chromatography using anti-hEGF antibody precolumn, *Anal. Sci.* **4:**313–316.

Hearn, M. T. W., 1991, Characterisation of the physicochemical relationships of displacer ions in the high performance ion exchange chromatography of proteins, *Anal. Sci.* **7:**1519–1523.

Hearn, M. T. W., Aguilar, M. I., Nguyen, T., and Fridman, M., 1988, High-performance liquid chromatography of amino acids, peptides and proteins LXXXIV. Application of derivative spectroscopy to the study of column residency effects in the reversed-phase and size-exclusion liquid chromatographic separation of proteins, *J. Chromatogr.* **435:**271–284.

Hemling, M. E., Roberts, G. D., Johnson, W., and Carr, S. A., 1990, Analysis of proteins and glycoproteins at the picomole level by on-line coupling of microbore high-performance liquid chromatography with flow fast atom bombardment and electrospray mass spectrometry: A comparative evaluation, *Biomed. Environ. Mass Spectrom.* **19:** 677–691.

Hjerten, S., Zelikman, I., Lindeberg, J., Liao, J.-I., Eriksson, K.-O., and Mohammad. J., 1989, High-performance adsorption chromatography of proteins on deformed non-porous agarose beads coated with insoluble metal compounds, *J. Chromatogr.* **481:** 175–186.

Hochuli, E., Dobeli, H., and Schacher, A., 1987, New metal chelate adsorbent selective for proteins and peptides containing neighbouring histidine residues, *J. Chromatogr.* **411:**177–184.

Iadarola, P., Zapponi, M. C., Minchiotti, L., Meloni, M. L., Galliano, M., and Ferri, G., 1990, Separation of fragments from human serum albumin and its charged variants by reversed-phase and cation-exchange high-performance liquid chromatography, *J. Chromatogr.* **512:**165–176.

Jacquot-Dourges, M. A., Zhou, F. L., Muller, D., and Jozefonvicz, J., 1991, Affinity chromatography of fibroblast growth factors on coated silica supports grafted with heparin, *J. Chromatogr.* **539:**417–424.

Josic, D., Hofmann, W., Habermann, R., and Reutter, W., 1988, High-performance concanavalin a affinity chromatography of liver and hepatoma membrane proteins, *J. Chromatogr.* **444:**29–39.

Kalghatgi, K., and Horváth, C., 1988, Rapid peptide mapping by high-performance liquid chromatography, *J. Chromatogr.* **443:**343–354.

Karger, B. L., and Blanco, R., 1989, The effect of on-column structural changes of proteins on their HPLC behaviour, *Talanta* **36:**243–248.

Kassel, D. B., Musselman, B. D., and Smith, J. A., 1991, Primary structure determination of peptides and enzymatically digested proteins using capillary liquid chromatography/mass spectrometry and rapid linked-scan techniques, *Anal. Chem.* **63:**1091–1097.

Kato, Y., Nakamura, K., and Hashimoto, T., 1983, New ion-exchanger for the separation of proteins and nucleic acids, *J. Chromatogr.* **266:**385–394.

Kato, Y., Kitamura, T., Nakamura, K., Mitsui, A., Yamasaki, Y., and Hashimoto, T., 1987, High-performance liquid chromatography of membrane proteins, *J. Chromatogr.* **391:**395–407.

Kato, Y., Nakatani, S., Kitamura, T., Yamasaki, Y., and Hashimoto, T., 1990, Reversed-phase high-performance liquid chromatography of proteins and peptides on a pellicular support based on hydrophillic resin, *J. Chromatogr.* **502:**416–422.

Kato, A., Kameyama, K., and Takagi, T., 1992, Molecular weight determination and compositional analysis of dextran-protein conjugates using low-angle laser light scattering technique combined with high-performance gel chromatography, *Biochim. Biophys. Acta* **1159:**22–28.

Katzenstein, G. E., Vrona, S. A., Wechsler, R. J., Steadman, B. L., Lewis, R. V., and Middaugh, C. R., 1986, Role of conformational changes in the elution of proteins from reversed-phase HPLC columns, *Proc. Natl. Sci. USA* **83:**4268–4272.

Kennedy, J. F., Rivera, Z. S., and White, C. A., 1989, The use of HPLC in biotechnology, *J. Biotechnol.* **9:**83–106.

Kennedy, R. T., and Jorgenson, J. W., 1990, Efficiency of packed microcolumns compared with large-bore packed columns in size-exclusion chromatography, *J. Microcol. Sep.* **2:**120–126.

Konishi, T., and Kamada, M., 1990, Evaluation of ammonium acetate as a volatile buffer for high-performance hydrophobic-interaction chromatography, *J. Chromatogr.* **515:**279–283.

Kumagaye, K. Y., Takai, M., Chino, N., Kimura, T., and Sakakibara, S., 1985, Comparison of reversed-phase and cation-exchange high-performance liquid chromatography for separating closely related peptides: Separation of Asp76-human parathyroid hormone(1-84) from Asn76-human parathyroid hormone(1-84), *J. Chromatogr.* **327:** 327–332.

Kunitani, M., and Kresin, L., 1993, High-performance liquid chromatographic analysis of carbohydrate mass composition in glycoproteins, *J. Chromatogr.* **632:**19–28.

Kunitani, M., Hirtzer, P., Johnson, D., Halenbeck, R., Boosman, A., and Koths, K., 1986, Reversed-phase chromatography of interleukin-2 muteins, *J. Chromatogr.* **359:** 391–402.

Kunitani, M. G., Cunico, R. L., and Staats, S. J., 1988, Reversible subunit dissociation of tumour necrosis factor during hydrophobic interaction chromatography, *J. Chromatogr.* **443:**205–220.

Kunitani, M., Dollinger, G., Johnson, D., and Kresin, L., 1991, On-line characterization of polyethylene glycol-modified proteins, *J. Chromatogr.* **588:**125–137.

Kurosu, Y., Sasaki, T., Takakuwa, T., Sakayanagi, N., Hibi, K., and Senda, M., 1990, Analysis of proteins by high-performance liquid chromatography with circular dichroism spectrophotometric detection, *J. Chromatogr.* **515:**407–414.

Lahm, H.-W., and Stein, S., 1985, Characterization of recombinant human interleukin-2 with micromethods, *J. Chromatogr.* **326:**357–361.

Lau, S. Y. M., Taneja, A. K., and Hodges, R. S., 1984, Effects of high-performance liquid chromatographic solvents and hydrophobic matrices on the secondary and quaternary structure of a model protein. Reversed-phase and size exclusion high-performance liquid chromatography, *J. Chromatogr.* **317:**129–140.

Liapis, A. I., and McCoy, M. A., 1992, Theory of perfusion chromatography, *J. Chromatogr.* **599:**87–104.

Lin, S., and Karger, B., 1990, Reversed-phase chromatographic behaviour of proteins in different unfolded states, *J. Chromatogr.* **499:**89–102.

Linde, S., and Welinder, B. S., 1991a, Reversed-phase chromatography of insulin and iodinated insulin, in: *High-Performance Liquid Chromatography of Peptides and Proteins* (C. T. Mant and R. S. Hodges, eds.), CRC Press, Boca Raton, FL, pp. 351–360.

Linde, S., and Welinder, B. S., 1991b, Silica versus polymer-based stationary phases for reversed-phase high-performance liquid chromatographic analyses of rat insulin biosynthesis. A comparison of resolution and recovery, *J. Chromatogr.* **548:**195–206.

Ling, V., Guzzetta, A. W., Canova-Davis, E., Stults, J. T., Hancock, W. S., Covey, T. R., and Shushan, B. I., 1991, Characterization of the tryptic map of recombinant DNA derived tissue plasminogen activator by high-performance liquid chromatography-electrospray ionization mass spectrometry, *Anal. Chem.* **63:**2909–2915.

Loetscher, P., Mottlau, L., and Hochuli, E., 1992, Immobilization of monoclonal antibodies for affinity chromatography using a chelating peptide, *J. Chromatogr.* **595:**113–119.

Lu, X. M., Benedek, K., and Karger, B. L., 1986, Conformational effects in the high-performance liquid chromatography of proteins. Further studies of the reversed-phase chromatographic behaviour of ribonuclease A, *J. Chromatogr.* **359:**19–29.

Maa,Y.-F., and Horváth, C., 1988, Rapid analysis of proteins and peptides by reversed-phase chromatography with polymeric micropellicular sorbents, *J. Chromatogr.* **445:** 71–86.

Malmquist, G., and Lundell, N., 1992, Characterization of the influence of displacing salts on retention in gradient elution ion-exchange chromatography of proteins and peptides, *J. Chromatogr.* **627:**107–124.

Mant, C. T., and Hodges, R. S. (eds.), 1991, *High-Performance Liquid Chromatography of Peptides and Proteins*, CRC Press, Boca Raton, FL.

Mant, C. T., Zhou, N. E., and Hodges, R. S., 1989, Correlation of protein retention times in reversed-phase chromatography with polypeptide chain length and hydrophobicity, *J. Chromatogr.* **476:**363–375.

McLeod, A., and Wood, S. P., 1984, High-performance liquid chromatography of insulin, *J. Chromatogr.* **285:**319–331.

Mhatre, R. M., and Krull, I. S., 1992, Interfacing gradient elution ion-exchange chromatography and low-angle laser light-scattering photometry for analysis of proteins, *J. Chromatogr.* **591:**139–148.

Miles, J., and Formosa, T., 1992, Protein affinity chromatography with purified yeast DNA polymerase α detects proteins that bind to DNA polymerase, *Proc. Natl. Acad. Sci. USA* **89:** 1276–1280.

Müller, W., 1990, New ion exchangers for the chromatography of biopolymers, *J. Chromatogr.* **510:**133–140.

Muszynska, G., Dobrowolska, G., Medin, A., Ekman, P., and Porath, J. O., 1992, Model studies on iron(III) ion affinity chromatography, *J. Chromatogr.* **604:**19–28.

Nakamura, H., Konishi, T., and Kamada, M., 1990, Use of volatile buffers in high performance hydrophobic interaction chromatography of proteins, *Anal. Sci.* **6:**137–138.

Nakazawa, H., 1988, Rapid characterization of natural and biotechnologically synthesized human growth hormones by fast atom bombardment mass spectrometry and high-performance liquid chromatography, *Chem. Pharm. Bull.* **36:**988–993.

Nau, D. R., 1989, Chromatographic methods for antibody purification and analysis, *BioChromatography* **4:**4–18.

Nelson, N. F., and Kitagawa, N., 1990, Biomolecule separations with two new HPLC ion-exchange columns, *J. Liq. Chromatogr.* **13:**4037–4050.

Nugent, K. D., Burton, W. G., Slattery, T. K., Johnson, B. F., and Snyder, L. R., 1988, Separation of proteins by reversed-phase high-performance liquid chromatography. II. Optimizing sample pretreatment and mobile phase conditions, *J. Chromatogr.* **443:** 381–397.

Ohgami, Y., Nagase, M., Nabeshima, S., Fukui, M., and Nakazawa, H., 1989, Characterization of recombinant DNA-derived human granulocyte macrophage colony stimulating factor by fast atom bombartment mass spectrometry, *J. Biotechnol.* **12:**219–230.

Ohlson, S., Hansson, L., Glad, M., Mosbach, K., and Larsson, P.-O., 1989, High performance liquid affinity chromatography: a new tool in biotechnology, *Trends Biotechnol.* **7:**179–186.

O'Keefe, D. O., Lee, A. L., and Yamazaki, S., 1992, Use of monobromobimane to resolve two recombinant proteins by reversed-phase high-performance liquid chromatography based on their cysteine content, *J. Chromatogr.* **627:**137–143.

Oroszlan, P., Wicar, S., Teshima, G., Wu, S.-L., Hancock, W. S., and Karger, B. L., 1992, Conformational effects in the reversed-phase chromatographic behaviour of recombinant human growth hormone (rhGH) and N-methionyl recombinant human growth hormone (Met-hGH), *Anal. Chem.* **64:**1623–1631.

Patel, K., and Borchardt, R. T., 1990, Deamination of asparaginyl residues in proteins: A potential pathway for chemical degradation of proteins in lyophilized dosage forms, *J. Parenter. Sci. Technol.* **44:**300–301.

Patrick, J. S., and Lagu, A. L., 1992, Determination of recombinant proinsulin fusion protein produced in *E. coli* using oxidative sulphitolysis and two-dimensional HPLC, *Anal. Chem.* **64:**507–511.

Philips, T. M., 1991, Theory and practical aspects of high-performance immunoaffinity chromatography in: *High-Performance Liquid Chromatography of Peptides and Proteins* (C. T. Mant and R. S. Hodges, eds.), CRC Press, Boca Raton, FL, pp. 507–515.

Poll, D. J., and Harding, D. R. K., 1989, Formic acid as a milder alternative to trifluoroacetic acid and phosphoric cid in two-dimensional peptide mapping, *J. Chromatogr.* **469:** 231–239.

Poll, D. J., and Harding, D. R. K., 1991, Column-friendly reversed-phase high performance liquid chromatography of peptides and proteins using formic acid with sodium chloride and dynamic column coating with crown ethers, *J. Chromatogr.* **539:**37–45.

Porath, J., 1988, High-performance immobilized-metal-ion affinity chromatography of peptides and proteins, *J. Chromatogr.* **443:**3–11.

Renlund, S., Klintrot, I.-M., Nunn, M., Schrimsher, J. L., Wernstedt, C., and Hellman, U., 1990, Peptide mapping of HIV polypeptides expressed in *E. coli*. Quality control of different batches and identification of tryptic fragments containing residues of aromatic amino acids or cysteine, *J. Chromatogr.* **512:**325–335.

Sadana, A., 1992, Inactivation of proteins and other biological macromolecules during chromatographic methods of bioseparation, *Bioseparation* **3:**145–165.

Sadler, J., Micanovic, R., Katzenstein, G. E., Lewis, R. V., and Middaugh, C. R., 1984, Protein conformation and reversed-phase high performance liquid chromatography, *J. Chromatogr.* **317:**93–101.

Scawen, M. D., 1991, Dye affinity chromatography, *Anal. Proc.* **28:**143–144.

Schafer, W. A., and Carr, P. W., 1991, Chromatographic characterization of a phosphate-modified zirconia support for bio-chromatographic applications, *J. Chromatogr.* **587:** 149–160.

Schröder, W., Dumas, M. L., and Klein, U., 1990, Rapid high-performance liquid chromatographic protein quantitation of purified recombinant Factor VIII containing interfering substances, *J. Chromatogr.* **512:**213–218.

Scouten, W. H., 1991, Affinity chromatography for protein isolation, *Curr. Opin. Biotechnol.* **2:**37–43.

Shalongo, W., Heid, P., and Stellwagen, E., 1993a, Kinetic analysis of the hydrodynamic transition accompanying protein folding using size exclusion chromatography. 1. Denaturant dependent baseline changes, *Biopolymers* **33:**127–134.

Shalongo, W., Jagannadham, M., and Stellwagen, E., 1993b, Kinetic analysis of the hydrodynamic transition accompanying protein folding using size-exclusion chromatography. 2. Comparison of spectral and chromatographic kinetic analyses, *Biopolymers* **33:**135–145.

Sing, Y. L. K., Kroviarski, Y., Cochet, S., Dhermy, D., and Bertrand, O., 1992, High-performance hydrophobic interaction chromatography of proteins on reversed-phase supports coated with non-ionic surfactants of polyoxyethylene type, *J. Chromatogr.* **598:**181–187.

Smith, J. B., Thévenon-Emeric, G., Smith, D. L., and Green, B., 1991, Elucidation of the primary structures of proteins by mass spectrometry, *Anal. Biochem.* **193:**118–124.

Snider, J., Neville, C., Yuan, L.-C., and Bullock, J., 1992, Characterization of the heterogeneity of polyethylene glycol-modified superoxide dismutase by chromatographic and electrophoretic techniques, *J. Chromatogr.* **599:**141–155.

Ståhlberg, J., Joensson, B., and Horvath, C., 1991, Theory for electrostatic interaction chromatography of proteins, *Anal. Chem.* **63:**1867–1874.

Ståhlberg, J., Joensson, B., and Horvath, C., 1992, Combined effect of coulombic and van der Waals interactions in the chromatography of proteins, *Anal. Chem.* **64:**3118–3124.

Stout, R. W., and DeStefano, J. J., 1985, A new, stabilized hydrophilic silica packing for the high-performance gel chromatography of macromolecules, *J. Chromatogr.* **326:** 63–78.

Stout, R. W., Sivakoff, S. I., Ricker, R. D., Palmer, H. C., Jackson, M. A., and Odiorne, T. J., 1986, New ion-exchange packings based on zirconium oxide surface-stabilized, diol-bonded, silica substrates, *J. Chromatogr.* **352:**381–397.

Strömqvist, M., Holgersson, J., and Samuelsson, B., 1991, Glycosylation of extracelluar superoxide dismutase studied by high-performance liquid chromatography and mass spectrometry, *J. Chromatogr.* **548:**293–301.

Tagaki, T., 1990, Application of low-angle laser light scattering detection in the field of biochemistry. Review of recent progress, *J. Chromatogr.* **506:**409–416.

Thévenon, G., and Regnier, F., 1989, Reversed-phase liquid chromatography of proteins with strong acids, *J. Chromatogr.* **476:**499–511.

Unger, K., 1983, The application of size-exclusion chromatography to the analysis of biopolymers, *Trends Anal. Chem.* **2:**271–274.

Unger, K., Anspach, B., and Giesche, H., 1984, Optimum support properties for protein separations by high-performance size-exclusion chromatography, *J. Pharm. Biomed. Anal.* **2:**139–151.

Utsumi, J., Yamazaki, S., Kawaguchi, K., Kimura, S., and Shimizu, H., 1989a, Stability of human interferon-β1: Oligomeric human interferon-β1 is inactive but is reactivated by monomerization, *Biochim. Biophys. Acta* **998:**167–172.

Utsumi, J., Yamamoto-Terasawa I., Yamazaki S., and Shimizu H., 1989b, Elimination of contaminating *Escherichia coli* peptides in the purification of *Escherichia coli*-derived recombinant human interferon-β1 by zinc chelate affinity chromatography, *J. Chromatogr.* **490:**193–197.

Vosters, A. F., Evans, D. B., Tarpley, W. G., and Sharma, S. K., 1992, On the engineering of rDNA proteins for purification by immobilized metal affinity chromatography: Applications to alternating histidine-containing chimeric proteins from recombinant *Escherichia coli*, *Protein Exp. Purif.* **3:**18–26.

Watson, E., and Kenney, W. C., 1988, High-performance size-exclusion chromatography of recombinant derived proteins and aggregated species, *J. Chromatogr.* **436:**289–298.

Weir, M. P., and Sparks, J., 1987, Purification and renaturation of recombinant human interleukin-2, *Biochem. J.* **245:**85–91.

Welinder, B. S., and Sørensen, H. H., 1991, Alternative mobile phases for the reversed-phase high-performance liquid chromatography of peptides and proteins, *J. Chromatogr.* **537:**181–199.

Welling, G. W., Geurts, T., van Gorkum, J., Damhof, R. A., Drijfhout, J. W., Bloemhoff, W., and Welling-Wester, S., 1990, Synthetic antibody fragment as ligand in immunoaffinity chromatography, *J. Chromatogr.* **512:**337–343.

Welling, G. W., Hiemstra, Y., Feijlbrief, M., Oervell, C., van Ede, J., and Welling-Wester, S., 1992, Comparison of detergents for extraction and ion-exchange high-performance liquid chromatography of Sendai virus membrane proteins, *J. Chromatogr.* **599:** 157–162.

Wirth, H. J., Unger, K. K., and Hearn, M. T. W., 1993, Influence of ligand density on the properties of metal-chelate affinity supports, *Anal. Biochem.* **208:**16–25.

Withka, J., Moncuse, P., Baziotis, A., and Maskiewicz, R., 1987, Use of high-performance size-exclusion, ion-exchange, and hydrophobic interaction chromatography for the measurement of protein conformational change and stability, *J. Chromatogr.* **398:** 175–202.

Wu, S.-L., Benedek, K., and Karger, B. L., 1986a, Thermal behaviour of proteins in high-performance hydrophobic-interaction chromatography, *J. Chromatogr.* **359:**3–17.

Wu, S.-L., Figueroa, A., and Karger, B. L., 1986b, Protein conformational effects in hydrophobic interaction chromatography. Retention characterization and the role of the mobile phase additives and stationary phase hydrophobicity, *J. Chromatogr.* **371:**3–27.

Yamada, T., Kato, K., Kawahara, K., and Nishimura, O., 1986, Separation of recombinant human interleukin-2 and methionyl interleukin-2 produced in *E. coli*, *Biochem. Biophys. Res. Commun.* **135:**837–843.

Yang, F., Zhu, D.-W., Wang, J.-Y., and Lin, S.-X., 1992, Rapid purification yielding highly active 17β-hydroxysteroid dehydrogenase: Application of hydrophic interaction and affinity fast protein liquid chromatography, *J. Chromatogr.* **582:**71–76.

Yang, Y., and Verzele, M., 1987, High-speed and high-performance size-exclusion chromatography of proteins on a new hydrophilic polystyrene-based resin, *J. Chromatogr.* **391:**383–393.

Yasukawa, K., Abe, F., Shida, N., Koizumi, Y., Uchida, T., Noguchi, K., Shima, K., 1992, High-performance affinity chromatography system for the rapid, efficient assay of glycosylated albumin, *J. Chromatogr.* **597:**271–275.

Zachariou, M., and Hearn, M. T. W., 1992, High-performance liquid chromatography of amino acids, peptides and proteins, *J. Chromatogr.* **599:**171–177.

Zhou, Z., and Smith, D. L., 1990, Location of disulphide bonds in antithrombin III, *Biomed. Environ. Mass Spec.* **19:**782–786.

8

Capillary Electrophoresis of Proteins

Tom A. A. M. van de Goor

1. GENERAL INTRODUCTION

Because of the developments in biotechnology, e.g., the introduction of recombinant protein technology, there is an increasing demand for suitable separation techniques for biomolecules. Up-scaling is required for preparative purposes such as purification or for subsequent structural analysis, while down-scaling to miniaturized dimensions is needed to obtain high resolution for purity control and other analytical purposes.

Electrophoresis has been a standard technique for separation of biomolecules due to the slow radial relaxation of biomolecules in chromatographic methods. Slab gel electrophoretic methods are routine methods for analysis of both native and denatured proteins.

Miniaturization is often a way toward higher resolution. The use of ultrathin slab gels has greatly enhanced separation efficiency, resolution, and speed. However, quantitation has been a difficult issue and it cannot be automated like high-performance liquid chromatography (HPLC). Capillary electrophoresis (CE) is another approach toward miniaturization and automation for analytical applications (Mikkers *et al.*, 1979a; Jorgenson and Lukacs, 1981, 1983; Jorgenson, 1987; Karger *et al.*, 1989). A fused silica capillary tube (25–100 μm inner diameter) is used as the separation column, allowing the application of high electric fields

---

Tom A. A. M. van de Goor • Eindhoven University of Technology, Laboratory of Instrumental Analysis, 5600 MB Eindhoven, The Netherlands; *present address*: Hewlett Packard Laboratories, Palo Alto, California 94303-0867.

Physical Methods to Characterize Pharmaceutical Proteins, edited by James N. Herron *et al.*, Plenum Press, New York, 1995.

without thermal problems and hence high-speed separations. Several modes of electrophoresis can be performed in such systems and levels of automation comparable to chromatographic methods can be reached. Being a relatively new separation technique, commercial instrumentation has only been available for a few years. Applications are still in the development stage, especially for protein separations. The search for suitable separation modes is still ongoing and a success-guaranteed approach has not yet been found. Several strategies, however, have proven to be successful for given applications and the results have been very encouraging (Jorgenson, 1986; Grossman *et al.*, 1989b; Novotny *et al.*, 1990; Hjerten, 1991; Chen *et al.*, 1991)

This chapter describes the basic principles of open tubular CE and its application in protein separations. It discusses the information that can be obtained and illustrates a number of applications. The use of capillary isotachophoresis in closed systems, a technique that has been around for over 20 years and for which instrumentation has been available for many years, has been discussed elsewhere (Everaerts *et al.*, 1976; Catsimpoolas, 1983) and is outside the scope of this chapter.

2. PRINCIPLES OF CAPILLARY ELECTROPHORESIS

2.1. Electrophoresis and Electrophoretic Mobility

As in all electrophoretic methods, CE is based on the different velocities (v in m/sec) charged species obtain when they encounter an electric field; v is proportional to the field strength (E in V/m) and the effective mobility (m_{eff} in m^2/Vsec).

$$v = m_{eff} {}^{*}E \qquad (1)$$

The mobility is scalar, being positive for cations and negative for anions, and dependent on the charge and size of the species and the environment in which the species are present. Buffer pH influences the charge and therefore the effective mobility. If there is a sufficient difference in effective mobility, two species can be separated in an electrophoretic experiment.

In classical electrophoresis a stabilizing medium, e.g., a gel, is necessary to prevent convective disturbances during the analysis. In CE, due to the narrow-bore capillary, convective disturbances can be minimized, allowing separations in free solution.

2.2. Electroosmosis and Electroosmotic Mobility

When a fused silica capillary tube is filled with the buffer solution, an electric double layer is formed at the inside surface. The surface usually becomes negatively charged with respect to the buffer solution because of polarization or ionization of, e.g., silanol groups at the silica surface or preferential adsorption of ions from the buffer. Ions with positive charge will order opposite to this negative surface to fulfill the requirement of electroneutrality. It is important to understand that the negative charge, which is part of the surface, is fixed, while the positive charge in the buffer is free to move. When an electric field is applied over this capillary, the positive charge starts to move toward the negative electrode, dragging the solvent along. This flow of electrolyte is called the *electroosmotic flow* (EOF). The velocity of the EOF (v_{eof} in m/sec) is proportional to the applied electric field (E in V/m) by a constant: the electroosmotic mobility (μ_{eof} in m^2/Vsec):

$$v_{eof} = \mu_{eof} {}^{*}E \tag{2}$$

This constant is dependent on the charge of the surface, characterized by the zeta potential (ζ in V) and solvent properties like viscosity (η in Nsec/m^2) and dielectric constant (ϵ in F/m) given by the Helmholtz-Smoluchovski equation:

$$\mu = \frac{\zeta^{*}\epsilon}{\eta} \tag{3}$$

If the direction of the flow is toward the negative electrode, we define the EOF to be positive. An interesting characteristic of the EOF is that the flow profile is pluglike. This has considerable advantages over the parabolic flow profile in the case of laminar flow, since dispersion of the zones is suppressed.

2.3. Electrophoresis and Electroosmosis

During a CE experiment both electrophoresis and electroosmosis will act on a species. Since both are vectorial, the sum of both velocities is the actual velocity with which the species is moving through the capillary tube. Figure 1 shows a schematic representation for the separation of three species: a cation, a neutral molecule, and an anion. The different vectors indicate the electrophoretic velocities of the species and the electroosmotic velocity of the buffer system in the tube. The result of the EOF is that, if sufficiently high, all species move in the same direction, making detection of all of them possible at one point of the tube.

2.4. Efficiency and Resolution

Efficiency in CE is often expressed in terms of plate numbers comparable to the chromatographic expression. Under ideal circumstances (no thermal effects or wall interaction), diffusion is the only major source of band broadening. For this situation, Jorgenson and Lukacs (1981) derived the number of theoretical plates.

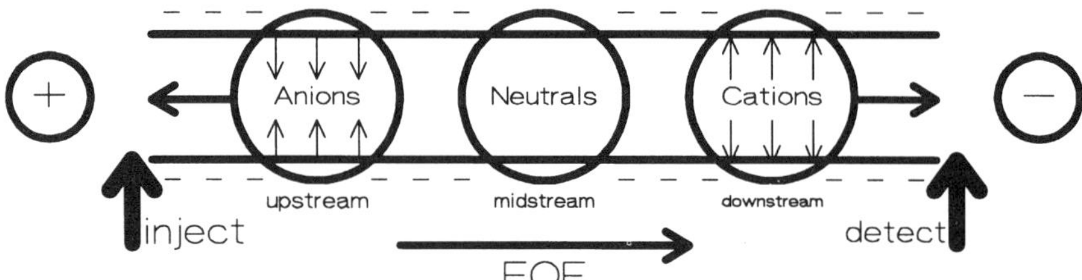

Figure 1. Principle of capillary electrophoresis with electroosmosis. If the EOF is sufficiently large, all ions move toward the detector. Ions with similar charge sign are repelled from the wall, while those with equal sign will be attracted.

$$N = \frac{[m_{eff} + \mu_{eof}]V}{2D} \tag{4}$$

In this equation, D is the diffusion coefficient (D in m^2/sec) and V is the applied voltage (V in V). This indicates that high voltages lead to increased resolution, as long as, e.g., temperature problems do not occur. For CE, field strengths of up to 500 V/cm are not unusual. Due to the tiny capillary dimensions, little Joule heat is generated, which can be removed at the capillary wall and often the system is externally thermostated.

Another possible conclusion from Eq. (4) is that high plate numbers are achieved when the osmotic velocity is high. This is indeed the case since time to diffuse is reduced. However, optimum resolution is obtained when the electrophoretic mobility counteracts electroosmotic mobility. This is, of course, at the expense of analysis time and plate numbers.

A third conclusion is that species with low diffusion coefficients will show high plate numbers, in contrast to chromatographic techniques.

2.5. Modes in Capillary Electrophoresis

Depending on the choice of the electrolyte system used, we can distinguish four types of open tubular electrophoretic techniques:

1. Zone electrophoresis, in which zones are migrating in a background electrolyte (Jorgenson and Lukacs, 1981).
2. Moving boundary electrophoresis, a mode not used for separations but important during injection (Beckers and Everaerts, 1990).

3. Isotachophoresis, in which the sample is stacked between a leading and a terminating electrolyte (Beckers *et al.*, 1991).
4. Isoelectric focusing, in which the sample is focused in a pH-gradient (Hjerten and Zhu, 1985a).

In case electrolyte additives are used or filled capillaries are applied, two more techniques can be added:

5. Micellar electrokinetic chromatography, in which a partitioning between migrating micelles and the electrolyte occurs (Terabe *et al.*, 1984).
6. Gel electrophoresis, in which sieving through a polymer network causes separation (Cohen and Karger, 1987).

The different modes and their first introduction in fused silica capillaries are discussed in the references stated with each mode.

2.6. Setup of Capillary Electrophoresis

Figure 2 shows the basic setup of an instrument for CE. The fused silica capillary tube is mounted between two electrolyte reservoirs containing the electrolytes. Electrodes in the reservoirs connected to a high-voltage power supply provide the electric field.

Prior to the analysis, the capillary is filled with the electrolyte. Injection into the column is performed by positioning the sample reservoir at the inlet end of the capillary. Applying pressure or vacuum forces a small amount (a few nanoliters) of sample into the tube. Another approach is to apply the electric field, which forces migration of ions onto the column. After injection, the electrolyte reservoir is replaced and the field applied. Sample components start to migrate toward the outlet. At a fixed point of the capillary detection is performed. A small part of the outer coating of the capillary is removed, creating a UV-transparent area. UV detection is the most common method of detection, although sensitivity can be enhanced using other strategies as on-column fluorescence detection (Green and Jorgenson, 1986), postcolumn fluorescence detection (Rose and Jorgenson, 1988), indirect fluorescence detection (Kuhr and Yeung, 1988), and laser-induced fluorescence detection (Swaile and Sepaniak, 1991). A time-based signal is the result, giving peaks similar to chromatographic methods, which can be quantified and stored. At the end of the analysis the capillary is rinsed and a new analysis can be performed. From measured data, mobility and plate numbers for the different components can be calculated. Substituting in the first equation, the electric field by the potential drop divided by the total capillary length (L in m) and the velocity by the length to the detector (l in m) divided by the migration time (t in sec), we obtain an expression for the total mobility:

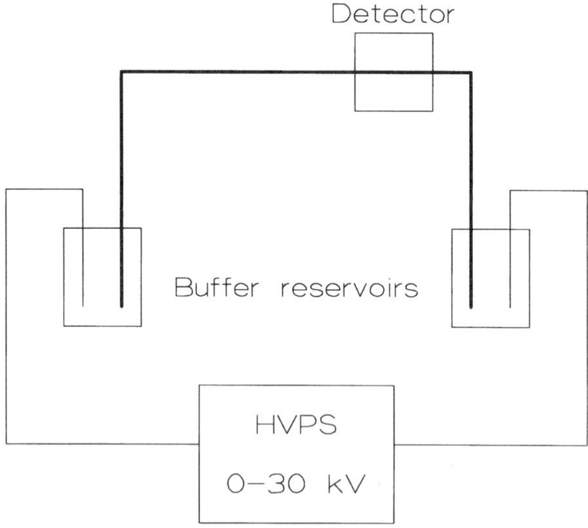

Figure 2. Basic setup for capillary electrophoresis.

$$m_{tot} = \frac{lL}{tV} \tag{5}$$

To obtain the effective mobility, one has to subtract the electroosmotic mobility, which can be measured by using the migration time for a neutral species ($m_{eff} = 0$). The number of theoretical plates N is often calculated using the expression for Gaussian peaks:

$$N = 5.54 \left[\frac{t_r}{w_{0.5}} \right]^2 \tag{6}$$

In this expression, t_r is the migration time of the compound (t_r in sec) and $w_{0.5}$ is the width of the peak at half height ($w_{0.5}$ in sec).

2.7. Advantages of Capillary Electrophoresis

CE offers possibilities for automation of the electrophoretic experiment. This leads to a high reproducibility. On-line detection methods and data storage allow easy quantification (Guzman and Hernandez, 1989). Using high electric fields, fast analyses with high separation efficiency can be performed. Reuse of the capillaries and on-line detection offer great time savings as compared to the gel approach. The use of free solution systems and the existence of EOF allow the simultaneous analysis of a wide isoelectric point (pI) range of compounds. Buffer additives can enhance resolution of compounds difficult to separate in classical systems and method development is simplified.

3. STRATEGIES FOR PROTEIN SEPARATIONS

As described earlier, under ideal CE condition diffusion is the only source of band broadening. The number of theoretical plates under these conditions can be several million per meter. However, for many biomolecules, especially proteins, interaction with the capillary wall is a serious problem, affecting the efficiency. To describe this wall interaction a chromatographic capacity factor (k') can be used. McManigill and Swedberg (1989) described an equation for open tubular electrophoresis as:

$$H = \frac{b^2}{12*L} + \frac{2*D}{v_{eof}} + \frac{k'*v_{eof}}{(1 + k')^2}\left[\frac{r^2*k'}{4*D} + \frac{2}{k_d}\right] \qquad (7)$$

In this equation the efficiency is expressed as a theoretical plate height (H in m), defined as the separation length divided by the number of theoretical plates; b represents the width of the injection band (b in m); r is the capillary radius (r in m); and k_d is the kinetic rate constant (k_d in m^{-1}). The first term represents the injection dispersion of the initial bandwidth, and the second term describes the axial dispersion. The third term describes the radial dispersion, and includes the capacity factor k', describing the axial relaxation, and a term describing adsorption–desorption kinetics at the surface. For a number of k' values, given typical values for the other parameters, the resulting plate height and consequent plate number are calculated (see Table I).

It is clear that a small chromatographic interaction has a large impact on the plate numbers (Maa *et al.*, 1991). One of the main reasons for wall interaction is often electrostatic effects. Recall from Section 2.2 that the fused silica surface is often negatively charged. When the electrolyte pH is below the pI value of the proteins, they will be positively charged, leading to interaction. Another effect can be hydrophobic interaction with silica material. This can especially be the case when unfolding of the protein occurs and hydrophobic sites become exposed.

Many strategies to overcome protein–wall interaction have focused on either selecting electrolyte conditions under which protein binding sites are reduced or treating the capillary surface to reduce active sites.

Table I. Calculation of Plate Height and Plate Numbers as a Function of the Capacity Factor for Proteins to a Capillary Surface

k'	Injection (μm)	Axial (μm)	Radial (μm)		Total (μm)	N
			Relax	Ads-des		
0	0.08	0.40	0	0.	0.48	2.1E6
0.001	0.08	0.40	0	0.10	0.58	1.7E6
0.005	0.08	0.40	0.02	0.50	1.00	1.0E6
0.010	0.08	0.40	0.08	0.098	1.54	6.5E5
0.050	0.08	0.40	1.77	4.53	6.78	1.5E5
0.100	0.08	0.40	6.46	8.26	15.68	6.6E4

3.1. Analysis at Extreme Electrolyte pH

The surface charge of the capillary tube is a function of pH. In free solution systems the point of zero charge for silica capillaries is around pH 2.5. To prevent electrostatic interaction, the protein charge and the surface charge should have the same sign. To reach this goal one could choose the pH to be extremely low (e.g., pH 2) (McCormick, 1988), to have a protonated surface, and to stay below the pI value of most of the proteins. Another area would be at extremely high pH (e.g., pH 11), as is demonstrated in Fig. 3 (Lauer and McManigill, 1986), where the surface carries a clear negative charge and pH is above most protein pI values.

Working at extreme pH conditions, however, can be problematic in finding suitable separation conditions. Often all titratable groups are either protonated or deprotonated, reducing differences in effective electrophoretic mobility. Estimation of protein mobility in free solution can help to check if separation can still be achieved (Deyl *et al.*, 1989).

At low pH the electroosmotic flow is negligible, making the separation slow. Since there is a fixed point of detection, one has to wait until even the slowest migrating ion has traveled the distance form injection to detection. At high pH the EOF in bare fused silica is generally high and separation can be insufficient.

Influencing the EOF is important in these cases. A change in pH during the run can have a beneficial effect in this respect (Foret *et al.*, 1990). Also, a change in the zeta potential using an external electric field is a solution (Wu *et al.*, 1992).

Other problems under extreme pH conditions include protein conformational changes. Unfolding or aggregation of proteins may occur, resulting in irreproducible data, often reduced efficiency, and loss of biological activity. Therefore, the use of this approach will be limited to a number of selected applications.

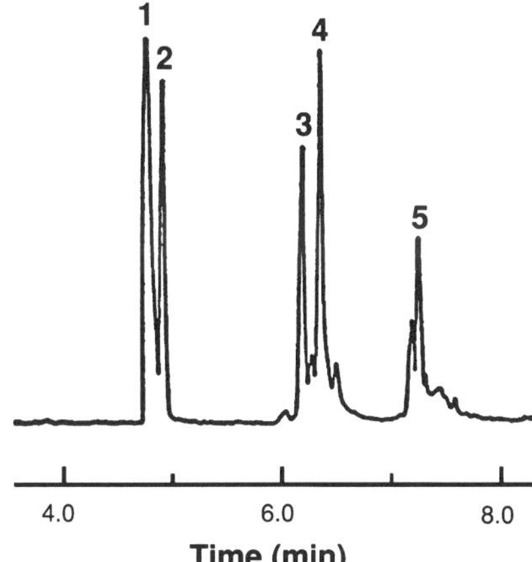

Figure 3. Typical protein separation at pH 11 in an uncoated fused silica tube. Elution order: 1 = lysozyme, 2 = mesityloxide, 3 = RNA, 4 = cytochrome *c*, and 5 = myoglobin. (From Lauer and McManigill, 1989, with permission.)

3.2. Modification of the Capillary Tube

To be able to work at moderate pH conditions, surface modification of the capillary tube can be helpful. It can decrease protein–surface interaction by reducing the surface charge of the silica material. Other interactions, like hydrophobic interactions, can also be diminished by selecting the proper surface material. Several strategies have been investigated that basically can be classified into three categories, although the boundaries between these approaches are not always clear-cut. In case of a chemical coating, the coating material is chemically attached to the silica surface. In case of a physical coating, the capillary is flushed with a substance that so strongly adheres to the surface that it forms a stable layer. However, the binding is based on pure physical effects. The third approach involves the addition of a substance to the buffer system that interacts with the surface and competes with the protein for binding sites. In this way the surface interaction is reduced. In the following sections these three approaches will be discussed.

3.2.1. CHEMICALLY BONDED COATINGS

Since the silica surface is a high energy surface, it generally needs pretreatment before a coating can be applied. This often involves surface activation by means of a hydroxide rinse and a subsequent attachment of a silanization reagent containing a reactive group for further modification. In a subsequent step mostly

Table II. Overview of Various Capillary Chemical Coatings

Coating	Reference
Methyl cellulose	Hjerten (1985)
Polyacrylamide Si-O-Si-C	Hjerten (1985)
Polyacrylamide SI-C	Cobb *et al.* (1990)
3-Glycidoxypropyltrimethoxy silane	Jorgenson and Lukacs (1983)
Epoxy-diol, maltose	Bruin *et al.* (1989)
Polyethylene glycol	Bruin *et al.* (1989)
Polyethylene imine	Towns and Regnier (1990)
Polyvinylpyrrolidone	McCormick (1988)
Arylpentafluoro(aminopropyltrimethoxy)silane	Swedberg (1990)
Alpha-lactalbumin	Maa *et al.* (1991)
Polyether	Nashabeh and El Rassi (1991a)
Hydrophilic, C1, C8, C18	Dougherty *et al.* (1991)
Ion exchangers	Kohr *et al.* (1991)
Nonionic surfactant	Towns and Regnier (1991)

hydrophilic polymers are attached, although the use of hydrophobic phases has been reported as well. Table II lists a number of phases that have been used for protein separations.

Two of the most important parameters in testing a column coating are the coating stability and the deactivation of the surface by the coating. Coating stability involves assessment of the working pH range under which the coating can be used, without being stripped from the surface. This also includes the resistance to certain capillary rinsing steps, e.g., hydroxide or acid rinses to clean the surface. It also takes into account the time the coating can be used effectively and after how many runs breakdown of the surface starts. Deactivation of the surface is often tested by measuring the EOF as a function of pH for the coated capillary. This parameter gives information about the rest surface charge on the column as a function of pH. As was discussed before, surface charge can be a cause of electrostatic interaction. Another approach is to run a variety of proteins on the column and measure the plate numbers for each of them. This gives an indication of retardation of acidic, basic, or hydrophobic proteins on the column. Measurement of streaming potential and frontal analysis is another approach (Wang and Hartwick, 1992). As an example, Fig. 4 shows a typical protein separation on a polyacrylamide coated capillary (Bruin *et al.*, 1989b).

For most of the coating materials it appears that certain proteins behave well in a given system, while others still show protein–wall interaction. Therefore, it is hard to classify any of the listed columns for their suitability in specific applications. However, classification of proteins in groups, like acidic proteins, basic

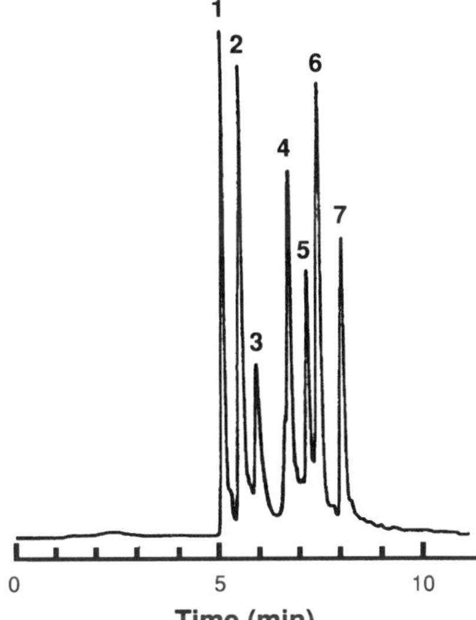

Figure 4. Typical protein separation using a PEG-coated capillary column and a run buffer at pH 3.8. 1 = Cytochrome *c*, 2 = lysozyme, 3 = myoglobin, 4 = trypsin, 5 = ribonuclease A, 6 = trypsinogen, and 7 = chymotrypsinogen. (From Bruin *et al.*, 1989b, with permission.)

proteins, and so on, can give an indication whether the attempt to use a certain column type will be successful or not.

3.2.2. PHYSICALLY BONDED COATING

To overcome at least the problem of the instability of the coating in time, the use of a physical coating rather than a chemical coating has been advocated. The idea is to choose a material that will strongly adhere to the surface, thus modifying its properties. This includes change of the surface charge to neutral, reversal of the surface charge by the formation of an additional double layer, or change in the surface viscosity, also changing the EOF. Although the coating may be removed in time, it can always be renewed by applying the dynamic coating procedure again. Table III lists a number of physical coatings.

Again, the coating should be tested for stability. It is necessary to determine when to reapply the coating procedure. Often addition of a very small amount of the coating material to the run buffer can extend the time between two subsequent coating procedures. Also, EOF measurements and standard protein tests can be applied to characterize its surface deactivation. There is a tendency that the run reproducibility with physical coatings is better than with the chemically bonded phase when the coating procedure is repeated regularly. An example of a physical

Table III. Overview of Various Capillary Physical Coatings

Coating	Reference
Polymers added to the buffer and/or sample	
Hydroxypropylmethylcellulose	Lindner *et al.* (1992)
Polyethylene glycol	Gordon *et al.* (1991)
Polyvinylalcohol	Motsch *et al.* (1992)
Nonionic surfactants used on a coated column	
Brij35	Dougherty *et al.* (1991)
Ephtaoxyethylenelaurylether	Hjerten *et al.* (1989)
Octylglycoside	Hjerten *et al.* (1989)
Buffer additives causing charge reversal	
FC134	Emmer *et al.* (1991)
CTAB	Tsuda (1987)
Polybrene	Wiktorowicz and Colburn (1990)

coating causing a charge reversal of the surface is shown in Fig. 5 (Emmer *et al.*, 1991).

3.2.3. BUFFER ADDITIVES

Another approach is to have a buffer composition containing species successfully competing with the protein for active sites on the surface. One way is to use high ionic strength buffers. The addition of a relatively high concentration of salt to the buffer helps to cover the active sites (Lauer and McManigill, 1986;

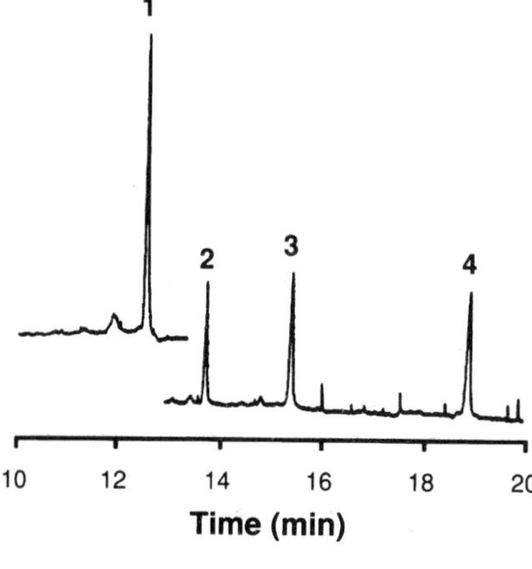

Figure 5. Typical protein separation using a surface charge reversal with FC-134 at pH 7. 1 = Myoglobin, 2 = ribonuclease A, 3 = cytochrome *c*, and 4 = lysozyme. Myoglobin injected separately. (From Emmer *et al.*, 1991, with permission.)

Green and Jorgenson, 1989). Another approach would be to increase the buffer concentration (Chen, 1991; Chen *et al.*, 1992). Apart from high ionic strength conditions, an additional advantage of this approach could be the improved buffer capacity of the system.

In general, there are several consequences to the use of high ionic strength buffers. The EOF decreases with an increase of ionic strength, as a result of a reduced zeta potential due to a collapse of the double layer. Detectability can be reduced by absorbance of the additive. High ionic strength may lead to higher conductivity of the buffer, causing extra Joule heating and possibly disturbing separation, although thermostating can help considerably (Rush *et al.*, 1991). Further reduction of capillary diameter can solve this problem, although this may increase problems with detection sensitivity. The addition of zwitterionic salts (Bushey and Jorgenson, 1989) is often a more suitable approach. They can be added up to concentrations of several moles to the buffer, thus greatly increasing ionic strength, but they do not contribute to the conductivity at their isoelectric point.

Some additives have surface-active properties such as amines. They interact with the capillary wall and reduce EOF. They also seem to have a positive influence on the separation of basic proteins when added to the run, as shown in Fig. 6 (Bullock and Yuan, 1991). Recently it was shown that the replacement of water in the buffer by deuterium oxide had a positive effect, too (Camillieri and Okafo, 1991).

3.3. Capillary Isoelectric Focusing

Isoelectric focusing is one of the major separation modes on classical gels. Also, in capillaries this mode can be used. It is important, however, to eliminate EOF, since it disturbs the formation of a stable pH gradient (Hjerten and Liao, 1986; Zhu *et al.*, 1989, 1990). The use of coated columns is in this case the most common approach (Bolger *et al.*, 1991), as shown in Fig. 7.

The capillary is filled with the ampholyte solution, which has the desired pH range. Voltage is applied, and after about 15 min the gradient is established. Often, the sample can be loaded together with the buffer or can be subsequently applied. Once the zones have focused, the separation is completed. To detect them, however, the gradient has to be moved through the detector, preferably without extensive loss in resolution. One way is to apply a small pressure from the inlet end, thus forcing the zones toward the detector. In this case, however, broadening is unavoidable. A change in the pH of the buffer reservoirs can also be used to move the gradient through the system. Recently a method was introduced to detect the bands on-column without moving the gradient by focusing a laser beam axially

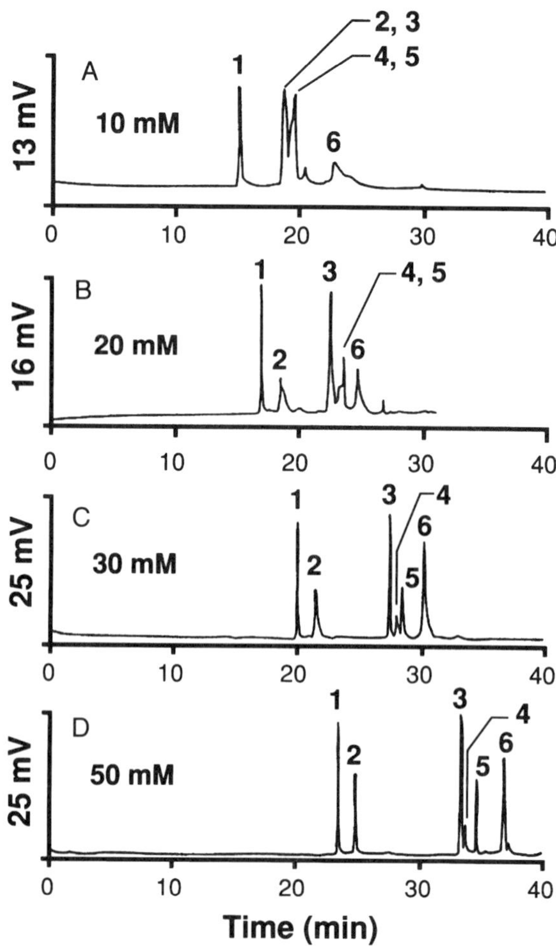

Figure 6. Typical protein separations using varying amounts of an 1,3-diaminopropane to reduce EOF. 1 = Lysozyme, 2 = cytochrome *c*, 3 = ribonuclease A, 4 = trypsinogen, 5 = chymotrypsinogen, and 6 = rhuIL-4. (From Bullock and Yuan, 1991, with permission.)

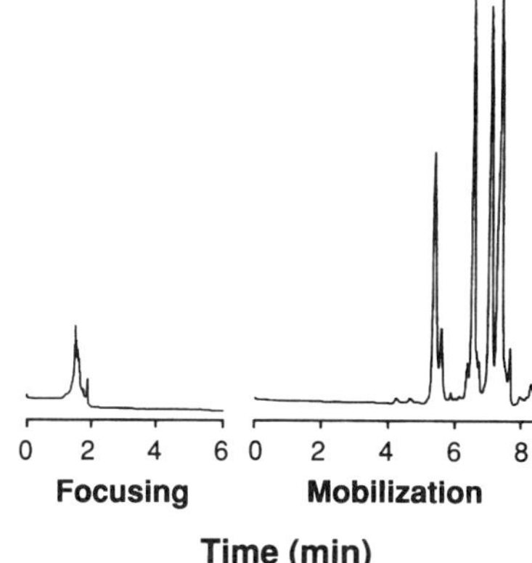

Figure 7. Focusing and mobilization signals of an isoelectric focusing experiment of hemoglobin variants in a coated capillary. (From Bolger *et al.*, 1991, with permission.)

in the capillary and radially monitoring the Schlierren effect (Wu and Pawliszyn, 1992).

3.4. Capillary Sodium Dodecyl Sulfate Gel Electrophoresis

Another method regularly used for protein analysis is a sodium dodecyl sulfate–polyacrylamide gel electrophoresis (SDS-PAGE) to obtain molecular weight information. A similar technique can be used in a capillary (Cohen and Karger, 1987), as shown in Fig. 8.

Polyacrylamide gels, however, are difficult to polymerize inside a capillary tube since they have the tendency to shrink during polymerization. This results in an expulsion of the gel from the tube due to EOF. Several approaches were studied to circumvent this problem, e.g., by polymerization at elevated pressures and the use of covalent linkage of the gel to the capillary surface (Cohen and Karger, 1987; Tsuji, 1991) or the use of linear non–cross-linked gels (Widhalm *et al.*, 1991). Another disadvantage of polyacrylamide is the detection interference at low UV, making protein sensitivity fairly low. The use of other gel types, e.g., dextrans (Hjerten *et al.*, 1989; Ganzler *et al.*, 1992), have been more successful and a reasonable size estimation up to 100 kDa can be accomplished.

4. INFORMATION FROM CAPILLARY ELECTROPHORESIS

When analyzing proteins one of the first classifications one makes is to distinguish between a native analysis and a denaturing analysis. Native analysis is

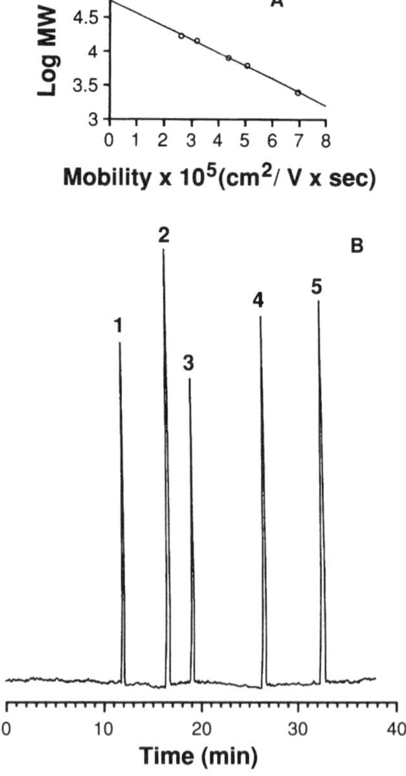

Figure 8. SDS-PAGE separation of myoglobin and several of its fragments, together with the Ferguson plot. 1 = Frag. III, mol. wt. 2510; 2 = Frag. II, mol. wt. 6210; 3 = Frag. I, mol. wt. 8160; 4 = Frag. I and II, mol. wt. 8160; and 5 = myoglobin, mol. wt. 17000. (From Cohen and Karger, 1987, with permission.)

the method of choice when information on the structure of the protein is required, while a denatured analysis is acceptable when one is only interested in protein, e.g., molecular weight or purity.

4.1. Analysis of Native Proteins

Relatively mild analytic conditions are required to retain the native structure of the species. This can sometimes interfere with the method chosen to prevent wall interaction. The use of extreme pH conditions may cause unfolding of the protein. High salt additives can generate Joule heating, leading to thermal denaturation. Other additives like surfactants can also destabilize the secondary structure. The use of isoelectric focusing can lead to a too high protein concentration, which leads to precipitation. Moderate pH conditions in free solution systems are therefore usually required and the use of surface-modified capillaries is usually the recommended approach.

Sample preparation is critically important when analyzing native proteins. Some proteins require surfactants to stay in solution, and some need special precautions not to become oxidized. Adjustment of pH of the sample can prevent loss of sample due to a locally altered EOF and pH, which can cause back-migration into the inlet reservoir. A safe option in this respect is to keep the sample pH a little bit under the run buffer pH. Matching of the sample buffer conductivity to the run buffer conductivity is also important. After injection of a zone with too low conductivity, local concentration and temperature may exceed the stability limits (Vinther *et al.*, 1992).

Following the above-presented approach, native analysis may be feasible and structural information on the protein may be obtained. This includes information on microheterogeneity, such as amidation (Vinther *et al.*, 1991), phosphorylation (Yannoukakos *et al.*, 1991), charge (Wu *et al.*, 1990), glycosilation (Josic *et al.*, 1990), isoelectric types (Compton, 1991), oxidation and reduction (Senda *et al.*, 1991), and carbohydrate contents (Tran *et al.*, 1991). Other examples are listed in Section 4.6.

4.2. Analysis of Denatured Proteins

To run under denatured conditions, the protein sample has to be pretreated. Urea, guanidine HCl, SDS, mercaptoethanol, or the elevation of temperature are often used to unfold the protein or break disulfide bridges. It is important to choose conditions that completely unfold the protein. In case of coexistence of native and denatured forms or in case aggregates are formed, the electrophoretic analysis may show broad peaks or even multiple peaks for a single protein (Kenndler and Schmidt-Beiwl, 1991). After the sample pretreatment, steps have to be taken to prevent refolding, aggregation, or complexation during the analysis. Addition of denaturants such as a high concentration of urea, SDS, or the addition of organic solvents to the run buffer can often solve problems. An important drawback of adding urea is its reduction of UV sensitivity.

4.3. Micropreparative Analysis and Combined Methods

The use of micropreparative CE is somewhat limited due to the capillary dimensions. The use of wider-bore capillaries (Hjerten and Zhu, 1985b), the collection of the same fraction from several runs (Bergman *et al.*, 1991), and the blotting of the protein onto a membrane (Eriksson *et al.*, 1992) are a few ways to get around this problem. Some examples of micropreparative CE are the subsequent analysis of a protein fraction for kinetic studies (Banke *et al.*, 1991) or fraction collection for amino acid sequencing (Hecht *et al.*, 1989b). Usually, however, when CE is used in combination with another analytical technique, it will be the second method after, e.g., gel permeation chromatography (GPC) (Yamamoto *et al.*, 1989), HPLC (Strickland, *et al.*, 1991), and HPEC (ABI model 230A micropreparative electrophoresis system) (Rosenblum, 1991).

4.4. Peptide Mapping

Another application of CE is peptide mapping to characterize a protein based on the sites, where it can be attacked by certain proteases. After digestion analysis is performed mostly in free solution under a variety of pH conditions to try to resolve as many peaks as possible (Cobb and Novotny, 1989, 1992; Young and Merion, 1990) (Fig. 9).

Solute–wall interaction is, in this case, usually no problem since small peptides tend to behave well in fused silica capillaries. A two-dimensional approach using both CE and reversed-phase HPLC to analyze the peptide digest often results in complementary information. A rule of thumb seems to be that peptides that are hard to separate in HPLC do well in CE and vice versa. When the primary structure of the peptide is known, a prediction of suitable buffer pH for optimized CE resolution can be made using the pK_a values of the amino acids in the peptide chains. A good screening parameter appears to be charge over molecular weight to the two-third power (Grossman *et al.*, 1989a; Rickard *et al.*, 1991). Peptide-oligosaccharide mapping is another peptide mapping approach (Nashabeh and El Rassi, 1991b).

4.5. Mass Spectrometry Interfacing

Interfacing CE to a mass spectrometer has been demonstrated using different interfaces. For protein applications the use of an electrospray interface seems to be the most promising (Smith *et al.*, 1989). In this case analytes leaving the capillary get multiply charged and are subsequently analyzed in the mass spectrometer. Since a mass spectrometer screens on mass over charge, the upper molecular mass limit suitable for analysis can be extended when the charge on the species increases (Fig. 10).

The ability of 60 positive charges on a single molecule have been demonstrated. This expands the operational molecular weight range of the mass spec-

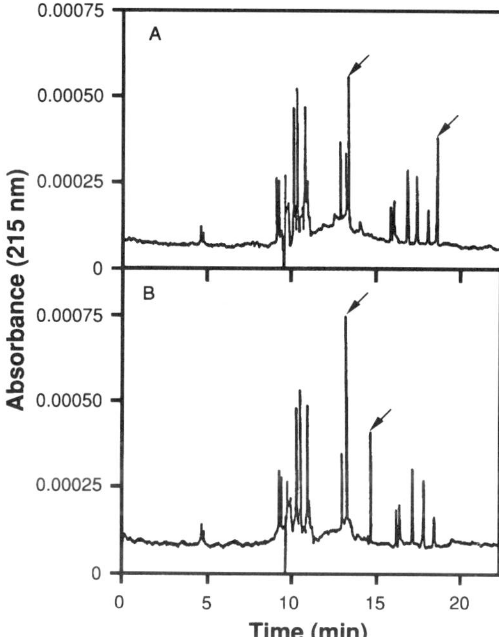

Figure 9. Comparison of a tryptic digest of (A) phosphorylated and (B) dephosphorylated β-casein. (From Cobb and Novotny, 1989, with permission.)

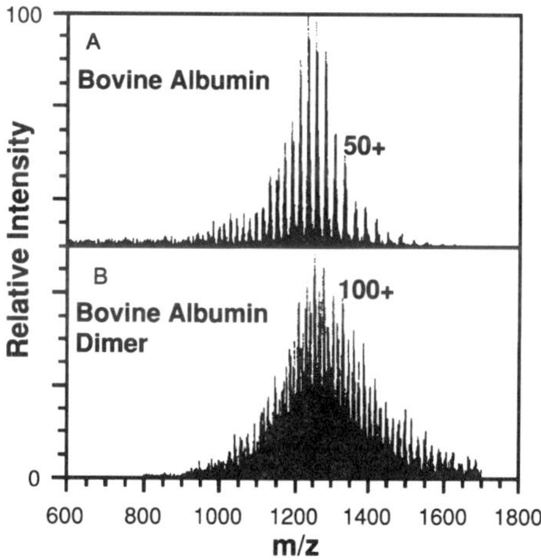

Figure 10. Mass spectrum of bovine serum albumin, molecular weight 66300, and the native dimer, molecular weight 133000, detectable due to 120 positive charges. (From Smith *et al.*, 1989, with permission.)

trometer well into the class of proteins with molecular weights over 10,000. The use of tandem mass spectrometry can give structural information based on subsequent fragmentation (Hunt *et al.*, 1990). Addition of an ion trap can turn the mass spectrometer into a powerful tool in which even protein sequencing could be performed.

4.6. Applications of CE for the Analysis of Proteins

A number of applications of CE to protein analysis, which were not discussed in the earlier sections of this chapter, are listed:

- In the pharmaceutical industry confirmation of purity of recombinant proteins and identification of derivatized products is a key issue. Complementarity of CE to other methods such as HPLC or SDS-PAGE has shown the usefulness of the technique (Wenish *et al.*, 1990; Rejman *et al.*, 1991; Nielsen and Rickard, 1990; Nielsen *et al.*, 1989; Frenz *et al.*, 1989) (Fig. 11).
- Studies on protein binding either to enzymes (Harrington *et al.*, 1991; Krueger *et al.*, 1991; Kajiwara *et al.*, 1991) or to metal ions (Kajiwara, 1991).

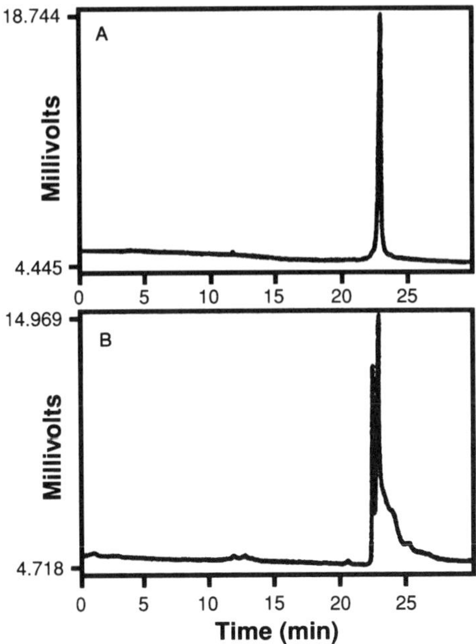

Figure 11. Electropherograms of (A) recombinant and (B) pituitary-derived human growth hormone using a coated capillary at pH 2.5. (From Frenz *et al.*, 1989, with permission.)

- Identification of proteins in a complex mixture to monitor health or disease state (Lai *et al.*, 1992; Guzman *et al.*, 1990)
- Degradation studies of proteins and peptides (Nyberg *et al.*, 1988).
- Separation of the native protein from its site-directed mutants in rDNA technology (Wiktorowicz *et al.*, 1991).
- Determination of protein mobility and diffusion coefficients in free solution is possible since under ideal conditions diffusion is the only source of band broadening (Wahlbroehl and Jorgenson, 1989).
- Analysis of protein polymorphism (Bell *et al.*, 1990).

5. CONCLUSIONS

Capillary electrophoresis is a relatively new separation technique using narrow capillary tubes as its separation medium. This allows free solution separations using high electric fields resulting in high resolution and fast analysis. Several separation modes can be selected, resulting in measurement of different compound characteristics.

Wall interaction is the most serious limitation for its use in protein analysis. Reduction of this interaction by means of buffer selection and capillary wall modification has resulted in introduction of CE into the biotechnology field. CE provides information other techniques cannot easily provide. Although the potential of CE has not been fully explored yet, it is clear that in the foreseeable future CE will be recognized as a valuable and indispensable technique for the analysis of peptides and proteins.

REFERENCES

Banke, N., Hansen, K., and Diers, I. J., 1991, Detection of enzyme activity in fractions collected from free solution capillary electrophoresis of complex samples, *J. Chromatogr.* **559:**325–335.

Beckers, J. L., and Everaerts, F. M., 1990, Isotachophoresis with two leading ions and migration behaviour in capillary zone electrophoresis, *J. Chromatogr.* **508:**19–26.

Beckers, J. L., Everaerts, F. M., and Ackermans, M. T., 1991, Isotachophoresis with electroosmotic flow: Open versus closed systems, *J. Chromatogr.* **537:**429–442.

Bell, K., McKenzie, H. A., and Shaw, D. C., 1990, Hemoglobin, serum albumin and transferrin variants of Bali (Banteng) cattle, Bos (Bibos) javanicus, *Comp. Biochem. Physiol.* **95B:**825–832.

Bergman, T., Agerberth, B., and Joernvall, H., 1991, Direct analysis of peptides and amino acids from capillary electrophoresis, *FEBS Lett.* **283:**100–103.

Bolger, C. A., Zhu, M., Rodriguez, R., and Wehr, T., 1991, Performance of uncoated and coated capillaries in free zone electrophoreses and isoelectric focusing of proteins, *J. Liq. Chromatogr.* **14:**895–906.

Bruin, G. J. M., Huisden, R., Kraak, J. C., and Poppe, H., 1989a, Performance of carbohydrate modified fused silica capillaries for the separation of proteins by zone electrophoresis, *J. Chromatogr.* **480:**339–349.

Bruin, G. J. M., Chang, J. P., Kuhlman, R. H., Zegers, K., Kraak, J. C., and Poppe, H., 1989b, Capillary zone electrophoretic separations of proteins in polyethylene glycol modified capillaries, *J. Chromatogr.* **471:**429–436.

Bullock, J. A., and Yuan, L. C., 1991, Free solution capillary electrophoresis of basic proteins in uncoated fused silica capillary tubing, *J. Microcolumn Sep.* **3:**241–248.

Bushey, M. M., and Jorgenson, J. W., 1989, Capillary electrophoresis of proteins in buffers containing high concentrations of zwitterionic salts, *J. Chromatogr.* **480:**301–310.

Camilleri, P., and Okafo, G. N., 1991, Replacement of water by deuterium oxide in capillary zone electrophoresis can increase the resolution of peptides and proteins, *J. Chem. Soc. Chem. Commun.* **3:**196–198.

Catsimpoolas, N., 1983, *Proteins*, Elsevier, Amsterdam.

Chen, F. T. A., 1991, Rapid protein analysis by capillary electrophoresis, *J. Chromatogr.* **559:**445–453.

Chen, F. A., Kelly, L., Palmieri, R., Biehler, R., and Schwartz, H. E., 1992, Use of high ionic strength buffers for the separation of proteins and peptides with capillary electrophoresis, *J. Liq. Chromatogr.* **15:**1143–1161.

Chen, F. T. A., Liu, C. M., Hsieh, Y. Z., and Sternberg, J. C., 1991, Capillary electrophoresis: A new clinical tool, *Clin. Chem.* **37:**14–19.

Cobb, K. A., and Novotny, M. V., 1989, High sensitivity peptide mapping by capillary zone electrophoresis and microcolumn liquid chromatography using immobilized trypsin for protein digestion, *Anal. Chem.* **61:**2226–2231.

Cobb, K. A., and Novotny, M. V., 1992, Peptide mapping of complex proteins at the low picomole level with capillary electrophoretic separations, *Anal. Chem.* **64:**879–886.

Cobb, K. A., Dolnik, V., and Novotny, M., 1990, Electrophoretic separations of proteins in capillaries with hydrolytically stable surface structures, *Anal. Chem.* **62:**2478–2483.

Cohen, A. S., and Karger, B. L., 1987, High performance sodium dodecyl sulfate polyacrylamide gel capillary electrophoresis of peptides and proteins, *J. Chromatogr.* **397:**409–417.

Compton, B. J., 1991, Electrophoretic modeling of proteins in free solution capillary electrophoresis and its application to monoclonal antibody microheterogeneity analysis, *J. Chromatogr.* **559:**357–366.

Deyl, Z., Rohlicek, V., and Struzinsky, R., 1989, Some rules applicable to capillary zone electrophoresis of peptides and proteins, *J. Liq. Chromatogr.* **12:**2515–2526.

Dougherty, A. M., Woolley, C. L., Williams, D. L., Swaile, D. F., Cole, R. O., and Sepaniak, M. J., 1991, Stable phases for capillary electrophoresis, *J. Liq. Chromatogr.* **14:** 907–921.

Emmer, A., Jansson, M., and Roeraade, J., 1991, Improved capillary zone electrophoretic separation of basic proteins, using a fluorosurfactant buffer additive, *J. Chromatogr.* **547:**544–550.

Eriksson, K. O., Palm, A., and Hjerten, S., 1992, Preparative capillary electrophoresis based on adsorption of the solutes (proteins) onto a moving blotting membrane as they migrate out of the capillary. *Anal. Biochem.* **210:**211–215.

Everaerts, F. M., Beckers, J. L., and Verheggen, Th. P. E. M., 1976, *Isotachophoresis: Theory, Instrumentation and Applications*, Elsevier, Amsterdam.

Foret, F., Fanali, S., and Bocek, P., 1990, Applicability of dynamic change of pH in the capillary zone electrophoresis of proteins, *J. Chromatogr.* **516:**219–222.

Frenz, J., Wu, S. L., and Hancock, W. S., 1989, Characterization of human growth hormone by capillary electrophoresis, *J. Chromatogr.* **480:**379–391.

Ganzler, K., Greve, K. S., Cohen, A. S., Karger, B. L., Guttman, A., and Cooke, N. C., 1992, High-performance capillary electrophoresis of SDS-protein complexes using UV-transparent polymer networks, *Anal. Chem.* **64:**2665–2671.

Gilges, M., Husmann, H., Kleemisch, M. H., Motsch, S. R., and Schomburg, G., 1992, CZE separations of basic proteins at low pH in fused silica capillaries with surfaces modified by silane derivatization and/or adsorption of polar polymers, *J. High Resolut. Chromatogr.* **15:**452–457.

Gordon, M. J., Lee, K. J., and Zare, R. N., 1991, Protocol for resolving protein mixtures in capillary zone electrophoresis, *Anal. Chem.* **63:**69–72.

Green, J. S., and Jorgenson, J. W., 1986, Variable wavelength on-column fluorescence detector for open tubular zone electrophoresis, *J. Chromatogr.* **352:**337–343.

Green, J. S., and Jorgenson, J. W., 1989, Minimizing adsorption of proteins on fused silica in capillary zone electrophoresis by the addition of alkali metal salts to the buffers, *J. Chromatogr.* **478:**63–70.

Grossman, P. D., Colburn, J. C., and Lauer, H. H., 1989a, A semiemperical model for the electrophoretic mobilities of peptides in free solution capillary electrophoresis, *Anal. Biochem.* **179:**28–33.

Grossman, P. D., Colburn, J. C., Lauer, H. H., Nielsen, R. G., Riggin, R. M., Sittampalam, G. S., and Rickard, E. C., 1989b, Application of free solution capillary electrophoresis to the analytical scale separation of proteins and peptides, *Anal. Chem.* **61:**1186–1194.

Guzman, N. A., and Hernandez, L., 1989, A rapid procedure for the quantitative analysis of monoclonal antibodies by high performance capillary electrophoresis, in: *Techniques in Protein Chemistry* (T. E. Hugli, ed.), Academic Press, San Diego, California, pp. 456–467.

Guzman, N. A., Hernandez, L., and Advis, J. P., 1990, Analysis of brain constituents by capillary electrophoresis, *Curr. Res. Protein Chem. Tech. Struct. Funct.* **89:**203–216.

Harrington, S. J., Varro, R., and Li, T. M., 1991, High performance capillary electrophoresis as a fast in-process control method for enzyme-labeled monoclonal antibody conjugates, *J. Chromatogr.* **559:**385–390.

Hecht, R. I., Morris, J. C., Stover, F. S., Fossey, L., and Demarest, C., 1989a, Capillary zone electrophoresis separation of recombinant human interleukin 3 and related proteins, *Prep. Biochem.* **19:**201–207.

Hecht, R. I., Coleman, J. F., Morris, J. C., Stover, F. S., and Demarest, C., 1989b, Micro-preparative capillary zone electrophoresis of recombinant human interleukin 3, *Prep. Biochem.* **19:**363–366.

Hjerten, S., 1985, High performance electrophoresis. Elimination of electroendosmosis and solute adsorption, *J. Chromatogr.* **347:**191–198.

Hjerten, S., 1991, Separation of proteins and peptides by high performance capillary electrophoresis: A versatile analytical and micropreparative method, in: *High Performance Liquid Chromatography of Proteins, Peptides and Polynucleotides* (M. T. Hearn, ed.), VCH, New York, pp. 737–770.

Hjerten, S., and Liao, J. L., 1986, Rapid separation of proteins by isoelectric focusing in the high performance electrophoresis apparatus, *Protides Biol. Fluids* **34:**727–730.

Hjerten, S., and Zhu, M., 1985a, Adaptation of the equipment for high performance electrophoresis to isoelectric focusing, *J. Chromatogr.* **346:**265–270.

Hjerten, S., and Zhu, M., 1985b, Micropreparative version of high performance electrophoresis. The electrophoretic counterpart of narrow bore high performance liquid chromatography, *J. Chromatogr.* **327:**157–164.

Hjerten, S., Valtcheva, L., Elenbring, K., and Eaker, D., 1989, High performance electrophoresis of acidic and basic low molecular weight compounds and of proteins in the presence of polymers and neutral surfactants, *J. Liq. Chromatogr.* **12:**2471–2499.

Hunt, D. F., Shabanowitz, J., Moseley, M. A., McCormack, A. L., Michel, H., Martino, P. A., Tomer, K. B., and Jorgenson, J. W., 1990, Protein and peptide sequence analysis by tandem mass spectrometry in combination with either capillary electrophoresis or microcapillary HPLC, in: *Methods Protein Sequence Analysis* (H. Jörnvall, and J. Höög, eds.), pp. 257–266.

Jorgenson, J. W., 1986, Electrophoresis, *Anal. Chem.* **58:**743A–758A.

Jorgenson, J. W., 1987, Capillary zone electrophoresis, *ACS Symp. Ser.* **335:**182–198.

Jorgenson, J. W., and Lukacs, K. D., 1981, Zone electrophoresis in open-tubular glass capillaries, *Anal. Chem.* **53:**1298–1302.

Jorgenson, J. W., and Lukacs, K. D., 1983, Capillary zone electrophoresis, *Science* **222:**266–272.

Josic, D., Zeilinger, K., Reutter, W., Boettcher, A., and Schmitz, G., 1990, High performance capillary electrophoresis of hydrophobic membrane proteins, *J. Chromatogr.* **516:**89–98.

Kajiwara, H., 1991, Application of high performance capillary electrophoresis to the analysis of conformation and interaction of metal binding proteins, *J. Chromatogr.* **559:**345–356.

Kajiwara, H., Hirano, H., and Oono, K., 1991, Binding shift assay of parvalbumin, calmodulin and carbonic anhydrase by high performance capillary electrophoresis, *J. Biochem. Biophys. Methods* **22:**263–268.

Karger, B. L., Cohen, A. S., and Guttman, A., 1989, High performance capillary electrophoresis in the biological sciences, *J. Chromatogr.* **492:**585–614.

Kenndler, E., and Schmidt-Beiwl, K., 1991, Effect of sodium dodecyl sulfate in protein samples on separation with free capillary zone electrophoresis, *J. Chromatogr.* **545:**397–402.

Kohr, J., and Engelhardt, H., 1991, Capillary electrophoresis with surface coated capillaries, *J. Microcolumn Sep.* **3:**491–495.

Krueger, R. J., Hobbs, T. R., Mihal, K. A., Tehrani, J., and Zeece, M. G., 1991, Analysis of endoproteinase Arg C action on adrenocorticotropic hormone by capillary electrophoresis and reversed phase high performance liquid chromatography, *J. Chromatogr.* **543:**451–461.

Kuhr, W. G., and Yeung, E. S., 1988, Optimization of sensitivity and separation in capillary zone electrophoresis with indirect fluorescence detection, *Anal. Chem.* **60:**2642–2646.

Lai, K., Xu, L., Colburn, J., Hong, A. L., and Pollock, J. J., 1992, The use of capillary electrophoresis to identify cationic proteins in human parotid saliva, *Arch. Oral Biol.* **37:**7–13.

Lauer, H. H., and McManigill, D., 1986, Capillary zone electrophoresis of proteins in untreated fused silica tubing, *Anal. Chem.* **58:**166–170.

Lindner, H., Helliger, W., Dirschlmayer, A., Jaquemar, M., and Puschendorf, B., 1992, High performance capillary electrophoresis of core histones and their acetylated modified derivatives, *J. Biochem.* **283:**467–471.

Maa, Y. F., Hyver, K. J., and Swedberg, S. A., 1991, Impact of wall modifications on protein elution in high performance capillary zone electrophoresis, *J. High Resolut. Chromatogr.* **14:**65–67.

McCormick, R. M., 1988, Capillary zone electrophoretic separation of peptides and proteins using low pH buffers in modified silica capillaries, *Anal. Chem.* **60:**2322–2328.

McManigill, D., and Swedberg, S. A., 1989, Factors affecting plate height in high performance zonal capillary electrophoresis (HPZCE), in: *Techniques in Protein Chemistry* (T. E. Hugli, ed.), Academic Press, San Diego, California, pp. 468–478.

Mikkers, F. E. P., Everaerts, F. M., and Verheggen, Th. P. E. M., 1979a, Concentration distributions in free zone electrophoresis, *J. Chromatogr.* **169:**1–10.

Mikkers, F. E. P., Everaerts, F. M., and Verheggen, Th. P. E. M., 1979b, High performance zone electrophoresis, *J. Chromatogr.* **169:**11–20.

Nashabeh, W., and El Rassi, Z., 1991a, Capillary zone electrophoresis of proteins with hydrophylic fused silica capillaries, *J. Chromatogr.* **559:**367–383.

Nashabeh, W., and El Rassi, Z., 1991b, Capillary zone electrophoresis of alpha-1 acid glycoprotein fragments from trypsin and endoglycosidase digestions, *J. Chromatogr.* **536:**31–42.

Nielsen, R. G., and Rickard, E. C., 1990, Applications of capillary zone electrophoresis to quality control, *ACS Symp. Ser.* **434:**36–49.

Nielsen, R. G., Sittampalam, G. S., and Rickard, E. C., 1989, Capillary zone electrophoresis of insulin and growth hormone, *Anal. Biochem.* **177:**20–26.

Novotny, M. V., Cobb, K. A., and Liu, J., 1990, Recent advances in capillary electrophoresis of proteins, peptides and amino acids, *Electrophoresis* **11:**735–749.

Nyberg, F., Zhu, M. D., Liao, J. L., and Hjerten, S., 1988, High performance electrophoresis in studies of substance P degradation, in: *Electrophoresis '88* VCH, New York, pp. 141–150.

Rejman, J., Landers, J., Goldberger, A., McCormick, D. J., Gosse, B., and Spelsberg, T. C., 1991, Purification of a nuclear protein associated with the chromatin acceptor sites for the avian oviduct progesteron receptor, *J. Protein Chem.* **10:**651–667.

Rickard, E. C., Strohl, M. M., and Nielsen, R. G., 1991, Correlation of electrophoretic mobilities from capillary electrophoresis with physicochemical properties of proteins and peptides, *Anal. Biochem.* **197:**197–207.

Rose, D. J., and Jorgenson, J. W., 1988, Postcapillary fluorescence detection in capillary zone electrophoresis using *o*-phthaldialdehyde, *J. Chromatogr.* **447:**117–131.

Rosenblum, B. B., 1991, Rapid analysis of protein fractions from the HPEC using capillary electrophoresis, *J. Liq. Chromatogr.* **14:**1017–1024.

Rush, R. S., Cohen, A. S., and Karger, B. L., 1991, Influence of column temperature on the electrophoretic behaviour of myoglobine and alpha-lactalbumine in high performance capillary electrophoresis, *Anal. Chem.* **63:**1346–1350.

Senda, M., Sasaki, T., Nakamoto, H., and Hiyama, T., 1991, Applications of capillary electrophoresis to redox proteins and nucleotides in photosynthetic system, *Anal. Sci.* **7:**1545–1548.

Smith, R. D., Loo, J. A., Barinaga, C. J., Edmonds, C. G., and Udseth, H. R., 1989, Capillary zone electrophoresis and isotachophoresis mass spectrometry of polypeptides and proteins based upon an electrospray ionization interface, *J. Chromatogr.* **480:**211–232.

Strickland, M., 1991, Use of capillary electrophoresis to screen peptide fractions from micro- and narrow bore reversed phase columns, *Am. Lab.* **23:**70–74.

Swaile, D. F., and Sepaniak, M. J., 1991, Laser based fluorometric detection schemes for the analysis of proteins by capillary zone electrophoresis, *J. Liq. Chromatogr.* **14:** 869–893.

Swedberg, S. A., 1990, Characterization of protein behaviour in high performance capillary electrophoresis using a novel capillary system, *Anal. Biochem.* **185:**51–56.

Terabe, S., Otsuka, K., Ichikawa, K., Tsuchija, A., and Ando, T., 1984, Electrokinetic separations with micellar solutions and open tubular capillaries, *Anal. Chem.* **56:**111–113.

Towns, J. K., and Regnier, F. E., 1990, Polyethyleneimine bonded phases in the separation of proteins by capillary electrophoresis, *J. Chromatogr.* **516:**69–78.

Towns, J. K., and Regnier, F. E., 1991, Capillary electrophoretic separations of proteins using non-ionic surfactants, *Anal. Chem.* **63:**1126–1132.

Tran, A. D., Park, S., Lisi, P. J., Huynh, O. T., Ryall, R. R., and Lane, P. A., 1991, Separation of carbohydrate mediated microheterogeneity of human erythropoietin by free solution capillary electrophoresis, *J. Chromatogr.* **542:**459–476.

Tsuda, T., 1987, Modification of electroosmotic flow with cetyltrimethyl-ammonium bromide in capillary zone electrophoresis, *J. High Resolut. Chromatogr.* **1987:**622–624.

Tsuji, K., 1991, High performance capillary electrophoresis of proteins. Sodium dodecyl sulfate polyacrylamide gel filled capillary column for the determination of recombinant biotechnology derived proteins, *J. Chromatogr.* **550:**823–830.

Vinther, A., Soeberg, H., Nielsen, L., Pedersen, J., and Biedermann, K., 1992, Thermal degradation of a thermolabile *Serratia marcescens* nuclease using capillary electrophoresis with stacking conditions, *Anal. Chem.* **64:**187–191.

Vinther, A., Soeberg, H., Soerensen, H., Holmegaard, J., and Munk, A., 1991, A practical approach to high performance capillary electrophoresis with biosynthetic human growth hormone as a model compound, *Talanta* **38:**1369–1379.

Walbroehl, Y., and Jorgenson, J. W., 1989, Capillary zone electrophoresis for the determination of electrophoretic mobilities and diffusion coefficients of proteins, *J. Microcolumn Sep.* **1:**41–45.

Wang, T., and Hartwick, R. A., 1992, Capillary modification and evaluation using streaming potential and frontal chromatography for protein analysis in capillary electrophoresis, *J. Chromatogr.* **594:**325–334.

Wenisch, E., Tauer, C., Jungbauer, A., Katinger, H., Faupel, M., and Righetti, P. G., 1990, Capillary zone electrophoresis for monitoring recombinant DNA protein purification in multi compartment electrolyzers with immobiline membranes, *J. Chromatogr.* **516:**133–146.

Widhalm, A., Schwer, C., Blaas, D., and Kenndler, E., 1991, Capillary zone electrophoresis with a linear, non-crosslinked polyacrylamide gel: Separation of proteins according to molecular mass, *J. Chromatogr.* **549:**446–451.

Wiktorowicz, J. E., and Colburn, J. C., 1990, Separation of cationic proteins via charge reversal in capillary electrophoresis, *Electrophoresis* **11:**769–773.

Wiktorowicz, J. E., Wilson, K. J., and Shirley, B. A., 1991, Structural analysis of RNase T1 native and site directed mutants by capillary electrophoresis, *Tech. Protein Chem.* **2:**325–333.

Wu, C. T., Lopes, T., Patel, B., and Lee, C. S., 1992, Effect of direct control of electroosmosis on peptide and protein separations in capillary electrophoresis, *Anal. Chem.* **64:**886–891.

Wu, J., and Pawliszyn, J., 1992, Universal detection for capillary isoelectric focusing without mobilization using concentration gradient imaging system, *Anal. Chem.* **64:** 224–227.

Wu, S. L., Teshima, G., Cacia, J., and Hancock, W. S., 1990, Use of high performance capillary electrophoresis to monitor charge heterogeneity in recombinant DNA derived proteins, *J. Chromatogr.* **516:**115–122.

Yamamoto, H., Manabe, T., and Okuyama, T., 1989, Gel permeation chromatography combined with capillary electrophoresis for microanalysis of proteins, *J. Chromatogr.* **480:**277–283.

Yannoukakos, D., Meyer, H. E., Vasseur, C., Driancourt, C., Wajcman, H., and Bursaux, E., 1991, Three regions of erythrocyte band 3 protein are phosphorylated on tyrosines: Characterization of the phosphorylation sites by solid phase sequencing combined with capillary electrophoresis, *Biochim. Biophys. Acta* **1066:**70–76.

Young, P. M., and Merion, M., 1990, Capillary electrophoresis analysis of species variations in the tryptic maps of cytochrome *c*, *Curr. Res. Protein Chem.* **89:**217–232.

Zhu, M. D., Hansen, D. L., Burd, S., and Gannon, F., 1989, Factors affecting free zone electrophoresis and isoelectric focusing in capillary electrophoresis, *J. Chromatogr.* **480:**311–319.

Zhu, M. D., Rodriguez, R., Hansen, D., and Wehr, T., 1990, Capillary electrophoresis of proteins under alkaline conditions, *J. Chromatogr.* **516:**123–131.

9

Applying Genetic Engineering to the Structural Analysis of Proteins

Paul T. Hamilton

1. INTRODUCTION

Developments in biochemistry and recombinant DNA technology make it possible to genetically clone, isolate, characterize, and modify any protein of interest. Current techniques in protein purification permit N-terminal sequence analysis of picomolar quantities of proteins. From this amino acid sequence information, DNA probes can be designed and used to clone the gene sequence encoding that protein. The gene can be expressed at high levels in the gram-negative bacterium *Escherichia coli* using specialized plasmid vectors. The overexpressed protein can often be purified in sufficient quantities to allow biophysical characterization. Mutagenesis can be used to analyze the structure–function relationships of the protein and to generate new and novel proteins for therapeutic and diagnostic applications.

This chapter will provide an overview of the molecular biology techniques involved in cloning, expressing, and mutating a DNA sequence. I will attempt to highlight some of the advantages and disadvantages of the various techniques. I will limit my discussion to the use of *E. coli* host–vector systems. A detailed account of many of the techniques can be found in *Current Protocols in Molecular*

Paul T. Hamilton • Becton Dickinson Research Center, Research Triangle Park, North Carolina 27709.

Physical Methods to Characterize Pharmaceutical Proteins, edited by James N. Herron *et al.*, Plenum Press, New York, 1995.

Biology (Ausubel *et al.*, 1987) and *Molecular Cloning: A Laboratory Manual* (Sambrook *et al.*, 1989).

The first and most critical step in the genetic engineering of a protein is the cloning of the DNA sequence that encodes that protein. Many methods of gene cloning require some sequence information about the gene. Amino acid sequence analysis of the protein of interest is often used as the starting point for gene cloning. Protein purification and sequence analysis techniques, however, are beyond the scope of this chapter (see Deutscher, 1990; Matsudaira, 1989, for protein purification and sequencing techniques).

2. MOLECULAR CLONING AND EXPRESSION

2.1. DNA Cloning

Two general approaches to isolate a gene sequence from an organism involve either the screening of a gene library by various methods, such as hybridization or immunodetection, or the direct amplification of the gene sequence using the polymerase chain reaction.

2.1.1. RECOMBINANT DNA LIBRARIES

Genomic libraries consist of segments of cellular DNA inserted into a cloning vector. Libraries of complementary DNA (cDNA) sequences can be generated from cellular messenger RNA (mRNA). The number of clones that make up the library must be large enough to ensure that the gene of interest is represented at least once (see Ausubel *et al.*, 1987; Sambrook *et al.*, 1989, for detailed protocols on library construction).

For genomic DNA libraries, chromosomal DNA is released from cells using enzymes and detergents. Contaminating proteins, RNA, and other macromolecules are removed by treatment with chemicals or enzymes. The purified DNA is fragmented to a size compatible with the cloning vector using either mechanical shearing or restriction enzyme digestion. Typically, genomic DNA libraries are constructed in bacteriophage lambda or cosmid vectors. These vectors have a high cloning efficiency and can accommodate relatively large DNA inserts. Genomic DNA libraries are generally screened by hybridization with a nucleic acid probe.

For cDNA libraries, total RNA or poly-A^{+}-containing mRNA is isolated from cells. To produce a good cDNA library, special care must be used to ensure that high-quality, full-length mRNA is obtained. The mRNA is converted into cDNA using reverse transcriptase. The cDNA is inserted into a cloning vector.

The cloning vector used is dependent on the method of screening. cDNA libraries are usually screened by either nucleic acid hybridization or immunological detection of an expressed antigen. Lambda gt10 is a typical cloning vector used for screening cDNA libraries by hybridization (Huynh *et al.*, 1985); λgt11 or λZAP are typical cloning vectors used with immunological detection (Huynh *et al.*, 1985; Short *et al.*, 1988).

2.1.2. ISOLATING GENES USING THE POLYMERASE CHAIN REACTION

The polymerase chain reaction (PCR) is based on the enzymatic amplification of a DNA fragment that is flanked by two oligonucleotide primers (Saiki *et al.*, 1985, 1988). The two primers hybridize to complementary strands at the opposite ends of the double-stranded DNA target sequence. Each amplification cycle of PCR involves thermal denaturation of the double-stranded DNA molecule, annealing of the primers to the target sequence, and extension of the annealed primers by a thermal-stable DNA polymerase. Repeated cycles result in the geometric amplification of the target sequence delimited by the primers.

PCR is a very sensitive method for isolating a target DNA sequence. The combination of the specificity of the primers for the target sequence along with the exponential amplification of that target sequence allow for the isolation of a single-copy gene out of an entire genome. The method is also capable of amplifying a single target sequence out of a complex mixture of RNA or DNA (Saiki *et al.*, 1988).

Since the PCR primers are incorporated into the DNA sequence of the amplified product, specific restriction sites can be included in the primers to aid in the cloning of the final PCR product (Higuchi, 1989). Similarly, transcription and translation control signals can be added to a gene sequence through PCR to facilitate the expression or analysis of a cloned gene (MacFerrin *et al.*, 1989; Stoflet *et al.*, 1988).

In order to isolate the gene encoding a protein of interest using PCR, it is necessary to have some sequence information. Degenerate PCR primers based on the reverse translation of an amino acid sequence are sufficient. Moremen (1989) designed degenerate PCR primers based on two regions of protein sequence from the enzyme mannosidase. The two regions were 42 kDa apart. PCR on cDNA from rat liver using these degenerate primers yielded a PCR product whose sequence matched protein sequence data. Cooper and Isola (1990) paired a degenerate primer based on N-terminal protein sequence information with a "universal" primer that was complementary to the 3′ end of poly-A$^+$ RNA to amplify and clone full-length cDNA.

It is also possible to use PCR to screen cDNA libraries. Using N-terminal amino acid sequence from urate oxidase, degenerate PCR primers were designed and used to amplify the 5′ end of the urate oxidase gene (Lee *et al.*, 1988). This PCR product was used to screen a cDNA library to isolate the full-length urate oxidase cDNA sequence. Tung *et al.* (1989) used a primer based on N-terminal amino acid sequence data and a second primer complementary to the cloning vector to screen a cDNA library for the gene encoding a bat exocrine protein.

Oligonucleotides for PCR can also be designed using evolutionarily conserved regions within gene families. A new class of protein–tyrosine kinases was identified in this way (Wilks, 1989).

PCR is a powerful technique for amplifying and isolating gene sequences. Primers for PCR can be designed from only a limited amount of amino acid sequence information. Using PCR methodology, DNA hybridization techniques, or immunological techniques, it is possible to isolate almost any gene sequence of interest.

2.2. Expression

Once the gene sequence has been isolated, it can be cloned into a plasmid expression vector for overproduction in *E. coli*. The primary elements controlling the level of expression in *E. coli* are transcription and translation control sequences. Also of importance are the stability of the mRNA and the protein product within the host strain. High expression in *E. coli*, however, is not a given. Each new sequence must be assessed on a case-by-case basis to determine how to maximize expression.

2.2.1. TRANSCRIPTION

DNA-dependent RNA polymerase binds and initiates mRNA synthesis at a site termed the promoter (McClure, 1985). The most common promoter sequences used in expression plasmids are derived from either natural *E. coli* sequences such as *lac* (Yanisch-Perron *et al.*, 1985), *trp* (Nichols and Yanofsky, 1983; Latta *et al.*, 1990), or *lpp* (Dufford *et al.*, 1987) promoters, phage promoters such as P_L or P_R from lambda (Remaut *et al.*, 1981; Elvin *et al.*, 1990) or T7 gene 10 (Tabor and Richardson, 1985; Studier *et al.*, 1990), or hybrid promoters such as *tac* (Amann, 1983; De Boer *et al.*, 1983). For these expression plasmids, transcription from the promoter is regulated by an adjacent operator sequence that binds a repressor protein. Transcription is initiated by an induction protocol that renders the repressor inactive. For promoters containing the *lac* operator, induction is accomplished by the addition of isopropyl β-D-thiogalactopyranoside (IPTG). IPTG binds to the *lac* repressor, making it unable to bind to the operator sequence, and blocks transcription. Transcription of the lambda promoters P_L and P_R is controlled by the lambda repressor *cI*, which binds to an adjacent operator sequence. Using a temperature-sensitive allele of *cI* (for example, *cI857*), induction of transcription can be accomplished by a temperature shift.

A given gene product, when expressed in *E. coli*, can be toxic to the cell, particularly when expressed in large quantities. Such toxic gene products should be cloned and maintained in the repressed state until overproduction is desired. For some gene products, even very low levels of expression may be deleterious to *E. coli*. These proteins are best expressed using very tightly regulated promoter systems, such as P_L or T7.

A cascade of regulation can be used to provide very tight regulation of expression. An example is the use of a T7 promoter in conjunction with a *lac* operator. The *lac* operator is placed downstream of the start site of a T7 promoter and binding of the *lac* repressor at the operator site blocks transcription (Studier *et al.*, 1990). The T7 RNA polymerase (T7 gene 1) is under the control of the *lac* operator–promoter. The gene of interest is cloned under the control of the T7 gene 10 promoter. In the presence of *lac* repressor, the expression of the T7 polymerase is down-regulated and expression of the gene product is blocked. To express the gene product, IPTG is added and the two promoters are simultaneously derepressed. This cascade of regulation can lead to very tight control of the expression of toxic gene products, as long as sufficient *lac* repressor is present. The use of such tightly regulated promoters allows the recombinant bacteria to be grown to a sufficient cell density prior to the expression of the toxic gene product.

2.2.2. TRANSLATION

The initiation of translation in *E. coli* involves sequences of the 5′ end of the mRNA called the ribosome-binding site (RBS) (Gold, 1988). The RBS is composed of a sequence just upstream from the translation initiation codon, termed the Shine-Dalgarno sequence and sequences around the translation initiation codon (ATG). The Shine-Dalgarno sequence is involved in the direct base-pairing of the mRNA to the 16S rRNA in the ribosome. Efficient translation initiation appears to require the following elements (Stormo, 1986). The optimum Shine-Dalgarno sequence contains at least four nucleotides of the sequence AGGAGGT, positioned between seven and nine bases upstream from the ATG initiation codon. For optimum expression, the spacer region between the Shine-Dalgarno sequence and the ATG should be rich in A and T nucleotides. The ribosome binding site should not contain sequences that can form secondary structures. The sequence following the ATG codon also influences the efficiency of translation. Many commonly used expression vectors provide good translation initiation signals. However, for cloned genes that lack optimal translation initiation sequences, site-directed mutagenesis techniques can be employed to convert the existing translation initiation sequences to more optimal sequences.

Alternatively, optimal translation sequences can be defined by cloning a translation sequence cassette (Bucheler *et al.*, 1992). The cassette contains GGAG for a Shine-Dalgarno sequence and is flanked by random sequence. The cassette is positioned upstream from the gene of interest, transformed into *E. coli*, and the expression level of the mutants is evaluated. Sequences flanking the Shine-Dalgarno bases that specifically enhance translation of the cloned sequence are therefore identified.

The translation of a gene into a protein requires the cell to decode the DNA sequence into an amino acid sequence. The genetic code is degenerate, in that most amino acids may be specified by more than one codon. The frequency with which each of the codons is used varies from one organism to another (Andersson and Kurland, 1990; Ikemura, 1985). The frequency of codon usage in *E. coli* appears to be correlated with the relative abundance of the cognate tRNA species. There also appears to be a correlation between the pattern of codon usage for a protein and its abundance within the cell (Ikemura, 1981). Highly expressed proteins appear to use codons for the abundant tRNA species and avoid the "rare" codons. Examples of codons that are rarely used in highly expressed genes in *E. coli* are the arginine codons AGA and AGG. Maximizing protein expression in *E. coli*, particularly for heterologous protein expression, might require the conversion of rare codons to the preferred codons.

2.2.3. PROTEIN STABILITY

A final area that can have a large impact on the expression of proteins is the stability of the gene product within *E. coli*. Various approaches have been applied to help prevent the loss of an overexpressed gene product due to proteolytic degradation. One approach is the use of protease-deficient strains of *E. coli* (Gottesman, 1990). The *lon* gene encodes an ATP-dependent protease (Chung and Goldberg, 1981; Gottesman, 1989). The Lon protease appears to be involved in the degradation of unstable proteins and abnormal proteins. Proteolysis in *E. coli* is greater at higher growth temperatures (Goff *et al.*, 1984). A number of proteases, including Lon, appear to be induced as part of the heat-shock response (Baker *et al.*, 1984; Goff *et al.*, 1984). Induction of these proteases is controlled by the *htpR* gene product. Expressing cloned gene products in *lon* and *htpR* mutants and in *lon htpR* double mutants has increased the accumulation of some recombinant proteins (Buell *et al.*, 1985).

Talmadge and Gilbert (1982) showed that the half-life of proinsulin expressed in *E. coli* is increased 10 times by secreting the protein into the periplasmic space. The proteases present in the cytoplasm are different from the proteases present in the periplasm or associated with the cell membrane (Swamy and Goldberg, 1981; Miller, 1987). A recombinant protein that is unstable in the cytoplasm, therefore, may be stabilized by secretion into the periplasm. A number of secreted proteases have been isolated, such as DegP (Strauch and Beckwith, 1988), OmpT (Sugimura and Nishihara, 1988), and protease III (Cheng and Zipser, 1979). Baneyx and Georgiou (1991) have shown that some recombinant proteins are more stable when expressed in strains deficient in two or all three of these periplasmic proteases. The differences in stability between a protein expressed in the cytoplasm and a protein expressed in the periplasm might also be influenced by the different conformations a protein might attain in each environment. The periplasmic space of *E. coli* has a redox potential that allows disulfide bonds to form.

An alternative approach to protease-deficient strains or secretion into the periplasm is the use of fusion proteins to stabilize a gene product of interest. In some cases, fusing a native *E. coli* protein sequence to the labile recombinant protein will help protect the recombinant protein from degradation. N-terminal fusion of an unstable protein to β-galactosidase has generated stable fusion products (Germino *et al.*, 1983). Shen (1984) found that fusing multiple copies of the proinsulin domain in tandem generated a stably expressed protein in *E. coli*.

2.3. Gene Fusions for Protein Purification

As noted above, gene fusions may be useful in protecting the gene product of interest from being degraded in *E. coli*. Another useful role for gene fusions is as an aid in protein purification (Sassenfeld, 1990). The modification of a cloned gene to facilitate protein purification is usually accomplished by adding a DNA sequence to either the 5′ or 3′ end of the gene of interest. This additional DNA sequence can code for an entire protein or a short polypeptide sequence. The resulting fusion protein contains the protein of interest and a tag that can be used to purify the fusion protein.

2.3.1. PROTEIN FUSIONS

There are numerous examples of gene fusions for protein purification. β-galactosidase fusion proteins are often resistant to proteolysis and can be purified by *p*-amino-phenyl-β-D-thiogalactoside affinity chromatography (Germino and Bastia, 1984). Affinity purification of β-galactosidase fusion proteins, like many enzyme fusion proteins, requires that the enzyme purification tag be soluble and active. *Staphylococcus* protein A binds to the constant region of immunoglobulins with high affinity. Gene fusions of the protein A gene or a DNA sequence encoding a synthetic immunoglobulin-binding protein derived from protein A to a gene of interest have been expressed in *E. coli* and purified using IgG-Sepharose (Moks *et al.*, 1987; Nilsson *et al.*, 1987). Protein A fusions must be secreted into the periplasm of *E. coli* to be active. Another purification tag is based on the *E. coli* maltose-binding protein (MBP). Proteins have been fused to the C-terminus of MBP and can be purified based on MBP's affinity to cross-linked amylose (Guan *et al.*, 1988; Maina *et al.*, 1988). The MBP fusion proteins can be purified in one step from a bacterial lysate and are eluted from the amylose affinity resin under nondenaturing conditions by the addition of maltose. The enzyme glutathione S-transferase (GST) has also been used as an affinity purification tag (Smith and Johnson, 1988). Gene fusions to GST have been expressed in *E. coli*. The GST fusion proteins are generally stable and soluble and can be purified under nondenaturing conditions by affinity chromatography on immobilized glutathione. An *E. coli* expression–purification vector has also been constructed using streptavidin (Sano and Cantor, 1991). Streptavidin fusion proteins expressed in *E. coli* can be purified based on the very high affinity binding of streptavidin to biotin.

2.3.2. PEPTIDE FUSIONS

An alternative to protein fusions as purification tags is the use of short polypeptide sequences with binding characteristics as tags. Hopp *et al.* (1988) used an octapeptide sequence (Asp-Tyr-Lys-Asp-Asp-Asp-Asp-Lys) engineered onto the N-terminus of recombinant proteins as an identification and purification tag. A monoclonal antibody specific to the first four animo acids of the octapeptide was generated and used for immunoaffinity purification. The binding of the peptide to the antibody was found to be calcium-dependent and reversible with EDTA. Other polypeptide purification tags are based on the selective interaction of some amino acids, particularly histidine or tryptophan, with immobilized transition metal ions (Smith *et al.*, 1988; Yip *et al.*, 1989). Polyhistidine sequences engineered at either the N-terminus or the C-terminus of a recombinant protein will bind the fusion protein to immobilized nickel (Hochuli *et al.*, 1988). The length of the histidine tail affects the conditions used for binding and elution of the fusion protein from the immobilized nickel affinity resin. A dihistidine tag allowed binding and elution of the recombinant protein in physiological buffers, whereas a hexahistidine tag allowed binding and elution of the recombinant protein in the presence of 6 M guanidine hydrochloride. Insoluble recombinant proteins, therefore, could be purified in the presence of a denaturant. Recently, polyhistidine tails have been used for metal affinity precipitation (Lilius *et al.*, 1991). A pentahistidine tail was added to galactose dehydrogenase. Addition of the *bis*-metal chelate $EGTA(Zn)_2$ resulted in the precipitation of galactose dehydrogenase $(His)_5$. Polyarginine tails have also been added to recombinant proteins to aid in purification. A penta-arginine tail was added to urogastrone. The fusion protein was expressed in *E. coli* and purified on a cation exchange resin (Sassenfeld and Brewer, 1984). Charge-based purification tags may yield variable results, however, depending upon the presence of interacting ionic components such as nucleic acids or denaturants (Sassenfeld, 1990).

2.3.3. CLEAVAGE OF FUSION PROTEINS

For structural and functional analysis of the protein of interest, it is often necessary to remove the purification tag from the fusion protein. To aid in the removal of the tag, a cleavage site is often engineered into the fusion protein between the purification tag and the protein of interest. The cleavage site is an amino acid sequence that is preferentially removed by chemical treatment or a site-specific protease. In both cases, the cleavage site must be absent from the protein of interest. For chemical cleavage, the following combinations have been used: cyanogen bromide cleavage after methionine residues (Nilsson *et al.*, 1987), hydroxylamine cleavage between asparagine–glycine residues (Moks *et al.*, 1987), and acid treatment to cleave between aspartic acid–proline residues (Nilsson *et al.*, 1985).

For enzymatic cleavage, the following proteases have been used: renin (Haffey *et al.*, 1987), collagenase (Germino and Bastia, 1984), enterokinase (Hopp *et al.*, 1988), thrombin (Smith and Johnson, 1988), trypsin (Shine *et al.*, 1980), and factor Xa (Maina *et al.*, 1988). For specific removal of polypeptide affinity tags that are fused to the carboxy-terminus of a protein, carboxypeptidase A and B have been used. Carboxypeptidase B, which specifically digests C-terminal arginine and lysine residues from proteins, was used to remove a C-terminal polyarginine purification tag on urogastrone (Sassenfeld and Brewer, 1984). Carboxypeptidase A was used to remove a carboxy-terminus polyhistidine tag (Hochuli *et al.*, 1988).

In general, the disadvantages of chemical cleavage methods are the relatively harsh reaction conditions (acidic or basic solutions sometimes with heat) and the relatively frequent occurrence of the cleavage sites in proteins. Enzymatic cleavage methods, on the other hand, are much more specific but are sometimes inefficient.

2.4. Mutagenesis of DNA Sequences

Mutagenesis is a powerful tool of molecular biology for studying the structure and function of proteins and for altering sequences in macromolecules. Two general approaches exist for the mutagenesis of a cloned gene. Site-specific mutagenesis targets a single region for change, deletion, or insertion. This approach is useful for testing the function of a given amino acid in a protein sequence or for altering a DNA sequence for easier subcloning. Random mutagenesis, on the other hand, does not rely on any prior structural or functional knowledge of the target protein. Rather, it can be used to generate a library of variant molecules that can be screened and analyzed to identify key regions in the target protein. The critical element for successful use of random mutagenesis methods is the quality of the selection or screening method used to sift through the variants.

2.4.1. SITE-SPECIFIC MUTAGENESIS

Oligonucleotide-directed mutagenesis is a commonly used strategy for site-specific mutagenesis of a cloned DNA segment (Zoller and Smith, 1983). The DNA to be mutated is cloned into a vector from which single-stranded DNA can be generated (for example, M13 or a plasmid containing an M13 origin of replication). A mutagenic oligonucleotide containing the desired mutation is annealed to the single-stranded DNA; the oligonucleotide serves as a primer for DNA polymerization. A complementary strand is synthesized and DNA ligase joins the ends of the newly synthesized strand. The resulting double-stranded DNA molecule with a mismatch at the site of the mutation is transformed into an *E. coli* host strain yielding both mutant and wild-type progeny. In theory, the number of mutant progeny should represent 50% of the total; however, in practice, the percentage of mutants is usually much lower. Initially, screening of the progeny for mutants was done by either hybridization (Zoller and Smith, 1983) or DNA sequencing (Duilio *et al.*, 1988). However, now a number of selection methods have been developed that increase the frequency of mutants. Kunkel and co-workers (1991) have utilized a *dut*$^-$ *ung*$^-$ (*dut* is deoxyuridinetriphosphatase and *ung* is uracil–DNA–glycosylase) mutant of *E. coli* that inserts a small number of uracil residues into the DNA sequence in place of the normal DNA component thymine. Vectors containing the DNA sequence to be mutated can be grown on a *dut*$^-$ *ung*$^-$ mutant of *E. coli* and uracil-containing single-stranded DNA prepared. Oligonucleotide-directed mutagenesis is performed as outlined above. The double-stranded DNA contains uracil in the nonmutant strand, while the mutant strand does not. On transformation into an *ung*$^+$ *E. coli* host strain, the nonmutant strand is selectively destroyed, resulting in mostly mutant progeny.

Eckstein and his colleagues (Sayers and Eckstein, 1988; Sayers *et al.*, 1988) have developed a method of eliminating the nonmutant strand based on the observation that DNA-containing nucleoside phosphorothioates are resistant to cleavage by certain restriction enzymes. In their oligonucleotide-directed mutagenesis procedure, the complementary strand is synthesized using a phosphorothioate analogue of deoxycytidine. The double-stranded DNA is treated with the restriction enzyme Nci I, which nicks the nonmutant strand but cannot cleave the phosphorothioate-containing mutant strand. The nonmutant strand is further degraded using an exonuclease and then the mutant strand is used as the template for repolymerization.

An alternate procedure to select for mutants in oligonucleotide-directed mutagenesis experiments uses two mutagenic primers to anneal to the single-stranded DNA. One primer contains the mutation of interest and the second primer produces a selectable mutation. Lewis and Thompson (1990) describe a coupled-primer mutagenesis method where the second primer corrected a defect in the ampicillin-resistance gene of their vector, whereby successful mutagenesis yields ampicillin-resistant progeny. Deng and Nickoloff (1992) describe a coupled-primer mutagenesis procedure in which the second primer contains a mutation that eliminates a unique nonessential restriction site. After mutagenesis, the DNA is digested with a restriction enzyme that recognizes that nonessential restriction site. Linearized nonmutant DNA transforms *E. coli* less efficiently than the undigested mutant DNA. Both coupled-primer mutagenesis procedures require two transformation steps. The first transformation uses an *E. coli* host that is defective in mismatch repair (*mutS*). All the methods for increasing the frequency of mutations from oligonucleotide-directed mutagenesis yield at least 50% mutant progeny; often the frequencies are between 60 and 90%.

PCR can also be used to generate specific site mutations (Higuchi *et al.*, 1988; Vallette *et al.*, 1989; Herlitze and Koenen, 1990). PCR primers can be designed to contain base substitutions, insertions, or deletions. These alterations in primer sequence become permanent in the amplified DNA fragment. Similarly, restriction enzyme cleavage sites can be added on to the 5′-end of the PCR primer to facilitate the cloning of the mutated DNA sequence. PCR-mediated site-specific mutagenesis is a highly efficient process that does not require any special strains of *E. coli*. The concern with PCR-mediated mutagenesis is the low fidelity of *Taq* polymerase and the creation of unintended mutations during amplification (Eckert and Kunkel, 1991). However, proper choice of PCR conditions and the use of thermal-stable polymerases with lower error rates than *Taq* polymerase should limit unintended mutations.

2.4.2. RANDOM MUTAGENESIS

Random mutagenesis techniques attempt to introduce mutations throughout the entire DNA molecule. Random mutagenesis can be achieved through the use of chemical modification of DNA *in vitro*. Sodium bisulfite (Shortle and Botstein, 1983; Pine and Huang, 1987) and methoxylamine (Kadonaga and Knowles, 1985) have been used to generate random GC to AT transition mutations in DNA. Both compounds react with cytosine residues in single-stranded DNA, sodium bisulfite catalyzes the deamination of cytosine residues to uracil, and methoxylamine reacts with cytosine residues to give N^4-methoxycytosine. On copying the mutated DNA by DNA polymerase, the modified base is paired with an adenosine residue to yield a GC to AT transition mutation. A disadvantage of chemical mutagenesis is that only a limited spectrum of mutations are generated.

Another method of randomly mutating a DNA sequence *in vitro* involves the misincorporation of nucleotides by DNA polymerase. The general principle of enzymatic misincorporation involves isolating the gene to be mutagenized as a single-stranded DNA molecule, hybridizing an oligonucleotide primer next to the target region, and extending DNA synthesis across the target region using DNA polymerase and imbalanced ratios of nucleotides. Liao and Wise (1990) describe a procedure that uses polymerases that lack proofreading activity, reverse transcriptase, and a mutant T7 polymerase. The extension reactions from the primer are done using a high ratio of one nucleotide to the other three. Lehtovaara *et al.* (1988) used limited elongation to extend the primer into the target region, thereby producing a population of DNA molecules that ended at different locations in the target region but at known bases. The target DNA was subjected to misincorporation using reverse transcriptase in the presence of three nucleotides. Using this so-called forced nucleotide misincorporation mutagenesis, the mutations appear random and all types of base substitutions are possible. The region of mutagenesis, however, appears to be limited to approximately 200 to 300 nucleotides from the primer.

Strains of *E. coli* that have mutations in mismatch repair (*mutH*, *mutL*, *mutS*, *mutT*, and *mutY*) or in the proofreading activity of DNA polymerase (*mutD*/*dnaQ*) have increased mutation rates compared with wild-type *E. coli* (Foster, 1991). These mutator strains can be used to generate random mutations in a gene sequence. A plasmid containing the gene of interest is transformed into a mutator strain and propagated. The plasmid DNA is isolated and transformed into another strain of *E. coli* for mutation analysis. Various types of mutations are generated by the different mutator strains (Foster, 1991; Wu *et al.*, 1990).

Selective regions within a gene can be targeted for mutagenesis by using an oligonucleotide cassette. The region to be mutated must be flanked by unique restriction enzyme sites. These sites can be naturally occurring in the sequence or can be generated by site-directed mutagenesis. An oligonucleotide with terminal sequences compatible to the restriction sites and containing a randomized sequence is synthesized to span the target region. The complementary strand of the mutagenic oligonucleotide is generated either synthetically or enzymatically. The mutagenic cassette is cloned into the target region, replacing the wild-type sequence and the mutants analyzed. Cassette mutagenesis has been applied to the enzyme β-lactamase (Dube and Loeb, 1989; Oliphant and Struhl; 1989). A random sequence was used to replace the sequence encoding the active site amino acids of the enzyme. Mutants were screened for β-lactamase activity. Mutant β-lactamases were isolated with altered substrate specificity and temperature-dependent activities. Amino acids responsible for these altered activities were defined. Cassette mutagenesis has the potential to generate a wide variety of sequences for a given target region, including sequences containing multiple amino acid changes. The target region, however, is limited to a relatively short segment defined by the length of the oligonucleotide. The method, therefore, requires some prior knowledge of functionally important regions within the protein of interest.

2.4.3. SPIKED OLIGONUCLEOTIDE MUTAGENESIS

A final method of mutagenesis that I would like to describe combines the technique of oligonucleotide-directed mutagenesis with the idea of random mutagenesis. In this method, "spiked" or "doped" oligonucleotides are produced by a protocol that randomly misincorporates bases during synthesis. The number, position, and type of mutation can be influenced during synthesis of the spiked oligonucleotides. Hermes *et al.* (1989) used spiked primer mutagenesis to generate a random mutant library spanning the entire gene encoding a crippled triosephosphate isomerase. From this mutant library, they were able to isolate second-site mutations that increased the activity of the starting enzyme.

In conclusion, mutagenesis techniques are powerful tools for exploring structural and functional relationships within a protein. They also increase the ease and speed with which gene sequences can be manipulated. Site-directed mutagenesis procedures allow you to probe the importance of a single amino acid in a protein by changing it or deleting it. They also allow you to customize a sequence for cloning or expression. Random mutagenesis procedures allow you to identify possible mutations that satisfy a particular functional assay. Mutagenesis in general has the potential of generating interesting functional sequences that are not found in nature.

3. APPLYING GENETIC ENGINEERING: PHAGE DISPLAY TECHNOLOGY

As an example of a new technique in molecular biology, I will describe the use of filamentous phage to display peptide or protein ligands on the bacteriophage surface. Mutagenesis and affinity selection of the displayed ligands can provide information about structure–function relationships and generate new therapeutic and diagnostic agents.

The filamentous bacteriophages, for example, M13, fd, and f1, are single-stranded DNA viruses that infect *E. coli*, having an F pilus (Marvin and Hohn, 1969; Gundling, 1992). Virus particles are about 900 by 6 nm and composed of five different proteins (Smith, 1987). The major coat protein of the virus is encoded by gene VIII. The virus particle contains approximately 3000 gene VIII protein subunits. One of the minor viral coat proteins is the gene III protein. Three to five copies of the gene III protein are found at one end of the virus particle. On infection, the single-stranded viral DNA is converted to a double-stranded replicative form (RF DNA). The RF DNA is replicated and accumulates in the cell. The virus then shifts to production of single-stranded DNA copies that are packaged into newly formed viral particles. Double-stranded DNA plasmids that have a filamentous phage origin of replication can also be converted to single-stranded DNA molecules and packaged into bacteriophage particles. The filamentous bacteriophages do not lyse or kill their bacterial host. The phages replicate inside the host cells and are extruded from the host. Infected cells grow more slowly than uninfected cells. Cultures of infected cells can produce a large number of bacteriophage particles, titers of 10^{11} to 10^{13} phage per milliliter are present in the culture supernatant.

Expression vectors have been constructed that allow a peptide or protein domain to be cloned as a fusion to a filamentous phage coat protein (pIII or pVIII). These fusion proteins are incorporated into the phage particle during its production in *E. coli*, and the peptide or protein domain is displayed on the surface of the phage particle. The gene encoding the fusion protein is packaged inside of the phage particle. A library of different peptide or protein sequences is generated and incorporated into the phage particles as fusion proteins. Each clone produces phage particles displaying the particular peptide or protein domain encoded by the gene fusion sequence it contains. Phage particles displaying a particular binding activity are isolated from the library by affinity selection or "panning." The number of different clones present in the library is on the order of 10^6 to 10^9. After multiple rounds of affinity selection, phages displaying a peptide or protein domain with the desired binding activity have been isolated. These selected phages are amplified and the DNA sequence of the cloned segment is determined. The amino acid sequence of the peptide or protein domain displayed by each phage is deduced from the DNA sequence.

This general method has been applied to random peptides, antibody fragments, and proteins. Several different groups have generated random peptide libraries in filamentous phage vectors. Devlin *et al.* (1990) constructed a peptide library comprised of about 2×10^7 different 15-amino-acid-residue sequences expressed on the surface of filamentous phages. The peptides were fused to the minor coat protein, pIII. The library was affinity-selected for streptavidin-binding peptide sequences. Phages encoding nine different streptavidin-binding sequences were isolated and the core consensus sequence of each was His-Pro-Gln.

Cwirla *et al.* (1990) constructed a library of random hexapeptides expressed at the N-terminus of pIII. The library was screened for binding to a monoclonal antibody specific for the N-terminus of β-endorphins (Tyr-Gly-Gly-Phe). They isolated and sequenced 51 clones after three rounds of affinity purification. All 51 clones contained an N-terminal tyrosine, and 48 of the clones had glycine as the second amino acid residue. They were also able to devise an affinity-enrichment procedure that selected for phages expressing peptides with high affinity for the monoclonal antibody.

Antibody fragments, either single-chain antibodies or Fab fragments, have also been displayed on the surface of phage particles (Barbas *et al.*, 1991; McCafferty *et al.*, 1990). Initial work with "phage antibodies" demonstrated that the antibody displayed on the phage particle retained the antigen-binding specificity of the parent monoclonal antibody and that the antibody-bearing phage could be selectively enriched out of a population of phage particles (McCafferty *et al.*, 1990). Subsequent work has shown that combinatorial libraries of antibody heavy-chain and light-chain variable domains can be generated from either mice or humans and displayed on the surface of phages (Clackson *et al.*, 1991; Marks *et al.*, 1991; Burton *et al.*, 1991).

Marks *et al.* (1991) isolated the immunoglobulin heavy- and light-chain V-genes from human peripheral blood lymphocytes, randomly combined them as single-chain Fv fragments, and generated a library of antibody molecules displayed on phage particles. The library was affinity-selected for antibodies binding to protein antigens (turkey egg-white lysozyme and bovine serum albumin) and a hapten (2-phenyloxazol-5-one). Phage antibodies isolated from this library exhibited affinity constants for their antigen in the range of 10^6 to 10^7 M^{-1}. Using one of the antibodies directed against the hapten 2-phenyloxazol-5-one (phOx) isolated from the phage display library, they were able to increase the affinity of the antibody for phOx by a process called *chain shuffling* (Marks *et al.*, 1992). Chain shuffling involved taking the heavy-chain V gene of the phOx-binding antibody and pairing it with a library of light-chain V genes. The best phOx-binding phage antibody was affinity-selected. The light-chain V gene and the heavy-chain third hypervariable loop of this antibody were then paired with a repertoire of V gene sequences encoding the first two hypervariable loops of the heavy chain. Again, the phage library was affinity-selected on phOx. An antibody was isolated that had a 320-fold improvement in its affinity for phOx compared with the original phage-selected, phOx-binding antibody.

The affinity and specificity of antibodies displayed on the surface of phages has also been altered through mutagenesis of the antibody followed by affinity selection. Gram *et al.* (1992) isolated a progesterone-binding antibody from combinatorial antibody library generated from nonimmunized mice. The progesterone-binding antibody was then randomly mutated using error-prone PCR, and a subsequent clone with 30-fold higher affinity was isolated. Barbas *et al.* (1992) started with an antibody that specifically bound tetanus toxoid. The DNA sequence encoding the heavy-chain third hypervariable loop from this antibody was randomized and the resulting mutated antibody molecules were displayed on the surface of phages. The library was selected for specificity toward fluorescein. Antibodies were isolated with affinities for fluorescein in the range of 10^7 M^{-1} and decreased binding specificity for tetanus toxoid.

Analyzing and altering the binding properties of almost any protein is possible using phage display technology. Wells and his colleagues at Genentech (Bass *et al.*, 1990; Lowman *et al.*, 1991) have applied phage display to the analysis of the binding of human growth hormone (hGH) to its receptor. hGH was cloned as a fusion to pIII. Libraries of variant hGH molecules were generated by randomizing four residues on either of two of the four helixes in hGH. The amino acid positions that were varied had previously been determined to be at or near residues involved in the binding of hGH to its receptor. From the affinity-selected library, the consensus residues for hGH binding were mapped at some positions. Variants were isolated with both increased and decreased affinity for the hGH receptor; the best variants bound approximately eightfold stronger than wild-type hGH. hGH not only binds to its own receptor but also binds to the prolactin receptor. Variants were isolated out of the mutant hGH libraries that had slightly increased affinity for the hGH receptor but greatly decreased affinity for the prolactin receptor. The changes in binding preference ranged from 1000- to 4000-fold compared with the wild-type hGH molecule.

Libraries of peptide or protein sequences displayed on the surface of filamentous phages can be used to isolate, alter, and characterize a variety of proteins of diagnostic and therapeutic interest. Peptide libraries can be used to map epitopes on protein antigens and to generate mimitopes of discontinuous epitopes. They also can be used to generate peptide mimics of interactions where one or both binding partners is nonproteinaceous. By screening the libraries with receptors or binding proteins of clinical interest, it should be possible to isolate peptides that modulate the activity of natural ligands. These peptides could be used to develop novel therapeutic agents. Combinatorial antibody libraries displayed on phages can be used to isolate antibodies against any antigen, thereby eliminating immunization and hybridoma antibody technology for the generation of diagnostic or therapeutic antibodies. Antibodies displayed on phage particles can be altered through mutagenesis to increase affinity, change specificity, or possess any characteristic for which a selection scheme can be designed. Protein ligands displayed on phages can be mutated and variants with modified binding specificity profiles isolated. Phage display technology offers a very powerful method of isolating, altering, and analyzing peptide or protein ligands.

4. SUMMARY

Genetic engineering offers many techniques that can be applied to the structural analysis of proteins. These techniques aid in the characterization of the protein but also can be applied to the generation of completely new diagnostic and therapeutic agents.

ACKNOWLEDGMENT. I would like to thank Wayne Beyer for helpful discussions and comments on this manuscript.

REFERENCES

Amann, E., Brosius, J., and Ptashne, M., 1983, Vectors bearing a hybrid *trp-lac* promoter useful for regulated expression of cloned genes in *Escherichia coli, Gene* **25:**167–178.

Andersson, S. G. E., and Kurland, C. G., 1990, Codon preferences in free-living microorganisms, *Microbiol. Rev.* **54:**198–210.

Ausubel, F. M., Brent, R., Kingston, R. E., Moore, D. D., Seidan, J. G., Smith, J. A., and Struhl, K. (eds.), 1987, *Current Protocols in Molecular Biology*, John Wiley, New York.

Baker, T. A., Grossman, A. D., and Gross, C. A., 1984, A gene regulating the heat shock response in *Escherichia coli* also affects proteolysis, *Proc. Natl. Acad. Sci. USA* **81:**6779–6783.

Baneyx, F., and Georgiou, G., 1991, Construction and characterization of *Escherichia coli* strains deficient in multiple secreted proteases: Protease III degrades high-molecular-weight substrates *in vivo*, *J. Bacteriol.* **173:**2696–2703.

Barbas, C. F., Kang, A. S., Lerner, R. A., and Benkovic, S. J., 1991, Assembly of combinatorial antibody libraries on phage surfaces: The gene III site, *Proc. Natl. Acad. Sci. USA* **88:**7978–7982.

Barbas, C. F., Bain, J. D., Hoekstra, D. M., and Lerner, R. A., 1992, Semisynthetic combinatorial antibody libraries: A chemical solution to the diversity problem, *Proc. Natl. Acad. Sci. USA* **89:**4457–4461.

Bass, S., Greene, R., and Wells, J. A., 1990, Hormone phage: An enrichment method for variant proteins with altered binding properties, *Proteins* **8:**309–314.

Bucheler, U. S., Werner, D., and Schirmer, R. H., 1992, Generating compatible translation initiation regions for heterologous gene expression in *Escherichia coli* by exhaustive periShine-Dalgarno mutagenesis. Human glutathione reductase cDNA as a model, *Nucleic Acids Res.* **20:**3127–3133.

Buell, G., Schulz, M. F., Selzer, G., Chollet, A., Movva, N. R., Semon, D., Escanez, S., and Kawashima, E., 1985, Optimizing the expression in *E. coli* of a synthetic gene encoding somatomedin-C (IGF-I), *Nucleic Acids Res.* **13:**1923–1938.

Burton, D. R., Barbas, C. F., Persson, M. A. A., Koenig, S., Chanock, R. M., and Lerner, R. A., 1991, A large array of human monoclonal antibodies to type 1 human immunodeficiency virus from combinatorial libraries of asymptomatic seropositive individuals, *Proc. Natl. Acad. Sci. USA* **88:**10134–10137.

Cheng, Y. S. E., and Zipser, D., 1979, Purification and characterization of protease III from *Escherichia coli, J. Biol. Chem.* **254:**4698–4706.

Chung, C. H., and Goldberg, A. L., 1981, The product of the *lon (capR)* gene in *Escherichia coli* is the ATP-dependent protease, protease La, *Proc. Natl. Acad. Sci. USA* **78:**4931–4935.

Clackson, T., Hoogenboom, H. R., Griffiths, A. D., and Winter, G., 1991, Making antibody fragments using phage display libraries, *Nature* **352:**624–628.

Cooper, D. L., and Isola, N., 1990, Full-length cDNA cloning utilizing the polymerase chain reaction, a degenerate oligonucleotide sequence and a universal mRNA primer, *BioTechniques* **9:**60–65.

Cwirla, S. E., Peters, E. A., Barrett, R. W., and Dower, W. J., 1990, Peptides on phage: A vast library of peptides for identifying ligands, *Proc. Natl. Acad. Sci. USA* **87:**6378–6382.

De Boer, H. A., Comstock, L. J., and Vasser, M., 1983, The *tac* promoter: A functional hybrid derived from the *trp* and *lac* promoters, *Proc. Natl. Acad. Sci. USA* **80:**21–25.

Deng, W. P., and Nickoloff, J. A., 1992, Site-directed mutagenesis of virtually any plasmid by eliminating a unique site, *Anal. Biochem.* **200:**81–88.

Deutscher, M. P. (ed.), 1990, *Methods in Enzymology: Guide to Protein Purification*, Vol. 182, Academic Press, San Diego, CA.

Devlin, J. J., Panganiban, L. C., and Devlin, P. E., 1990, Random peptide libraries: A source of specific protein binding molecules, *Science* **249:**404–406.

Dube, D. K., and Loeb, L. A., 1989, Mutants generated by the insertion of random oligonucleotides into the active site of the β-lactamase gene, *Biochemistry* **28:**5703–5707.

Duffard, G. D., March, P. E., and Inouye, M., 1987, Expression and secretion of foreign proteins in *Escherichia coli*, in: *Methods in Enzymology*, Vol. 153 (R. Wu and L. Grossman eds.), Academic Press, New York, pp. 492–507.

Duilio, A., Cimino, F., and Russo, T., 1988, Oligonucleotide-directed mutagenesis: A sequence-based screening, *Anal. Biochem.* **174:**613–622.

Eckert, K. A., and Kunkel, T. A., 1991, DNA polymerase fidelity and the polymerase chain reaction, *PCR Methods Applic.* **1:**17–24.

Elvin, C. M., Thompson, P. R., Argall, M. E., Hendry, P., Stamford, P. J., Lilley, P. E., and Dixon, N. E., 1990, Modified bacteriophage lambda promoter vectors for overproduction of proteins in *Escherichia coli*, *Gene* **87:**123–126.

Foster, P. L., 1991, *In vitro* mutagenesis, in: *Methods in Enzymology*, Vol. 204 (J. F. Miller, ed.), Academic Press, New York. pp. 114–125.

Germino, J., Gray, J. G., Charbonneau, H., Vanaman, T., and Bastia, D., 1983, Use of gene fusions and protein–protein interactions in the isolation of a biologically active regulatory protein: The replication initiator protein of plasmid R6K, *Proc. Natl. Acad. Sci. USA* **80:**6848–6852.

Germino, J., and Bastia, D., 1984, Rapid purification of a cloned gene product by genetic fusion and site-specific proteolysis, *Proc. Natl. Acad. Sci. USA* **81:**4692–4696.

Goff, S. A., Casson, L. P., and Golberg, A. L., 1984, Heat shock regulatory gene *htpR* influences rates of protein degradation and expression of the *lon* gene in *Escherichia coli*, *Proc. Natl. Acad. Sci. USA* **81:**6647–6651.

Gold, L., 1988, Posttranscriptional regulatory mechanisms in *Escherichia coli*, *Annu. Rev. Biochem.* **57:**199–233.

Gottesman, S., 1989, Genetics of proteolysis in *Escherichia coli*, *Annu. Rev. Genet.* **23:** 163–198.

Gottesman, S., 1990, Minimizing proteolysis in *Escherichia coli*: Genetic solutions, in:

Methods in Enzymology, Vol. 185 (D. V. Goeddel, ed.), Academic Press, New York, pp. 119–129.

Gram, H., Marconi, L. A., Barbas, C. F., Collet, T. A., Lerner, R. A., and Kang, A. S., 1992, *In vitro* selection and affinity maturation of antibodies from a naive combinatorial immunoglobulin library, *Proc. Natl. Acad. Sci. USA* **89:**3576–3580.

Guan, C. D., Li, P., Riggs, P. D., and Inouye, H., 1988, Vectors that facilitate the expression and purification of foreign peptides in *Escherichia coli* by fusion to maltose-binding protein, *Gene* **67:**21–30.

Gundling, G. J., 1992, Antibody libraries on filamentous bacteriophage, *J. Clin. Immunoassay* **15:**31–34.

Haffey, M. L., Lehman, D., and Boger, J., 1987, Site-specific cleavage of a fusion protein by renin, *DNA* **6:**565–571.

Herlitze, S., and Koenen, M., 1990, A general and rapid mutagenesis method using polymerase chain reaction, *Gene* **91:**143–147.

Hermes, J. D., Parekh, S. M., Blacklow, S. C., Koster, H., and Knowles, J. R., 1989, A reliable method for random mutagenesis: The generation of mutant libraries using spiked oligodeoxyribonucleotide primers, *Gene* **84:**143–151.

Higuchi, R., 1989, Using PCR to engineer DNA, in: *PCR Technology: Principles and Applications for DNA Amplification* (H. A. Erlich, ed.), Stockton Press, New York, pp. 61–70.

Higuchi, R., Krummel, B., and Saiki, R. K., 1988, A general method of *in vitro* preparation and specific mutagenesis of DNA fragments: Study of protein and DNA interactions, *Nucleic Acids Res.* **16:**7351–7367.

Hochuli, E., Bannwarth, W., Dobeli, H., Gentz, R., and Stuber, D., 1988, Genetic approach to facilitate purification of recombinant proteins with a novel metal chelate adsorbent, *Bio/Technology* **6:**1321–1325.

Hopp, T. P., Prickett, K. S., Price, V. L., Libby, R. T., March, C. J., Cerretti, D. P., Urdal, D. L., and Conlon, P. J., 1988, A short polypeptide marker sequence useful for recombinant protein identification and purification, *Bio/Technology* **6:**1204–1210.

Huynh, T. V., Young, R. A., and R. W. Davis, 1985, Constructing and screening cDNA libraries in λgt10 and λgt11, in: *DNA Cloning: A Practical Approach*, Vol. 1 (D. M. Glover, ed.), IRL Press, Oxford, pp. 49–78.

Ikemura, T., 1981, Correlation between the abundance of *Escherichia coli* transfer RNAs and the occurrence of the respective codons in its protein genes: A proposal for a synonymous codon choice that is optimal for the *E. coli* translation system, *J. Mol. Biol.* **151:**389–409.

Ikemura, T., 1985, Codon usage and tRNA content in unicellular and multicellular organisms, *Mol. Biol. Evol.* **2:**13–34.

Kadonaga, J. T., and Knowles, J. R., 1985, A simple and efficient method for chemical mutagenesis of DNA, *Nucleic Acids Res.* **13:**1733–1745.

Kunkel, T. A., Bebenek, K., and McClary, J., 1991, Efficient site-directed mutagenesis using uracil-containing DNA, in: *Methods in Enzymology*, Vol. 204 (J. F. Miller, ed.), Academic Press, New York, pp. 125–139.

Latta, M., Philit, M., Maury, I., Soubrier, F., Denefle, P., and Mayaux, J. F., 1990, Tryptophan promoter derivatives on multicopy plasmids: A comparative analysis of expression plasmids in *Escherichia coli*, *DNA Cell Biol.* **9:**129–137.

Lee, C. C., Wu, X., Gibbs, R. A., Cook, R. G., Muzny, D. M., and Caskey, C. T., 1988, Generation of cDNA probes directed by amino acid sequence: Cloning of urate oxidase, *Science* **239:**1288–1291.

Lehtovaara, P. M., Koivula, A. K., Bamford, J., and Knowles, J. K. C., 1988, A new method for random mutagenesis of complete genes: Enzymatic generation of mutant libraries *in vitro*, *Protein Eng.* **2:**63–68.

Lewis, M. K., and Thompson, D. V., 1990, Efficient site directed *in vitro* mutagenesis using ampicillin selection, *Nucleic Acids Res.* **18:**3439–3443.

Liao, X., and Wise, J. A., 1990, A simple high-efficiency method for random mutagenesis of cloned genes using forced nucleotide misincorporation, *Gene* **88:**107–111.

Lilius, G., Persson, M., Bulow, L., and Mosbach, K., 1991, Metal affinity precipitation of proteins carrying genetically attached polyhistidine affinity tails, *Eur. J. Biochem.* **198:**499–504.

Lowman, H. B., Bass, S. H., Simpson, N., and Wells, J. A., 1991, Selecting high-affinity binding proteins by monovalent phage display, *Biochemistry* **30:**10832–10838.

MacFerrin, K. D., Terranova, M. P., Schreiber, S. L., and Verdine, G. L., 1989, Overproduction and dissection of proteins by the expression-cassette polymerase chain reaction, *Proc. Natl. Acad. Sci. USA* **87:**1937–1941.

Maina, C. V., Riggs, P. D., Grandea, A. G., Slatko, B. E., Moran, L. S., Tagliamonte, J. A., McReynolds, and Guan, C., 1988, An *Escherichia coli* vector to express and purify foreign proteins by fusion to and separation from maltose-binding protein, *Gene* **74:**365–373.

Marks, J. D., Hoogenboom, H. R., Bonnert, T. P., McCafferty, J., Griffiths, A. D., and Winter, G., 1991, By-passing immunization: Human antibodies from V-gene libraries displayed on phage, *J. Mol. Biol.* **222:**581–597.

Marks, J. D., Griffiths, A. D., Malmqvist, M., Clackson, T. P., Bye, J. M., and Winter, G., 1992, By-passing immunization: Building high affinity human antibodies by chain shuffling, *Bio/Technology* **10:**779–783.

Marvin, D. A., and Hohn, B., 1969, Filamentous bacterial viruses, *Bacteriol. Rev.* **33:** 172–209.

Matsudaira, P. T. (ed.), 1989, *A Practical Guide to Protein and Peptide Purification for Microsequencing*, Academic Press, San Diego.

McCafferty, J., Griffiths, A. D., Winter, G., and Chiswell, D. J., 1990, Phage antibodies: Filamentous phage displaying antibody variable domains, *Nature* **348:**552–554.

McClure, W. R., 1985, Mechanism and control of transcription initiation in prokaryotes. *Annu. Rev. Biochem.* **54:**171–204.

Miller, C. G., 1987, Protein degradation and proteolytic modification, in: *Escherichia coli and Salmonella typhimurium* (F. C. Neidhardt, ed.), American Society for Microbiology, Washington, D.C., pp. 680–691.

Moks, T., Abrahmsen, L., Holmgren, E., Bilich, M., Olsson, A., Uhlen, M., Pohl, G., Sterky, C., Hultberg, H., Josephson, S., Holmgren, A., Jornvall, H., and Nilsson, B., 1987, Expression of human insulin-like growth factor in bacteria: Use of optimized gene fusion vectors to facilitate protein purification, *Biochemistry* **26:**5239–5244.

Moremen, K. W., 1989, Isolation of a rat liver Golgi mannosidase II clone by mixed oligonucleotide-primed amplification of cDNA, *Proc. Natl. Acad. Sci. USA* **86:**5276–5280.

Nichols, B. P., and Yanofsky, C., 1983, Plasmids containing the *trp* promoters of *Escherichia coli* and *Serratia marcescens* and their use in expressing cloned genes, in: *Methods in Enzymology*, Vol. 101 (R. Wu, L. Grossmann, and K. Moldave, eds.), Academic Press, New York, pp. 155–164.

Nilsson, B., Holmgren, E., Josephson, S., Gatenbeck, S., Philipson, L., and Uhlen, M., 1985, Efficient secretion and purification of human insulin-like growth factor I with a gene fusion vector in staphylococci, *Nucleic Acids Res.* **13:**1151–1162.

Nilsson, B., Moks, T., Jansson, B., Abrahmsen, L., Elmblad, A., Holmgren, E., Henrichson, C., Jones, T. A., and Uhlen, M., 1987, A synthetic IgG-binding domain based on staphylococcal protein A, *Protein Eng.* **1:**107–113.

Oliphant, A. R., and Struhl, K., 1989, an efficient method for generating proteins with altered enzymatic properties: Application to β-lactamase, *Proc. Natl. Acad. Sci. USA* **86:**9094–9098.

Pine, R., and Huang, P. C., 1987, An improved method to obtain a large number of mutants in a defined region of DNA, in: *Methods in Enzymology*, Vol. 154 (R. Wu and L. Grossman, eds.), Academic Press, New York, pp. 415–430.

Remaut, E., Stanssens, P., and Fiers, W., 1981, Plasmid vectors for high-efficiency expression controlled by the P_L promoter of coliphage lambda, *Gene* **15:**81–93.

Saiki, R. K., Scharf, S., Faloona, F., Mullis, K. B., Horn, G. T., Erlich, H. A., and Arnheim, N., 1985, Enzymatic amplification of β-globin genomic sequences and restriction site analysis for diagnosis of sickle cell anemia, *Science* **230:**1350–1354.

Saiki, R. K., Gelfand, D. H., Stoffel, S., Scharf, S. J., Higuchi, R., Horn, G. T., Mullis, K. B., and Erlich, H. A., 1988, Primer-directed enzymatic amplification of DNA with a thermostable DNA polymerase, *Science* **239:**487–491.

Sambrook, J., Fritsch, E. F., and Mantiatis, T., 1989, *Molecular Cloning: A Laboratory Manual*, 2nd ed., Cold Spring Harbor Laboratory Press, Cold Spring Harbor.

Sano, T., and Cantor, C. R., 1991, Expression vectors for streptavidin-containing chimeric proteins, *Biochem. Biophys. Res. Commun.* **176:**571–577.

Sassenfeld, H. M., 1990, Engineering proteins for purification, *Trends Biotechnol.* **8:**88–93.

Sassenfeld, H. M., and Brewer, S. J., 1984, A polypeptide fusion designed for the purification of recombinant proteins, *Bio/Technology* **2:**76–81.

Sayers, J. R., and Eckstein, F., 1988, Phosphorothioate-based oligonucleotide-directed mutagenesis, in: *Genetic Engineering: Principles and Methods*, Vol. 10, (J. K. Setlow and A. Holleander, eds.), Plenum Press, New York, pp. 109–122.

Sayers, J. R., Schmidt, W., and Eckstein, F., 1988, Phosphorothioate-based site-directed mutagenesis: Extending the basic methodology, *Nucleosides Nucleotides* **7:**625–628.

Shen, S., 1984, Multiple joined genes prevent product degradation in *Escherichia coli*, *Proc. Natl. Acad. Sci. USA* **81:**4627–4631.

Shine, J., Fettes, I., Lan, N. C. Y., Roberts, J. L., and Baxter, J. D., 1980, Expression of cloned β-endorphin gene sequences by *Escherichia coli*, *Nature* **285:**456–461.

Short, J. M., Fernandez, J. M., Sorge, J. A., and Huse, W. D., 1988, λZAP: A bacteriophage λ expression vector with *in vivo* excision properties, *Nucleic Acids Res.* **16:**7583–7600.

Shortle, D., and Botstein, D., 1983, Directed mutagenesis with sodium bisulfite, in: *Methods in Enzymology*, Vol. 100B (R. Wu, L. Grossman, and K. Moldave, eds.), Academic Press, New York, pp. 457–468.

Smith, D. B., and Johnson, K. S., 1988, Single-step purification of polypeptides expressed in *Escherichia coli* as fusions with glutathione S-transferase, *Gene* **67:**31–40.

Smith, G. P., 1987, Filamentous phages as cloning vectors, in: *Vectors: A Survey of Molecular Cloning Vectors and Their Uses* (R. L. Rodriguez and D. T. Denhardt, eds.), Butterworths, Boston, pp. 61–83.

Smith, M. C., Furman, T. C., Ingolia, T. D., and Pidgeon, C., 1988, Chelating peptide-immobilized metal ion affinity chromatography, *J. Biol. Chem.* **263:**7211–7215.

Stoflet, E. S., Koeberl, D. D., Sarkar, G., and Sommer, S. S., 1988, Genomic amplification with transcript sequencing, *Science* **239:**491–494.

Stormo, G., 1986, Translation initiation, in: *Maximizing Gene Expression* (W. Reznikoff and L. Gold, eds.), Butterworths, Boston, pp. 195–224.

Strauch, K. L., and Beckwith, J., 1988, An *Escherichia coli* mutation preventing degradation of abnormal periplasmic proteins, *Proc. Natl. Acad. Sci. USA* **85:**1576–1580.

Studier, W. F., Rosenberg, A. H., Dunn, J. J., and Dubendorff, J. W., 1990, Use of T7 RNA polymerase to direct expression of cloned genes, in: *Methods in Enzymology*, Vol. 185 (D. V. Goeddel, ed.), Academic Press, San Diego, pp. 60–89.

Sugimura, K., and Nishihara, T., 1988, Purification, characterization, and primary structure of *Escherichia coli* protease III with specificity for paired basic residues: Identity of protease III and ompT, *J. Bacteriol.* **170:**5625–5632.

Swamy, K. H. S., and Goldberg, A. L., 1981, *E. coli* contains eight soluble proteolytic activities, one being ATP dependent, *Nature* **292:**652–654.

Tabor, S., and Richardson, C. C., 1985, A bacteriophage T7 RNA polymerase/promoter system for controlled exclusive expression of specific genes. *Proc. Natl. Acad. Sci. USA* **82:**1074–1078.

Talmadge, K., and Gilbert, W., 1982, Cellular location affects protein stability in *Escherichia coli, Proc. Natl. Acad. Sci. USA* **79:**1830–1833.

Tung, J. S., Daugherty, B. L., O'Neill, L., Law, S. W., Han, J., and Mark, G. E., 1989, PCR amplification of specific sequences from a cDNA library, in: *PCR Technology: Principles and Applications for DNA Amplification* (H. A. Erlich, ed.), Stockton Press, New York, pp. 99–104.

Vallette, F., Mege, E., Reiss, A., and Adesnik, M., 1989, Construction of mutant and chimeric genes using polymerase chain reaction, *Nucleic Acids Res.* **17:**723–733.

Wilks, A. F., 1989, Two putative protein-tyrosine kinases identified by application of the polymerase chain reaction, *Proc. Natl. Acad. Sci. USA* **86:**1603–1607.

Wu, T. H., Clarke, C. H., and Marinus, M. G., 1990, Specificity of *Escherichia coli mutD* and *mutL* mutator strains, *Gene* **87:**1–5.

Yanisch-Perron, C., Vieira, J., and Messing, J., 1985, Improved M13 phage cloning vectors and host strains: Nucleotide sequences of the M13mp18 and pUC19 vectors, *Gene* **33:**103–119.

Yip, T. T., Nakagawa, Y., and Porath, J., 1989, Evaluation of the interaction of peptides with Cu(II), Ni(II), and Zn(II) by high-performance immobilized metal ion affinity chromatography, *Anal. Biochem.* **183:**159–171.

Zoller, M. J., and Smith, M., 1983, Oligonucleotide-directed mutagenesis of DNA fragments cloned into M13 vectors, in: *Methods in Enzymology*, Vol. 100B (R. Wu, L. Grossman, and K. Moldave, eds.), Academic Press, New York, pp. 468–500.

Index

organic spec

w kemp £12.50

0468-623-288